INTELLIGENT RAILWAY ALIGNMENT DESIGN

Intelligent Railway ALIGNMENT DESIGN

HAO PU
Central South University
PAUL SCHONFELD
University of Maryland
WEI LI
Central South University
TAORAN SONG
Central South University
HONG ZHANG
East China Jiaotong University

ELSEVIER

ACADEMIC PRESS
An imprint of Elsevier

Academic Press is an imprint of Elsevier
125 London Wall, London EC2Y 5AS, United Kingdom
50 Hampshire Street, 5th Floor, Cambridge, MA 02139, United States

ISBN: 978-0-443-45923-8

For information on all Academic Press publications visit our website at https://www.elsevier.com/books-and-journals

Publisher: Matthew Deans
Acquisitions Editor: Glyn Jones
Editorial Project Manager: Shivani Barthwal
Production Project Manager: Bharatwaj Varatharajan
Cover Designer: Mark Rogers

Typeset by TNQ Tech

Contents

Foreword

Railway alignment design is a fundamental task that governs the overall planning of railway construction while also affecting future operations and life-cycle performance. It is influenced and constrained by a wide range of factors, including the natural, social, and economic environments along the proposed corridor. The design process faces a "three-many" dilemma—characterized by multidimensional complex environments, multiple coupled constraints, and multiple evaluation objectives—as well as compounded effects arising from the interaction of these factors within complex contexts. These challenges make railway alignment design a globally recognized NP-hard problem. Although the widespread adoption of computer-aided route selection technologies has significantly improved design efficiency, the generation and evaluation of alignment schemes still heavily depend on expert experience. Theoretically, in three-dimensional space, there are infinitely many potential alignment alternatives between the start and end points. Considering various complex influencing factors, identifying the optimal alignment requires considerable time and effort, and high-quality alternatives are often easily overlooked. Therefore, leveraging advanced intelligent optimization theories and technologies to enhance the efficiency and quality of alignment design has become a critical prerequisite for promoting the sustainable and effective development of the nation's railway system.

To address the aforementioned challenges, over the past 4 decades, the research team for Digital Design and Intelligent Optimization for Railway Alignments at Central South University, under the leadership of the author, and including a member from the University of Maryland, has collaborated extensively with China Railway Survey and Design Institutes as well as academic institutions. Drawing upon a series of major national and industry-level research initiatives—including general programs funded by the National Natural Science Foundation of China, a National Key R&D Program project, a major system-level project under the Science and Technology R&D Program of China State Railway Group Co., Ltd., and a major project supported by China Railway Group Limited—the team has conducted systematic research on the digital design and intelligent optimization of railway alignments. This work has addressed several key technical challenges, including the development of integrated geographic information models, baseline optimization frameworks for digital alignment design, risk-aware optimization models for typical geological hazards, intelligent corridor identification algorithms, automated generation of horizontal and vertical alignment alternatives, and the intelligent optimization of reconstruction processes for existing lines. These efforts have led to a series of theoretical and methodological advancements.

Due to the limitations of the author's knowledge and practical experience, there may be shortcomings and omissions in the book. Comments and corrections from readers are sincerely welcome.

The author
November 2025

Acknowledgments

In the course of this research, important contributions have been made by Zhu Ying, Chief Design Master of China Railway and a National Engineering Survey and Design Master. Significant support was also provided by Professorate Senior Engineers Hu Jianping, Hu Guangchang, and Liu Jiangtao from the China Railway Eryuan Engineering Group Co., Ltd. From the China Railway Siyuan Survey and Design Group Co., Ltd., valuable input was offered by Professorate Senior Engineers Song Wenxiang, Peng Xianbao, and Peng Lihui. Further contributions came from Wang Jie, Deputy Chief Engineer at the China Railway First Survey and Design Institute Group Co., Ltd., along with Professorate Senior Engineers Feng Wei and Meng Cunxi of the same institute. Finally, members of the research group at Central South University, including faculty member Song Zhanfeng and the graduate students supervised by the author—Fan Xiaomeng, Wang Guanghui, Fu Huidan, Wan Xinjie, Hu Ting, Zeng Qingxin, Ran Yang, Li Jiaheng, Yang Yuting, and Zhou Zhengwen—made important contributions to this work.

I would also like to extend my heartfelt gratitude to my esteemed mentor, Professor Zhan Zhenyan, a pioneering leader in China's computer-aided intelligent railway alignment design. Professor Zhan has made extensive foundational and groundbreaking contributions to key areas such as digital terrain modeling, integrated optimization of railway horizontal and vertical alignments, and 3D visualization of alignment design. Under his leadership, the research group developed several CAD software systems for railway and highway alignments, which have been widely adopted by design institutes across China. Moreover, Professor Zhan has nurtured numerous outstanding talents in the field, establishing a solid theoretical and practical foundation that underpins our team's ongoing research efforts.

CHAPTER ONE

Introduction

Abbreviations and corresponding full texts

3D-RRT-star 3D Rapidly-exploring Random Tree-star algorithm
AASHTO American Association of State Highway and Transportation Officials
AI Artificial Intelligence
AO Alignment Optimization
BIM Building Information Modeling
CPs Cutting Planes
CRITIC Criteria Importance Through Inter-criteria Correlation
CSU Central South University
DT Distance Transform
GA Genetic Algorithm
GIS Geographic Information System
KaRAD-KG Karst Alignment Design Knowledge Graph
MCTS Monte Carlo Tree Search
MH-QNF Multi-Haul Quasi Network Flow Model
PSO Particle Swarm Optimization
RAD-KG Railway Alignment Design Knowledge Graph
RRT Rapidly-exploring Random Tree algorithm

1.1 Research background

Railways represent a tremendous industrial achievement of modern civilization, evolving continuously alongside advancements in science and technology. On Sep 27, 1825, the 32-km standard-track railway from Stockton to Darlington in the United Kingdom was officially placed into operation, marking the birth of the world's first steam-powered railway. The rise of the railway industry constitutes a significant progress in the history of human civilization. *Encyclopedia Britannica* (1973–74 edition) praised the railways: "The development of railways is one of the greatest milestones in the progress of civilization." The 1978 international edition of the *Encyclopedia Britannica* acclaimed that "railways are one of the most important means for humanity to transport passengers and goods."

As a crucial national infrastructure and the main artery of the national economy, railways play an essential role in the development of the national social economy. Railway design is a massive systems engineering undertaking, involving more than 20 disciplines, including pre-construction

Intelligent Railway Alignment Design
ISBN: 978-0-443-45923-8
https://doi.org/10.1016/B978-0-443-45923-8.00015-7

disciplines such as surveying and mapping, geology, alignment, station yards, bridges, and tunnels, as well as post-construction professions such as communications, signaling, electrification, and rolling stock. Alignment design is the forerunner and foundation of railway construction, as well as the comprehensive work regarding all aspects of the entire railway development process. It is the "leading profession" among all specialties in railway design. Its basic tasks mainly include determining standards, layouts, alignment positions, and construction points. Specifically, according to the functional requirements of a project, combined with the natural, social, and economic environment of the regions it passes through, the basic direction of the alignment is planned, the main technical standards are decided, the spatial location of the alignment is determined, and the layouts of various structures are coordinated.

Alignment design faces the challenges of multi-dimensional environmental factors, coupling constraints, as well as hard-to-quantify and mutually-conflicting evaluation objectives:

First, the environmental factors affecting alignment design are multi-dimensional, including both natural factors such as terrain, geology, surface cover, hydrology, climate, floodplains and wetlands, as well as humanistic factors such as economy, society, politics, and military affairs. These factors are interrelated, coupled, and dynamically changing. The ultimate goal of alignment design is to achieve a comprehensive coordination and balance between the alignment and the multi-dimensional environmental factors. It is necessary to fully consider the temporal and spatial dynamic characteristics of the multi-dimensional environment. Spatially, it should adapt to the distribution characteristics of multi-dimensional influencing factors such as terrain, geology, hydrology, climate, economic activity centers, and road network planning. Temporally, it should fully consider the dynamic changes of the natural and humanistic environments, such as the evolution of geo-hazards, the planning and construction of economic activity centers, and the future development trends of economy and society.

Second, alignment design needs to satisfy the multiple complex coupling constraints between alignment-structures (such as subgrades, bridges, tunnels) and environment (such as natural and humanistic environmental factors) within a massive study area with a large temporal and spatial span. Meanwhile, this constraint coupling relation also presents dynamic characteristics, which will be dynamically created or updated with the generation of alignment alternatives. The above dynamic constraints must be handled in real-time. The coupling constraint relation among alignment-structures-environment is shown in Fig. 1.1.

Third, evaluating a "high-quality" alignment alternative requires comprehensive consideration of multiple objectives such as economy, safety, environmental protection, and comfort, in order to integrate alignment design with the sustainable development of nature, economy, and society. However, most of the above evaluation objectives are usually difficult to quantify and may conflict with each other. Therefore, the quantitative evaluation of multiple objectives along with the resolution and balance of conflicts among multiple objectives are urgent problems to be solved.

It is evident that quickly searching for an alignment alternative that meets multiple coupling constraints and achieves the optimal comprehensive performance among multiple objectives in a multi-dimensional environment is a huge challenge faced by alignment design. Especially in complex environments, the above-mentioned problems interact, yielding a universally recognized NP-hard problem. Meanwhile, there may be infinitely many potential feasible alternatives in a large search space. With manual alignment design, a time-consuming and labor-intensive process is required. Meanwhile, during the design, it is possible to miss some feasible alternatives. To address these problems, computer-aided digital

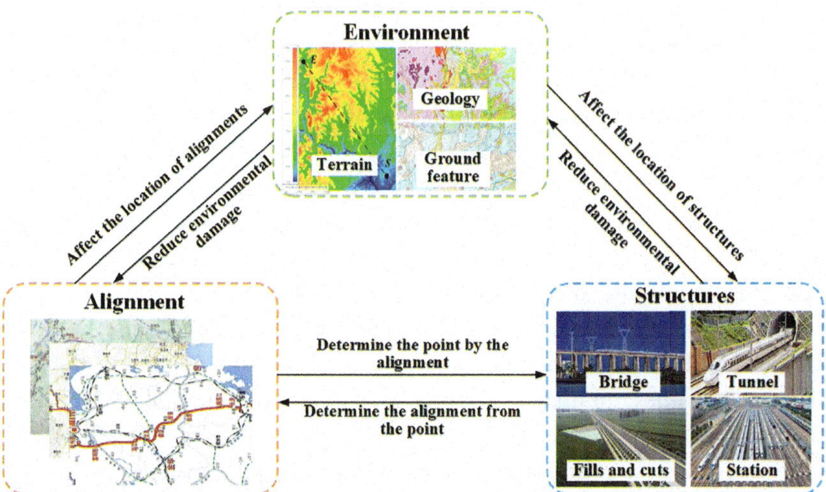

Figure 1.1 Coupling constraint relation among alignment-structures-environment.

alignment design emerged. Since the 1960s, information technology has triggered a new wave of industrial revolution. The rapid development of electronic computers and the interdisciplinary integration of various emerging disciplines have injected new vitality into the traditional alignment design. The development of aerial photogrammetry, remote sensing technology, and computer-aided design technology, as well as the integration of digital geographic models, virtual reality environment models, computers and their auxiliary equipment in survey and design, have gradually promoted the vigorous development of computerized railway alignment design.

In the early 1960s, computers were first applied to the design of railways and highways, saving much human labor and facilitating the comparison of multiple alternatives. In the 1970s, the Digital Terrain Model (DTM) began to be applied, and computer-aided design (CAD) technology could directly provide design and construction drawings, replacing manual drawing, and greatly enhancing alignment design. In the 1980s, many countries had established integrated systems consisting of aerial survey equipment, computers (including plotters, digitizers, and other external devices) and special software packages, which could realize data collection, establishment of DTM, and optimization technologies to make full set of computer calculations, drawing, and report generation. Examples include several typical systems, such as CANDID system developed by Louis Berger Inc. (in the United States), SICAD land information and graphics processing system from Siemens AG (Germany), as well as ROADCAD program system (Finland). In the 1990s, railway and highway CAD technologies had developed into integrated and systematic technologies integrating data collection and processing, design, analysis, and optimization. Typical railway (highway) CAD systems included: the German CARD/1 system, which emphasizes highway geometric design and drainage design; the Finnish ROADCAD road system, which integrates data collection, processing, and image output; the American INROADS system, which features advanced image processing and interactive design technologies; and the British MXRoad system, which can be used for the design of railways, highways, mines, drainage, airports, ports, and other civil engineering projects.

Since the 21st century, with the continuous increase in railway construction demand worldwide, particularly in China, the geographical and geological environments along railway alignments have become increasingly complex, and people have devoted increasing attention to environmental, landscape, and geological alignment design. As a result, BIM design software based on Building Information Modeling (BIM) technology has emerged, playing a pivotal role in the field of digital alignment design. The BIM software widely used in alignment design worldwide mainly includes OpenRoads Designer and OpenRail Designer from Bentley Systems, Inc. in the United States. Among them, OpenRoads Designer is a 3D road design software with an outstanding integrated solution in data collection, suitable for surveying, drainage, underground facilities, and highway design; OpenRail Designer is a 3D design software applicable to light rail, subway, heavy rail, high-speed railway, and maglev railway projects. It can unify design and construction from the concept stage to commissioning and completion. Meanwhile, it can display the results with excellent visual and animation effects. In addition, Autodesk's Civil 3D is also a commonly used software in civil engineering road design with high industry recognition. Its 3D dynamic engineering model helps in quickly completing road engineering, site, stormwater/sewage drainage system, and site planning design, enabling rapid evaluation of multiple design alternatives, decision-making, and visualization.

Although the rapid development of digital alignment design has significantly improved design efficiency compared with manual design, it is still limited to the expression stage of design alternatives. The most important stages of alternative formation, evaluation, adjustment, and decision-making still rely on human experience, resulting in relatively low design efficiency and the need for great human and material resources. Moreover, due to the uneven level of human experience, the quality of alternatives is difficult to guarantee. Additionally, the coupling effect of multi-dimensional complex environments, multiple coupling constraints, and multiple evaluation objectives has further compounded the difficulty of alignment design. Designers generally face difficulties such as high labor intensity, long design cycles, and lack of effective demonstration of alternatives. Furthermore, limited by time, energy, equipment, and human subjectivity, alignment engineers can usually only conduct detailed research on a small number of alternatives, thereby neglecting some potentially valuable alignment alternatives. This may affect very adversely the subsequent structural design, construction, operation, and maintenance.

To further solve the above problems, alignment optimization has become a cutting-edge issue and a research hotspot in the field, attracting the attention of a large number of researchers worldwide.

1.2 Railway alignment optimization

Artificial intelligence (AI) is the study of basic theories, methods, and technologies for applying computers to simulate certain human intelligent behaviors. At the Dartmouth Conference in 1956, John McCarthy et al. [1] proposed the concept of "artificial intelligence" for the first time, marking the birth of AI. AI is mainly classified into three major approaches: symbolism represented by expert systems and knowledge engineering; connectionism represented by artificial neural networks and deep learning; and evolutionism represented by bionic intelligent algorithms (e.g., genetic algorithms, particle swarm optimization algorithms). After more than 60 years of development, AI algorithms have been playing an increasingly important role, indicating that intelligence is an inevitable trend in the development of various fields in the future.

Alignment design requires, under specific natural, economic, and social environments, designing 3D spatial alignments that meet various constraints and are safe, economical, comfortable, and environmentally friendly. Meanwhile, the alignment must be coordinated with the layouts of structures such as bridges, tunnels, and stations. Essentially, alignment design is an optimization problem. The introduction of intelligent algorithms into the field of alignment optimization can be traced back to the 1960s.

The solution idea for alignment optimization problem can be roughly summarized as follows: an alignment optimization model is constructed by combining optimization algorithms with inputs on environmental factors such as terrain, land use, geology, and ecology. The optimization model includes: (1) a set of independent variables mainly based on spatial alignment control point information, (2) an optimization objective function that comprehensively evaluates engineering costs, and possibly other costs, e.g., for operations, maintenance, environmental impacts, safety and user time, as well as (3) constraint conditions composed of design specifications (based on safety and other operational requirements), resource availability (including funding) and environmental factors. Then, artificial intelligence or mathematical algorithms are used to solve this optimization model, generate alignment alternatives based on intelligent search strategies, and finally evaluate and select the best alternatives according to the objective function. With the continuous improvement of researchers' cognitive level and the advancement of computer technology, alignment optimization has developed from 2D alignment horizontal or longitudinal optimization to 3D spatial overall optimization of horizontal and longitudinal alignment.

The optimization models have expanded from early single-objective optimization focusing on construction costs to multi-objective optimization integrating construction costs, operating costs, environmental protection, and other factors. The search algorithms have gradually developed from early analytical mathematical methods to intelligent heuristic searches, with particle swarm optimization and genetic algorithms being prominent representatives. The research focus has also shifted from early artificially simulated numerical cases to actual flat and hilly terrain cases, and in recent years, active exploration has been carried out on alignment optimization problems in complex environments.

1.2.1 Alignment optimization model

In the field of railway alignment optimization, optimization models usually adopt means such as mathematical formulas to describe and characterize practical alignment engineering problems. Most railway alignment optimization follows the research idea of first establishing an optimization model and then solving it automatically. According to the number of objectives considered in the optimization model, alignment optimization may be categorized into single-objective and multi-objective studies. In single-objective alignment optimization, a cost-related function is the most fundamental kind of objective.

(1) Single-objective optimization model

Engineering construction investment is particularly important for each alignment design problem. For any highway or railway project, the cost must be reasonably controlled during the alignment design. So far, optimization research on cost investment has become relatively mature, and various cost-related objective functions have been proposed, including:

- Construction cost, such as the bridge and tunnel costs, intersection and interchange costs, earthwork costs, right-of-way costs, and other length-dependent costs (e.g., track engineering volume and power facilities).
- Operation cost, such as energy consumption costs, travel time costs, fuel and electricity costs, as well as operation management costs.
- Maintenance cost, such as that for structures maintenance (i.e., tracks, subgrades, bridges and tunnels), daily maintenance, and preventive maintenance.
- Penalty cost, which is mainly incurred on alignment sections traversing environmentally or historically sensitive regions.

(2) Multi-objective optimization model

In the actual alignment design process, apart from ensuring economic benefits during the construction and operation phases, it is more important to guarantee the safety of construction workers, operation and maintenance personnel, and passengers. Moreover, with the development and progress of society, people are paying increasing attention to ecological environment protection. Therefore, during the alignment design process, it is necessary to minimize the impact on the ecological environment as much as possible. However, objectives such as economy, safety, and ecology often conflict mutually. Thus, determining how to comprehensively consider, resolve, and balance the conflicts among various objectives and selecting the optimal alignment alternative is a key and difficult issue in the construction of multi-objective optimization models.

The degree of safety along alignments has long been a major concern of researchers. Initially, this factor was treated as an accident cost due to alignment characteristics and then combined into a total cost function. In later studies, it has become an independent objective. For example, Easa and Mehmood [2] constructed a collision frequency model for horizontal alignment optimization considering the geometric characteristic of highways. Then, the traffic accident rate [3] was considered by using a safety performance function proposed by AASTHO for highway alignment safety assessment. A safety performance function based on a so-called inertial speed method has also been applied [4] specifically for tunnel alignment design.

Besides, structural stability is also an important engineering safety factor for transportation infrastructures. In this regard, subgrade slope stability is the most frequently discussed topic for earthwork sections [5,6]. For such studies, the subgrade slope gradient and aspect are generally combined with alignment elevation and direction to estimate the stability of cuts and fills.

Moreover, geo-hazards are a major hidden danger affecting the safety of alignment construction and operation. When a geo-hazard occurs along a railway, it often causes significant losses of life and property. Therefore, in the alignment design stage, geo-hazards are also an important research aspect in safety evaluation, which have drawn attention from researchers at Central South University in recent years. The alignment optimization team of Central South University (hereinafter referred to as CSU AO team) [7] proposed a geo-hazard evaluation model of existing landslides for mountain railway alignment optimization by combining landslide susceptibility and structural vulnerability analyses. Meanwhile, the

Multi-criteria Tournament Decision method (MTD) with the Crowded Distance Calculation (CDC) were applied to conduct 3D optimization of horizontal and vertical alignments. Then, the model was [8] extended by considering potential landslide hazards in regions with no existing landslides. In addition, a probabilistic seismic risk model [9] was developed by integrating probabilistic seismic hazard, structural fragility and loss analyses for mountain railway alignment optimization. It should be noted that geological risks are fairly complicated but crucial in the optimization process. To this end, several related studies are still being conducted at Central South University [10,11].

Alignment design is highly intertwined with ecological influencing factors. Especially with the currently changing climate, ecological concern has become an increasingly important aspect for transportation planning and design to pursue a sustainable development. In this field, a region-based ecological impact analysis has been widely applied. As shown in Fig. 1.2, based on the GIS model, the areas of alignment-traversed noise sensitive regions, floodplains and wetlands, habitats as well as water source protection regions can be readily computed by accumulating GIS grids. This is a simple

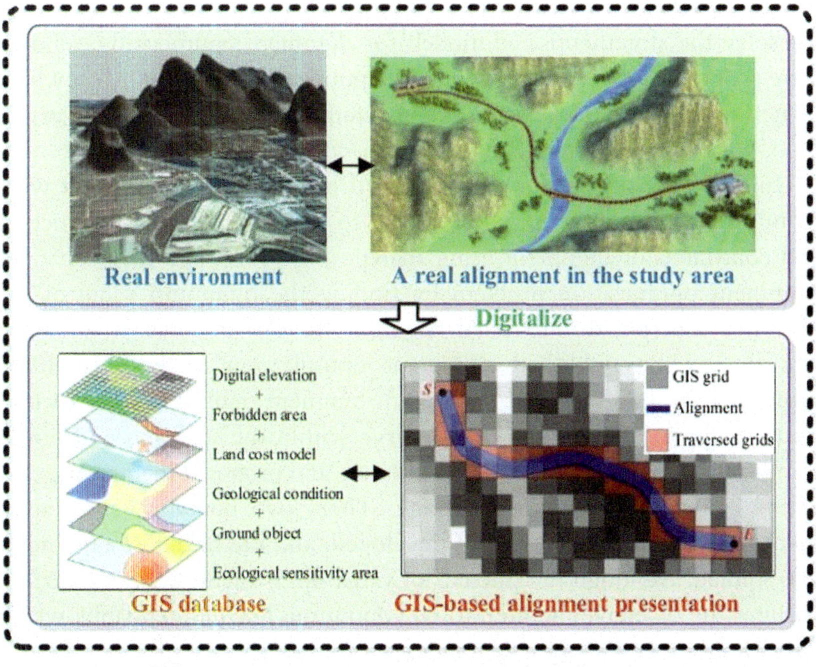

Figure 1.2 Grid-based GIS for alignment optimization.

but effective approach for estimating the effects of alignments within ecologically sensitive regions.

In addition, several specific ecological evaluation models have also been proposed. For example, a tri-level highway alignment optimization framework [12] was proposed. In the upper and lower levels, construction cost and traffic flow were optimized. In the intermediate level, environmental emissions related to highway alignment designs, such as air, noise and water pollution, were formulated. In another study, the coupled relations between road alignment location and animal migration [13] were considered. The model predicted different possible animal death rates based on dissimilar alignment alternatives. The CSU AO team [14] developed formulations for computing vegetation destruction and soil erosion within a railway alignment optimization. Most recently, a model for predicting carbon emission of earthwork constructions during a highway alignment design process [15] was developed. Given significantly increasing concerns for ecological protection in various countries, it seems likely that more ecological alignment optimization studies will be published in the future.

1.2.2 Alignment optimization methods

To solve the aforementioned models for alignment optimization, scholars have proposed a series of representative optimization methods. Meanwhile, related research has long been a popular and cutting-edge direction in this field, with over 20 different algorithms having been proposed.

Initially, the search algorithms applied in the field of alignment were mainly analytical methods. The idea was to express the objective function and constraints of alignment using analytical formulas, and then solve the alignment optimization problem by finding the minimum value of the objective function through analytic methods. The calculus of variations was the earliest published alignment optimization method. In 1968, Howard et al. [16] regarded horizontal alignment optimization as a variational problem of finding a spatial curve within the start and end points to minimize the integral value of a function, and proposed the optimal curvature principle for alignment search. Then, two numerical integration methods based on this were [17] developed, and the fitting of alignments was studied. Although the calculus of variations is theoretically optimal, it requires the objective function to be continuous and differentiable, which rarely holds in real-world alignment environments.

A numerical search algorithm is another typical algorithm. In 1970, a numerical search algorithm [18] was proposed to solve the vertical alignment optimization problem. In 1988, dynamic programming method and numerical search algorithm [19] were proposed for discrete and continuous search spaces. Subsequently, based on this work, an optimization model [20] was devised for 3D alignment in 1989.

Although analytical methods can theoretically yield the optimal solution of the objective function, they usually require the function to be continuous and differentiable. However, the real alignment environment and constraint conditions can hardly satisfy such requirements, which require the introduction of simplifying assumptions during the solution process. As a result, the obtained optimization results may reflect insufficiently the actual situation. Therefore, heuristic algorithms have begun to interest researchers.

Heuristic algorithms are a type of algorithm constructed based on intuition or experience. They can provide an approximately optimal feasible solution to a combinatorial optimization problem within an acceptable computation time and space. The degree of deviation between this feasible solution and the optimal solution is generally unpredictable. At this stage, heuristic algorithms are mainly nature-inspired, with typical ones including ant colony algorithms, genetic algorithms, simulated annealing algorithms, and neural networks. Currently, various heuristic algorithms and shortest path algorithms have been applied in the field of alignment design, among which the representative ones include grid optimization, dynamic programming, distance transform, mixed-integer programming, genetic algorithm, swarm intelligence, and mesh adaptive direct search algorithms.

The grid optimization method involves dividing the entire study area into a series of grids, each assigned with different geographical information attributes and cost information. It uses mature methods such as a shortest path algorithm to search for the optimal grid connection linking the start and end points, and then forms the final alignment by connecting the grids with arcs. This method was first proposed in 1971 [21] and applied to horizontal alignment optimization. Later, many scholars made improvements to this method. However, the alignment generated by the grid optimization method consists of a series of interconnected arcs, which is inaccurate in describing a horizontal alignment which must have smooth curves. Moreover, for longer alignments, enormous space is required to store grid data.

The dynamic programming method was proposed in the early 1950s by American mathematician R. E. Bellman [22] when studying the optimization problem of multistep decision processes. In 1973, the dynamic programming method was introduced into alignment optimization, when it was used to optimize vertical alignments with the goal of minimizing earthwork engineering costs [23]. Since then, researchers have made multiple improvements to this method, but it cannot guarantee the optimality of the alignment generated by fitting the optimal polygonal path. Moreover, the search range is limited to the nodes of each divided path stage, making it impossible to search in continuous space. In response to this, CSU AO team proposed a 3D dynamic programming method [24]. It divides the entire search space into a series of 3D grids in both horizontal and vertical directions, takes the vertices of the 3D grids as nodes, uses dynamic programming to generate 3D corridor lines, and finally uses the direction acceleration method to fit an alignment alternative that is closest to the optimal corridor line, which frees the alignment from fixed grid points. However, due to the huge amount of grid data and excessively long computation time, it is not suitable for railway alignment design in large spaces.

In 2006, the neighborhood heuristic search was first used to optimize horizontal alignment with the goal of minimizing engineering construction costs [25], followed by vertical alignment optimization based on mixed-integer programming. This method can generate horizontal alignments that satisfy horizontal curve constraints and incorporate transition curves. However, it considers the horizontal and vertical alignments separately rather than as an integrated optimization problem. In addition, to improve the accuracy and reliability of existing mixed-integer linear programming models, a new multi-haul quasi network flow model (MH-QNF) was proposed for vertical alignment optimization [26]. Experiments have shown that this method has high accuracy and robustness.

A Genetic Algorithm (GA) is a meta-heuristic algorithm based on the survival of the fittest rationale. The first GA for alignment optimization problems was pioneered in a Ph.D. dissertation [27] by Jong and advanced through critical collaborations with Schonfeld [28], whose contributions have laid the foundation of alignment design using GA. Schonfeld's team at the University of Maryland introduced an integration of GAs with Geographic Information Systems (GIS) for preliminary highway design, enabling the optimization of horizontal alignments by directly leveraging GIS databases to handle irregular geographic spaces, land-acquisition costs, environmental impacts, length-dependent costs, and user costs. This method addressed limitations of early mathematical methods that failed to handle non-smooth functions, minimum radius constraints, or explicit user costs [29]. Moreover, that team further developed comprehensive cost functions with detailed formulations for fuel consumption, travel time, and accident costs tied to alignment geometry and traffic conditions, which ensured realistic cost modeling. Based on these, Schonfeld' s team developed methods for 3D alignments optimization, devised a method allowing for backtracking to fit complex terrain and satisfying various design constraints (e.g., maximum gradient, minimum curve length). That team also promoted the integration of GAs with visualization techniques, enhancing stakeholder engagement and decision-making by visualizing alignment impacts [30]. Additionally, that team addressed intersection modeling, developing methods for optimizing intersections within the GA framework [31]. These efforts established GAs as the dominant approach in highway alignment optimization, with the models validated through various real-world case studies. By using a GA, alignment optimization problems could be formulated more realistically and solved more effectively compared to early mathematical methods. This has inspired researchers in developing further GA-related methods. Except for some GAs which employ specifically-designed solution search procedures, the first step of most GAs in this field is to set a series of cutting planes (CPs) along the straight line between two endpoints. These CPs are usually orthogonal to the straight line and distributed uniformly. Then, PIs can be generated on these CPs and, hence, alignments can be determined. Afterward, specific genetic operators are used to evolve the generated alignments by updating their PIs' locations on CPs. As shown in relevant summaries, up to now, GAs have become the most frequently applied algorithms in this field.

The PSO's rationale is the social communication mechanisms of insects, birds and fish. A representative PSO customized for alignment optimization [32] was proposed as an alternative of GA. Inspired by that research, the CSU AO team [33] proposed a hybrid PSO and GA method to combine the strengths of both algorithms for mountain railway alignment optimization. Specifically, a PSO formulation was developed for continuously searching 3D alignments; it was then refined with customized genetic operators to improve algorithm robustness. In recent years, the PSO-based heuristic method has been successively applied to more studies, such as multi-objective alignment optimization [14] as well as concurrent optimization of alignments and station locations [34].

However, in complex mountainous regions, alignments are usually very circuitous in order to overcome the drastically undulating terrain. In these conditions, the CPs are difficult to set for alignments. To solve this problem, DT-based algorithms have drawn researchers' interests. DT is a shortest path algorithm first introduced to alignment optimization by de Smith [35]. Its basic idea is to first scan every GIS grid to find the best local path within a neighboring mask, and then propagate as well as accumulate all the local paths to generate the best global path to the target point. Considering the complex constraints in a mountain railway design, the CSU AO team [36] developed a bi-directional scanning strategy for DT by defining the start and end points as two targets. The search spaces of this method have been enlarged to 3D spaces [37] to form a 3D-DT, which was then parallelized [38] using OpenMP library to speed up the search. It should be noted that DT-produced paths should be further refined into railway or road alignments. Thus, DT paths, which are piecewise linear, are generally deemed as centerlines of transportation corridors. Based on these corridors, a large study area can be narrowed and detailed alignment search can then be performed with other methods, such as GA [39] and PSO [40].

Additionally, numerous other alignment optimization methods have been proposed to address various different alignment design problems.

In 2015, an alignment optimization model was developed with the coordinates of horizontal intersection points and circular curve radii as independent variables, and the cost of cut and fill works as the objective function. The mesh adaptive direct search algorithm was also introduced into the field of alignment optimization, realizing the horizontal alignments optimization within a given alignment corridor [41]. However, this method has the disadvantage that a reasonable initial alignment position needs to be preset, and the initial alignment position has a significant impact on the optimization results. Therefore, this algorithm is suitable for further optimization of existing alignments.

Monte Carlo Tree Search (MCTS) also shows promise in alignment optimization. The CSU AO team improved a terrain-driven approach with MCTS for vertical railway alignments but fixed horizontal alignments [42]. Then, it also proposed 3D-MCTS [43], customizing selection with time-varying coefficient decay, expanding via dynamic feasible spaces, and simulating with global reward estimation. This method performs with lower costs and higher efficiency than DT, balancing exploration and exploitation effectively.

Faced with the complex terrain in mountainous regions, existing methods sometimes still struggle to adapt to the intricate topographical environment. To address this, several approaches based on environmental analysis have been proposed. Firstly, through the 3D environmental suitability analysis model, the study area is abstracted as a set of voxels. It integrates topographic factors (cost, stability) and geological factors (landslides, seismic impacts), and combines the CRITIC multi-criteria decision-making method to generate a suitability map, which provides a preliminary screening basis for alignment selection and significantly improves search efficiency [44]. For densely distributed obstacles and complex terrain, the 3D-RRT-star algorithm [45] rapidly generates obstacle-avoiding paths through heuristic sampling and dynamic constraint handling, achieving a 15–43 times efficiency improvement compared with traditional grid methods. For regions with the "isolated island effect" caused by terrain interception, a semantic topological map method [46] is proposed, which identifies favorable regions using DBSCAN clustering and simplifies path search by combining graph theory algorithms. In addition, the event tree-based distance transform algorithm [47] simultaneously optimizes alignment and station locations, increasing the diversity of solutions through multiple types of search operators. These methods break through the limitations of traditional alignment selection from the dimensions of environmental adaptation, obstacle handling, and collaborative optimization, providing new solutions for alignment optimization.

With the rapid advancement of computer science, a multitude of optimization algorithms within the domain of artificial intelligence (AI) have been successively proposed. These algorithms not only lay a solid theoretical and technical foundation for the intelligent evolution of modern society but also create favorable opportunities for the advancement of intelligent alignment design. In 2000, Yi et al. introduced the knowledge engineering of artificial intelligence into railway alignment design. They conducted systematic research on knowledge in the field of railway alignment, proposed an object-oriented rule-based knowledge representation model, and realized effective and flexible expression of railway alignment knowledge. Based on ontology theory, the rapid acquisition and utilization of various design data, e.g., about terrain, geology, hydrology, and human geography in the study area, was used to develop an auxiliary design system for railway alignment design, and the research results have been applied in multiple railway projects of the China Railway Eryuan Engineering Group Co., Ltd. In order to flexibly apply knowledge in alignment optimization, the CSU AO team proposed the a railway alignment design knowledge graph (RAD-KG) [48], which used a hierarchical semantic network with <condition-relation-treatment > triples to enable structured knowledge representation and efficient querying. For karst regions, it also developed a knowledge graph for alignment design (KaRAD-KG) [49] integrating hypergraph semantics to handle high-coupling relations, which decreased karst hazards by 18.73%.

Motion planning algorithms originate from the field of robotics and are used to find feasible paths for robots between given start and end points. Sushma et al. [50] made appropriate modifications to these algorithms, enabling their successful application to the horizontal alignment design of expressways. This method does not require optimization based on any initial alignment; instead, it effectively uses low-discrepancy point samples to explore the search space and generate the optimal horizontal alignment.

Deep reinforcement learning is a machine learning method, first proposed in 2013 [51] in which an agent learns how to perform actions through interaction with the environment. In 2021, a deep reinforcement learning [52] was introduced into the field of alignment optimization. By leveraging self-learning from human experience to guide the alignment optimization process, the automatic exploration and optimization of railway alignments have been initially realized. Moreover, the CSU AO team combined reinforcement learning with Monte Carlo tree search for railway optimization, improving autonomous decision-making in complex scenarios [53]. Meanwhile, deep learning was also introduced in alignment design. A voting framework for efficient railway major technical standards decision-making was proposed [54]. After that, a Transformer-driven method for alignment optimization in complex mountainous regions was devised, fusing multi-dimensional environmental data to enhance adaptability [55].

In summary, alignment optimization has long been a hot, difficult, and cutting-edge issue in the fields of transportation and civil engineering. Over the course of more than half a century of evolution, alignment optimization methods have undergone continuous innovation. This trajectory has yielded substantial methodological breakthroughs and garnered extensive practical applications. Notably, research objects have exhibited an increasingly diversified trend, optimization models have become more comprehensive, search algorithms have grown more problem-centric, and the problems addressed have become more complex while remaining closely aligned with engineering practice.

1.3 Main contents and features of this book

The CSU AO team has been conducting long-term and in-depth research on key digital alignment design technologies such as intelligent alignment, alignment CAD, 3D GIS, and visualization since the late 1970s. As one of the earliest research teams in China to engage in studies in this field, it has developed the digital alignment design software system that currently enjoys the widest application and the most comprehensive product portfolio in China's railway industry.

This book encapsulates the team's research heritage and academic accumulation spanning over 40 years. It represents the achievements attained through in-depth collaborations with numerous international experts in the field of alignment, such as Professors Paul Schonfeld and Harry Teng, as well as multifaceted development and research on new intelligent alignment technologies, including a comprehensive geographic information model [44], a railway alignment design knowledge graph [48], multi-task deep learning methods for determining railway major technical standards [54], a construction and operation cost optimization model for railway alignment [56], an alignment optimization model considering geological disaster risk assessment [7], an environmental model for railway alignment optimization [14,57], railway alignment corridor search methods [36], searching methods for detailed alignment alternatives [33,39], models and methods for the collaborative optimization of infrastructure structures and railway alignments [58], intelligent reconstruction optimization of existing alignments [59,60], BIM for railway alignment and station [34,47,61], as well as system development and applications. These achievements have addressed a multitude of key technical issues in alignment optimization.

The purpose of compiling this book is to summarize the theories and methods proposed in our existing research findings, facilitate exchanges with peers both domestically and internationally, with a view to further advancement.

This book is divided into 14 chapters, whose contents are presented in Table 1.1.

Table 1.1 Chapter summary.

Index	Title	Content
Chapter 1	Introduction	Introduces the background of its compilation and the main research contents; provides a detailed summary and review of the existing research achievements.
Chapter 2	Comprehensive geographic information model	Presents the alignment search space representation method, the construction of 3D environmental suitability analysis model, and alignment adaptability optimization under multi-modal environmental distribution characteristics.
Chapter 3	Railway alignment design knowledge graph	Introduces the modeling of railway alignment design knowledge graph and methods for rapid knowledge retrieval.
Chapter 4	Multi-task deep learning methods for determining railway major technical standards	Proposes a multi-task deep learning method for determining railway technical standards with a voting framework.
Chapter 5	Construction and operation cost optimization model for railway alignment	Demonstrates the development of the economic alignment optimization model considering construction and operation cost, which

Table 1.1 Chapter summary.—cont'd

Index	Title	Content
		consists of decision variables, constraints, and objective functions.
Chapter 6	Alignment optimization model considering geological disaster risk assessment	Introduces the development of geo-hazard optimization model for railway alignment, including the classification of typical geo-hazards, geo-hazards evaluation methods, spatial geo-hazard assessment, and knowledge-driven karst hazard assessment method.
Chapter 7	Environmental model for railway alignment optimization	Develops a dual-objective optimization model with traditional environmental protection indicators and cost, or carbon emission indicators and cost.
Chapter 8	Railway alignment corridor search methods	Elaborates on three railway corridor search methods: Presetting of dominating structures (i.e., high bridges and long tunnels), distance transformation, and Monte Carlo tree search.
Chapter 9	Searching methods for detailed alignment alternatives	Based on the generated railway corridor, introduces methods for further optimizing

(*Continued*)

Table 1.1 Chapter summary.—cont'd

Index	Title	Content
		railway alignment within a corridor, including improved genetic algorithm and particle swarm optimization with hybrid genetic operators.
Chapter 10	Models and methods for the collaborative optimization of infrastructure structures and railway alignments	Introduces several collaborative optimization models and methods for new railway-related structures: Concurrent optimization of railway alignments and station locations, railway alignments and large-scale auxiliary construction projects, and optimization of grade-separated road and railway crossings.
Chapter 11	Intelligent recreation optimization of existing Railways	Introduces several existing line reconstruction optimization models and methods step by step, including optimization model for horizontal and vertical alignment reconstruction of existing lines, overall oscillating iterative reconstruction and optimization method, adaptive reconstruction and optimization method,

Table 1.1 Chapter summary.—cont'd

Index	Title	Content
		and alignment reconstruction method for existing subways.
Chapter 12	Building information modeling (BIM) for railway alignment and station	Presents the 3D visualization methods for optimized alignment alternative and BIM design of railway stations.
Chapter 13	System development and applications	Introduces the main functions of railway alignment optimization software.
Chapter 14	Research prospects	Discusses the possible development directions of future intelligent alignment optimization technologies at the service stage.

References

[1] J. McCarthy, M.L. Minsky, N. Rochester, et al., A proposal for the dartmouth summer research project on artificial intelligence: August 31, 1955, AI Mag. 27 (2006) 12–14.

[2] S.M. Easa, A. Mehmood, Optimizing design of highway horizontal alignments: New substantive safety approach, Comput. Aided Civ. Infrastruct. Eng. 23 (7) (2008) 560–573.

[3] C. Li, L. Ding, B. Zhong, Highway planning and design in the qinghai-tibet plateau of China: A cost-safety balance perspective, Eng. 5 (3) (2019) 337–349.

[4] Z. Zhang, L. Lu, L. Zhu, et al., Preference information incorporation for decision making in the bi-objective alignment optimization problem of river-crossing tunnels, J. Transp. Eng., Part A: Syst. 146 (10) (2020) 04020121.

[5] S. Sadek, M. Bedran, I. Kaysi, GIS platform for multicriteria evaluation of route alignments, J. Transport. Eng. 125 (2) (1999) 144–151.

[6] M. Karlson, C.S.J. Karlsson, U. Mörtberg, et al., Design and evaluation of railway corridors based on spatial ecological and geological criteria, Transport. Res. Transport Environ. 46 (2016).

[7] T. Song, H. Pu, P. Schonfeld, et al., Mountain railway alignment optimization considering geological impacts: A cost-hazard bi-objective model, Comput. Aided Civ. Infrastruct. Eng. 35 (12) (2020) 1365–1386.

[8] H. Pu, J. Xie, P. Schonfeld, et al., Railway alignment optimization in mountainous regions considering spatial geological hazards: A sustainable safety perspective, Sustainability 13 (4) (2021) 1661.
[9] T. Song, H. Pu, P. Schonfeld, et al., Bi-objective mountain railway alignment optimization incorporating seismic risk assessment, Comput. Aided Civ. Infrastruct. Eng. 36 (2) (2020) 143–163.
[10] T. Song, H. Pu, P. Schonfeld, et al., Robust optimization method for mountain railway alignments considering preference uncertainty for costs and seismic risks, ASCE-ASME J. Risk Uncertain. Eng. Syst., Part A Civil Eng. 8 (1) (2022) 04021077.
[11] T. Song, H. Pu, T.Y. Yang, et al., Optimizing net present values of risk avoidance for mountain railway alignments with seismic performance evaluation, Comput. Aided Civ. Infrastruct. Eng. 39 (7) (2023) 944–962.
[12] S. Mishra, M.-W. Kang, M.K. Jha, Empirical model with environmental considerations in highway alignment optimization, J. Infrastruct. Syst. 20 (2014).
[13] N. Davey, S. Dunstall, S.K. Halgamuge, Optimal road design through ecologically sensitive areas considering animal migration dynamics, Transport. Res. C Emerg. Technol. 77 (2017) 478–494.
[14] H. Zhang, H. Pu, P. Schonfeld, et al., Multi-objective railway alignment optimization considering costs and environmental impacts, Appl. Soft Comput. 89 (2020) 106105.
[15] Y. Gao, T. Gao, Y. Wu, et al., Low-construction-emission cross-section optimization for mountainous highway alignment designs, Transport. Res. Transport Environ. 105 (Suppl. C) (2022) 103249.
[16] B.E. Howard, Z. Bramnick, J.F.B. Shaw, Optimum curvature principle in highway routing, J. Highw. Div. 94 (1) (1968) 61–82.
[17] F.B. Shaw, B.E. Howard, Expressway route optimization by OCP, Transp. Eng. J. 108 (3) (1982) 227–243.
[18] R.W. Hayman, Optimization of vertical alignment for highways through mathematical programming, Highw. Res. Rec. (1970) 1–9.
[19] C.J. Goh, E.P. Chew, T.F. Fwa, Discrete and continuous models for computation of optimal vertical highway alignment, Transp. Res. Part B Methodol. 22 (6) (1988) 399–409.
[20] E.P. Chew, C.J. Goh, T.F. Fwa, Simultaneous optimization of horizontal and vertical alignments for highways, Transp. Res. Part B Methodol. 23 (5) (1989) 315–329.
[21] A.K. Turner, R.D. Miles, A computer-assisted method of regional route location, Highw. Res. Rec. (1971) 1–15.
[22] R.E. Bellman, Dynamic Programming, Princeton University Press, Princeton, NJ, 2010.
[23] J.D. Murchland, Methods of vertical profile optimization for an improvement to an existing road, in: PTRC Seminar Proceedings on Cost Models and Optimization in Highways, 1973.
[24] W. Li, H. Pu, H. Zhao, et al., Approach for optimizing 3d highway alignments based on two-stage dynamic programming, J. Softw. 8 (11) (2013) 2967–2973.
[25] J.-F. Cheng, Y. Lee, Model for three-dimensional highway alignment, J. Transport. Eng. 132 (12) (2006) 913–920.
[26] V. Beiranvand, W. Hare, Y. Lucet, et al., Multi-haul quasi network flow model for vertical alignment optimization, Eng. Optim. 49 (10) (2017) 1777–1795.
[27] J.-C. Jong, Optimizing Highway Alignments with Genetic Algorithms, University of Maryland, College Park, MD, 1998.
[28] J.-C. Jong, P. Schonfeld, An evolutionary model for simultaneously optimizing three-dimensional highway alignments, Transp. Res. Part B 37 (2) (2003) 107–128.
[29] M.K. Jha, P. Schonfeld, A highway alignment optimization model using geographic information systems, Transport. Res. Pol. Pract. 38 (6) (2004) 455–481.

[30] E. Kim, M.K. Jha, D.J. Lovell, et al., Intersection modeling for highway alignment optimization, computer-aided civil and infrastructure, Eng. 19 (2) (2004) 119–129.
[31] M.K. Jha, C. McCall, P. Schonfeld, Using GIS, genetic algorithms, and visualization in highway development, Comput. Aided Civ. Infrastruct. Eng. 16 (6) (2001) 399–414.
[32] Y. Shafahi, M. Bagherian, A customized particle swarm method to solve highway alignment optimization problem, Comput. Aided Civ. Infrastruct. Eng. 28 (1) (2013) 52–67.
[33] H. Pu, T. Song, P. Schonfeld, et al., Mountain railway alignment optimization using stepwise & hybrid particle swarm optimization incorporating genetic operators, Appl. Soft Comput. 78 (C) (2019) 41–57.
[34] T. Song, H. Pu, P. Schonfeld, et al., Simultaneous optimization of 3D alignments and station locations for dedicated high-speed railways, Comput. Aided Civ. Infrastruct. Eng. 37 (4) (2021) 405–426.
[35] M.J. de Smith, Determination of gradient and curvature constrained optimal paths, Comput. Aided Civ. Infrastruct. Eng. 21 (1) (2006) 24–38.
[36] W. Li, H. Pu, P. Schonfeld, et al., Methodology for optimizing constrained 3-dimensional railway alignments in mountainous terrain, Transport. Res. C Emerg. Technol. 68 (2016) 549–565.
[37] H. Pu, T. Song, P. Schonfeld, et al., A three-dimensional distance transform for optimizing constrained mountain railway alignments, Comput. Aided Civ. Infrastruct. Eng. 34 (11) (2019) 972–990.
[38] T. Song, H. Pu, P. Schonfeld, et al., Parallel three-dimensional distance transform for railway alignment optimization using openMP, J. Transport. Eng., Part A: Systems 146 (5) (2020) 04020029.
[39] W. Li, H. Pu, P. Schonfeld, et al., Mountain railway alignment optimization with bidirectional distance transform and genetic algorithm, Comput. Aided Civ. Infrastruct. Eng. 32 (8) (2017) 691–709.
[40] T. Song, H. Pu, P. Schonfeld, et al., Mountain railway alignment optimization integrating layouts of large-scale auxiliary construction projects, Comput. Aided Civ. Infrastruct. Eng. 38 (4) (2022) 433–453.
[41] S. Mondal, Y. Lucet, W. Hare, Optimizing horizontal alignment of roads in a specified corridor, Comput. Oper. Res. 64 (0) (2015) 130–138.
[42] H. Zhang, T. Song, P. Schonfeld, et al., Vertical alignment optimization of mountain railways with terrain-driven greedy algorithm improved by monte carlo tree search, Comput. Aided Civ. Infrastruct. Eng. 38 (7) (2023) 873–891.
[43] G. Wang, H. Pu, T. Song, et al., A 3D monte carlo tree search method for railway alignment optimization, Appl. Soft Comput. 151 (2024) 111158.
[44] H. Pu, X. Wan, T. Song, et al., A geographic information model for 3-D environmental suitability analysis in railway alignment optimization, Integrated Comput. Aided Eng. 30 (1) (2022) 67–88.
[45] H. Pu, X. Wan, T. Song, et al., A 3D-RRT-star algorithm for optimizing constrained mountain railway alignments, Eng. Appl. Artif. Intell. 130 (2024) 107770.
[46] X. Wan, H. Pu, P. Schonfeld, et al., Railway alignment optimization in regions with densely-distributed obstacles based on semantic topological maps, Integrated Comput. Aided Eng. 31 (4) (2024) 421–437.
[47] X. Wan, H. Pu, T. Song, et al., An event tree-based distance transform algorithm for simultaneously determining mountain railway alignments and station locations, Expert Syst. Appl. 261 (2025) 125575.
[48] H. Pu, T. Hu, T. Song, et al., Modeling and application of a customized knowledge graph for railway alignment optimization, Expert Syst. Appl. 244 (2024) 122999.

[49] H. Pu, T. Hu, T. Song, et al., Knowledge graph-driven mountain railway alignment optimization integrating karst hazard assessment, Appl. Soft Comput. 167 (Part B) (2024) 112421.
[50] M.B. Sushma, A. Maji, A modified motion planning algorithm for horizontal highway alignment development, Comput. Aided Civ. Infrastruct. Eng. 35 (8) (2020) 818–831.
[51] V. Mnih, K. Kavukcuoglu, D. Silver, et al., Playing atari with deep reinforcement learning, 2013 arXiv preprint arXiv:1312.5602.
[52] T. Gao, Z. Li, Y. Gao, et al., A deep reinforcement learning approach to mountain railway alignment optimization, Comput. Aided Civ. Infrastruct. Eng. 37 (1) (2021) 73–92.
[53] G. Wang, H. Pu, T. Song, et al., Integrated method based on reinforcement learning and monte carlo tree search for railway alignment optimization, J. China Railw. 47 (5) (2025) 102–110, https://doi.org/10.3969/j.issn.1001-8360.2025.05.011.
[54] H. Pu, G. Wang, T. Song, et al., Multi-task deep learning methods for determining railway major technical standards, IEEE Trans. Intell. Transport. Syst. 25 (6) (2024) 5904–5918.
[55] G. Wang, H. Pu, T. Song, et al., A transformer-driven intelligent optimization method for railway alignment in mountainous multi-dimensional environments, J. China Railw. Sci. Eng. 22 (10) (2025) 4476–4488.
[56] H. Zhang, H. Pu, P. Schonfeld, et al., Railway alignment optimization considering life-cycle costs, IEEE Intell. Transp. Syst. Mag. 14 (5) (2021) 22–40.
[57] H. Pu, L. Cai, T. Song, et al., Minimizing costs and carbon emissions in railway alignment optimization: A bi-objective model, Transport. Res. Transport Environ. 116 (Suppl. C) (2023) 103615.
[58] H. Pu, Z. Liang, P. Schonfeld, et al., Optimization of grade-separated road and railway crossings based on a distance transform algorithm, Eng. Optim. 54 (2) (2021) 232–251.
[59] W. Li, H. Pu, P. Schonfeld, et al., A method for automatically recreating the horizontal alignment geometry of existing railways, Comput. Aided Civ. Infrastruct. Eng. 34 (1) (2019) 71–94.
[60] H. Pu, H. Fu, T. Song, et al., Recreating horizontal alignments of existing railways with a hybrid analytic and harmony search algorithm, Eng. Appl. Artif. Intell. 128 (2023) 107354.
[61] X. Fan, H. Pu, P. Schonfeld, et al., Microprism-based layered BIM modeling for railway station subgrade, Autom. Construct. 166 (2024) 105652.

CHAPTER TWO

Comprehensive geographic information model

Abbreviations and corresponding full texts

CGIM	Comprehensive Geographic Information Model
CRITIC	Criteria Importance Through Intercriteria Correlation
DEM	Digital Elevation Model
GIS	Geographic Information System
NDVI	Normalized Difference Vegetation Index

The comprehensive environmental data, such as on topography, geology, and ecology, within the study area, serve as the fundamental design basis for alignment design. In traditional manual alignment design, the interpretation of geographic information such as terrain and geological conditions relies heavily on the knowledge and experience of the designers. Therefore, the first step in intelligent alignment design is to "teach" the computer to "understand" various maps, such as topographic, geological, land use, and socio-economic maps, enabling it to automatically extract and organize comprehensive geographic data. To this end, Geographic Information Systems (GIS's) have been applied to intelligent alignment for over 30 years [1], providing powerful, efficient, and flexible tools for storing and managing the vast amounts of data required during the alignment optimization process.

From the late 20th century to the early 21st century, GIS models were primarily used to support cost estimation during the alignment optimization process, such as for estimating earthwork volumes [2]. During this period, the Digital Elevation Model (DEM) gradually evolved into a mature technology for characterizing the topographical features of the study area in alignment research.

2.1 Search space representation

It is desirable that a model for optimizing railway alignment should exploit a GIS database because massive amounts of spatial data such as on topography, soil conditions, forbidden areas, unit costs of earthwork, bridges, tunnels and right of way, are essential for railway alignment design.

Intelligent Railway Alignment Design
ISBN: 978-0-443-45923-8
https://doi.org/10.1016/B978-0-443-45923-8.00013-3

A Comprehensive Geographic Information Model (CGIM) is designed to store and manage these data [3]. Topography is often represented using a regular rectangular lattice of points or cells for which elevation values are provided. The data sets of this form are known as digital elevation model (DEM) data. Similarly to the DEM, the data structure of CGIM is also a regular rectangular lattice and has the same dimensions. All the needed geographic information is stored in the cells. Fig. 2.1 shows the CGIM structure.

For three-dimensional alignment spaces, a two-dimensional grid can be further extended into three-dimensional voxels, providing a more precise, comprehensive, and rich representation of geographic information data. In the context of alignment optimization, voxel dimensions are anisotropic: the dimensions in the horizontal x and y directions are generally the same. However, the z-direction dimension is much smaller than those in the horizontal plane because the maximum allowable gradient restricts vertical connections.

For example, assuming a maximum gradient $\gamma = 20‰$ and a vertical voxel height $z = 30\text{m}$, a horizontal distance of at least 1500m would be required to determine whether two adjacent voxels can be connected in the z-direction. Many voxels between these two would be unreachable by the alignment and thus skipped during the search space scanning process of intelligent alignment design, as their connection slope exceeds the permissible limit (as notable in Fig. 2.2). This significantly reduces the connection rate, efficiency, and the number of feasible local paths.

If the voxel size (in x, y, and z) increases, the number of voxels skipped during the scanning process will also increase [4]. Therefore, for alignment

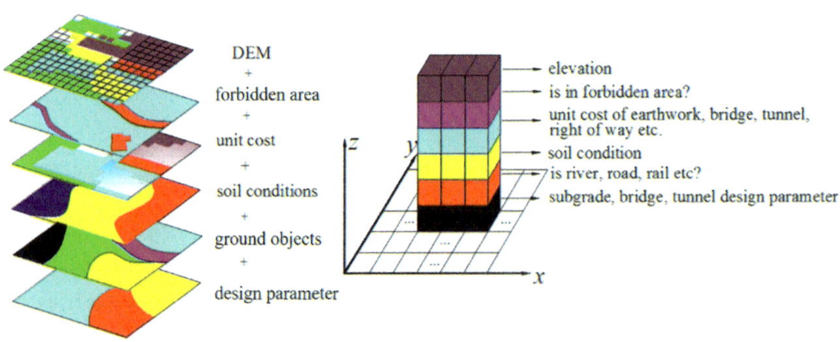

Figure 2.1 Comprehensive geographic information model.

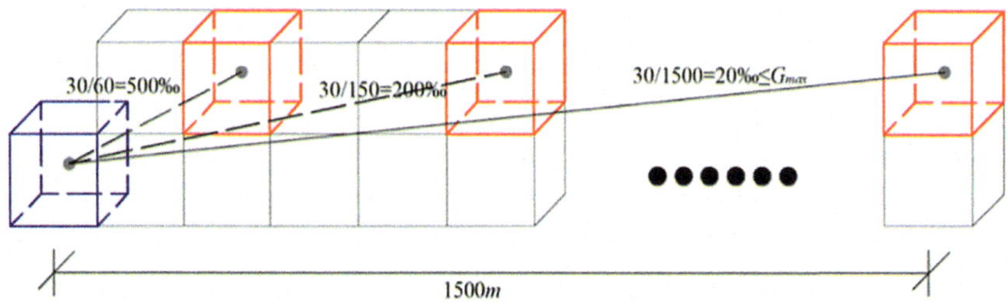

Figure 2.2 The maximum gradient limits the connectivity of isotropic voxels in the z-direction.

optimization problems, voxel dimensions should be anisotropic—for example, defining all voxels with sizes $x = y = 30$m and $z = 1$m.

2.2 Integrated geographic information encoding and storage

After constructing the CGIM model, interpolation based on these models is required to obtain terrain (elevation) and geological (slope, aspect, lithology, and attitude) information for each voxel. Surface cover, controlling factors (such as existing roads, railways, restricted zones, landslides, and nature reserves), economic indicators, and design standards are also essential geographic information data for alignment design. Therefore, the integrated geographic information required for alignment optimization primarily includes the following categories.

2.2.1 Topographic information

This primarily refers to the surface elevation data of the study area. For each voxel, in addition to storing its own elevation, the corresponding ground surface elevation derived from the DEM is also recorded.

2.2.2 Geological information

This mainly includes lithology and attitude, stratigraphic sequence, and mechanical properties of soil and rock, as well as unfavorable geological conditions such as faults, fracture zones, karst, weak interlayers, groundwater conditions, and geological hazards (e.g., landslides, debris flows, and collapses). Lithology is used to identify underground geological structures, while attitude describes the dip direction and dip angle of strata or rock layers, reflecting their horizontal distribution and inclination. Stratigraphic sequence indicates the vertical distribution of strata, which is essential for predicting excavation conditions. Mechanical properties reflect the strength, deformability, and permeability of the ground, directly influencing construction feasibility. Faults and fracture zones reveal structural weaknesses that may induce instability. Karst features highlight subsurface cavities and dissolution hazards that can threaten foundation stability. Weak interlayers identify potential slip surfaces within slopes or tunnels. Groundwater conditions determine the risk of water inflow, which may increase pore water pressure and chemical interactions with the surrounding rock, ultimately resulting in the softening of surrounding rock or soil. Geological hazards such as landslides and debris flows indicate external environmental threats to alignment safety and long-term operation.

2.2.3 Surface cover information

In mountainous areas with fragile ecosystems, railway construction should minimize damage to vegetation and soil to avoid soil erosion and maintain ecological stability. NDVI (Normalized Difference Vegetation Index), a widely used remote sensing index, can detect vegetation growth and coverage. NDVI values are discretized into each voxel of the integrated geographic information model as quantitative indicators of vegetation disturbance. NDVI values range from -1 to 1, where negative values indicate water, snow, clouds, while 0 indicates bare rock or soil, and positive values indicate vegetation coverage, increasing with coverage. The NDVI is computed as Eq. (2.1):

$$\mathrm{NDVI} = \frac{\rho_{\mathrm{NIR}} - \rho_{\mathrm{RED}}}{\rho_{\mathrm{NIR}} + \rho_{\mathrm{RED}}} \tag{2.1}$$

where ρ_{NIR} reflectance in the near-infrared band; ρ_{RED} reflectance in the red band.

2.2.4 Alignment design controlling factors

In addition to surface cover, controlling factors must be properly addressed in the alignment optimization process to ensure feasibility. These can be categorized into:

(1) Linear controlling factors: such as rivers, existing roads, and railways.

(2) Planar controlling factors: such as restricted zones, landslides, and nature reserves.

If a proposed alignment intersects linear features, clearances and intersection angles must satisfy regulatory requirements for safety and constructability. For planar features, the alignment should avoid restricted zones or, if necessary, traverse landslide, debris flow, or other geohazard areas with appropriate structures that satisfy clearance requirements.

2.2.5 Cost information

Cost indicators are a crucial part of alignment evaluation and include costs for subgrades, bridges, tunnels, land acquisition, and tracks. In special areas (e.g., seismic zones, forested lands, wetlands), these costs may differ from standard rates. Therefore, engineering unit costs should be recorded within the voxels of special zones. These data comprise the cost information layer of the integrated geographic information model.

2.2.6 Design standard information

Major design standards form the foundational constraints of alignment design and significantly influence its quality. These include design speed, minimum curve radius, maximum gradient, traction type, locomotive type, and arrival-departure track length. Within the study area, key technical standards may vary locally—for instance, a larger minimum curve radius may be adopted in flat terrain, while a smaller radius is permissible in difficult areas. Design standard information specifies the constraint conditions that each voxel must satisfy and serves as a basis for managing complex constraints in intelligent alignment optimization.

The integrated geographic information data used in alignment optimization are typically multi-source and heterogeneous. For example, NDVI data are commonly stored in formats such as.asc or.tiff, whereas controlling factor data are often stored in.dwg or.mdb formats. Due to the massive volume and diversity of such data, the regular voxel-based integrated geographic information model structure usually consumes a large amount of memory. Moreover, during the intelligent alignment optimization process, data access operations to the voxel model occur extremely frequently. Therefore, loading such a massive dataset entirely into memory for real-time reading and writing within the alignment optimization program can lead to serious problems such as memory exhaustion and inefficiency.

To address this, geographic information data are encoded and stored in a compressed form to reduce storage space and improve data access efficiency. Each voxel must store data including terrain, geology, NDVI, controlling factors, economic indicators, and design standards. Among these, terrain and geological information can be obtained through interpolation from the previously constructed digital terrain and geological models. NDVI data, stored in a regular grid, allows easy access to surface cover information for each voxel. However, region-based information, such as controlling factors, economic indicators, and design standards, is better suited for uniform compression and storage within voxels of localized study areas.

These data are eventually compressed and integrated into a structured format and stored in each three-dimensional voxel. The specific compression method for region-based information is illustrated using the treatment of controlling factors as an example. The data structure of a voxel is as follows:

```
typedef struct
{
float GridH; // Voxel elevation
float Slope; // Slope
float dSlope; // Aspect
double Lithology; // Lithology
double RockSoilType; // Rock-soil Category
double NDVI; // Normalized Difference Vegetation Index
bool isKarstArea; // Karst area (controlling factor)
double bridgeCost_unitLength; // Unit cost of bridges
double MaxGradient; // Maximum allowable gradient
...
} GridData; // Voxel information
```

The geometry of controlling factors can be expressed using control points, such as polyline nodes for linear and zonal features, or polygon vertices for area-based constraints. If control points of geological boundaries are stored directly within each voxel, substantial storage would be required. To avoid significantly increasing voxel memory usage, a distributed storage strategy is proposed.

With this strategy, the coordinates of control points of geological boundaries are stored in an external point set, while only their index references are recorded within each voxel. To further reduce storage requirements, an unsigned integer UINT and an integer array INTARRAY are used to store the presence and index information of controlling factors:

1) UINT(1, 2, ..., n) is a binary number of length n representing n types of controlling factors, where each bit denotes the presence (1) or absence (0) of a specific type.

2) INTARRAY(1, 2, ..., n, n+1) stores n+1 integers, with positions n and n+1 representing the start and end indices of the corresponding control points of geological boundaries in the external set.

The distributed storage strategy proceeds as follows (see Fig. 2.3):

Step 1. Initialize each bit of UINT in all voxels to 0. Let N = 1.

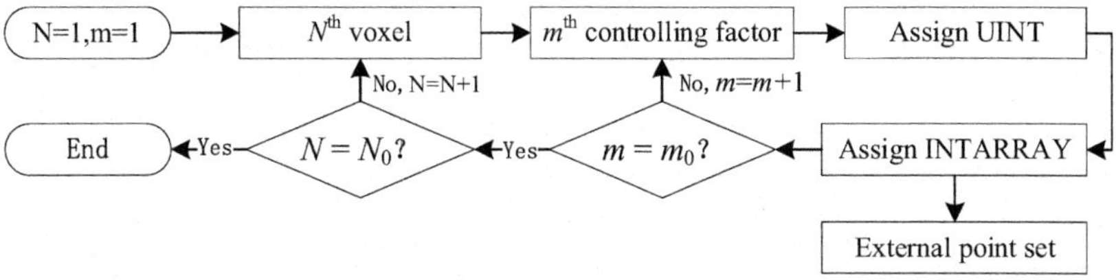

Figure 2.3 Distributed storage strategy.

Step 2. Locate all voxels intersecting the N-th type of controlling factor; let m = 1.

Step 3. For the m-th voxel, set the N-th bit in UINT to 1, and sequentially store the coordinates of the control points in the external point set.

Step 4. Store the start and end indices of these control points in the N and N+1 positions of the INTARRAY for the m-th voxel.

Step 5. Increment m and repeat steps (3)–(4) until all relevant voxels for the N-th controlling factor are processed.

Step 6. Increment N and repeat steps (2)–(5) until all types of controlling factors are assigned to their corresponding voxels.

At this point, the regular voxel-based integrated geographic information model is complete. In this framework, the study area is represented as a set of 3D voxels. Various geographic data—such as about terrain (elevation, slope, aspect), geology (lithology, attitude, mechanical properties of soil and rock, karst, landslide distribution), surface cover, controlling factors, economic indicators, and design standards—are discretized into these voxels to simulate the natural environmental features of the alignment corridor.

This model enables normalized management of massive, heterogeneous, multi-source geographic information, facilitating real-time, efficient retrieval of relevant data during the alignment search process.

To satisfy the data management requirements of intelligent alignment optimization, digital terrain and geological models are developed in this chapter. They subdivide the design space into 3D voxels, and interpolate key information such as elevation, geology, land cover, and controlling factors needed for alignment design. These data are then encoded and stored, forming the integrated geographic information model. This model serves as the foundation for intelligent alignment optimization, supporting both rapid solution evaluation (e.g., elevation interpolation for earthwork cost estimation) and intelligent optimization (e.g., constraint detection and mitigation) by providing essential geographic information services.

2.3 3-D environmental suitability analysis model

Based on the above CGIM, most existing intelligent alignment methods adopt a "generate alignments then evaluate their fitness" search strategy. That is, candidate alignment alternatives are first generated, and then the terrain, geological, and constraint information of the traversed voxels is retrieved to perform constraint checks. If an alignment fails to satisfy the constraints, it is regenerated through iterative adjustments by the optimization algorithm until all constraints are satisfied, thus providing a feasibility evaluation. Subsequently, the objective function is computed, and a progressive optimization mechanism is developed to minimize it, this provides a quality evaluation, iteratively improving the alignment until convergence.

This optimization strategy, which follows a trial-and-error paradigm of "search the environment via the candidate alignments," can work effectively when the types and numbers of constraints are limited and sparsely distributed. However, in densely constrained environments, where infeasible regions dominate the search space, this method tends to waste significant computational time and resources on blind exploration of infeasible regions. As a result, both search efficiency and optimization quality are greatly reduced, and in some cases, it may even be impossible to generate feasible solutions that satisfy all dense constraints.

In reference to the manual alignment process, designers often adopt the strategy of "evaluate environment and then design alignments." Based on professional experience or domain knowledge, they first assess the alignment environment to identify high-quality areas with high suitability for alignment passage. This allows the alignment work to be focused within specific high-potential regions or corridors, thereby reducing search time and improving design quality.

A deeper analysis reveals that the prevalent use of the "generate alignments then evaluate their fitness" and "search the environment via the candidate alignments" strategies in existing studies fundamentally stems from the close coupling between environmental suitability assessment and configurations of structures such as bridges, tunnels, and embankments. Suitability evaluation models must necessarily compute indicators such as structural stability, economic cost, and reliability, all of which depend on both the type of structure and the environmental conditions, in order to assess the service performance of the alignment and reflect its environmental suitability.

Thus, environmental suitability assessment must be based on the alignment-dependent distribution of structures. Consequently, most existing studies adopt a post-evaluation strategy, where alignment alternatives are first generated, the corresponding infrastructure layout is derived, and then the environmental suitability of the alignment is analyzed.

If an alternative "evaluate environment then design alignments" strategy is used to conduct a pre-evaluation of environmental suitability before generating the alignment, it becomes difficult to perform a comprehensive assessment due to the absence of structural layout information. This makes the analysis of the time-varying nature of environmental suitability and its adaptation to the alignment infeasible.

However, through extensive research and practical experience in railway alignment design, a critical property has been discovered by the CSU AO team: the study region exhibits a layered pattern of structure types. When the alignment environment is discretized into a regular 3D voxel grid, the structural type at any voxel through which an alignment may pass is isotropic in the horizontal plane—i.e., regardless of the direction in which the line crosses a voxel, the structure type (bridge, tunnel, or embankment) remains unchanged. It is solely determined by the vertical height difference between the voxel and the ground surface (Δh) and its relation to the threshold values that define transitions between embankments, bridges, and tunnels.

By traversing the voxel grid and analyzing for each voxel the relations between Δh and the threshold values for fill—bridge height (h_{fb}), general bridge—high bridge height (h_b), maximum allowable bridge height (h_{bmax}), cut - tunnel depth (d_{ct}), shallow buried or normal buried tunnel depth (d_{sn}) and normal buried or deep buried tunnel depth (d_{nd}), a layered map of structural types across the entire space can be constructed (see Fig. 2.4), typically categorized from top to bottom into inaccessible layers, bridge layers, embankment layers, tunnel layers, and other layer types.

This discovery breaks the traditional bottleneck where environmental suitability cannot be assessed in advance due to the unknown structure layout. It makes it possible to implement an "evaluate environment then design alignments" strategy and conduct a pre-evaluation of alignment suitability based purely on the environment.

Thus, a comprehensive evaluation model has been developed, incorporating multi-dimensional adaptability indicators, including construction cost, operational cost, risks of typical geological hazards (such as rockfalls, landslides, and debris flows), seismic risk, karst risk, structural stability, carbon emissions, soil erosion, and vegetation damage.

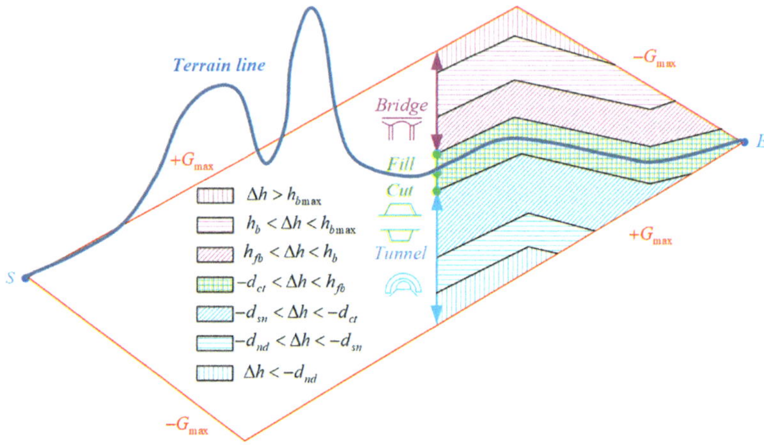

Figure 2.4 Division of the structural layers.

Afterward, a multi-criteria decision-making method, termed as CRITIC (Criteria Importance Through Intercriteria Correlation), is applied to integrate the above evaluation indicators [5]. This method determines the objective weight of each criterion by considering both the contrast intensity (i.e., the variability) of the criterion and the conflict or correlation between criteria. Criteria with higher variability and lower correlation with others are assigned greater importance, ensuring that the integrated evaluation reflects both the discriminative power and the uniqueness of each indicator. This strategy allows us to compute the environmental suitability value for each voxel within the alignment environment and to assemble these values into a comprehensive environmental suitability field (see Fig. 2.5).

2.4 Alignment adaptability optimization under multi-modal environmental distribution characteristics

The spatiotemporal distribution of high-suitability environments derived from the above environmental suitability analysis often exhibits multi-modal patterns, including:

(1) Connected patterns, where suitable zones form a continuous and uninterrupted corridor between the start and end points;

(2) Blocked patterns, where natural barriers such as rivers, mountains, or fault zones create complete disconnections;

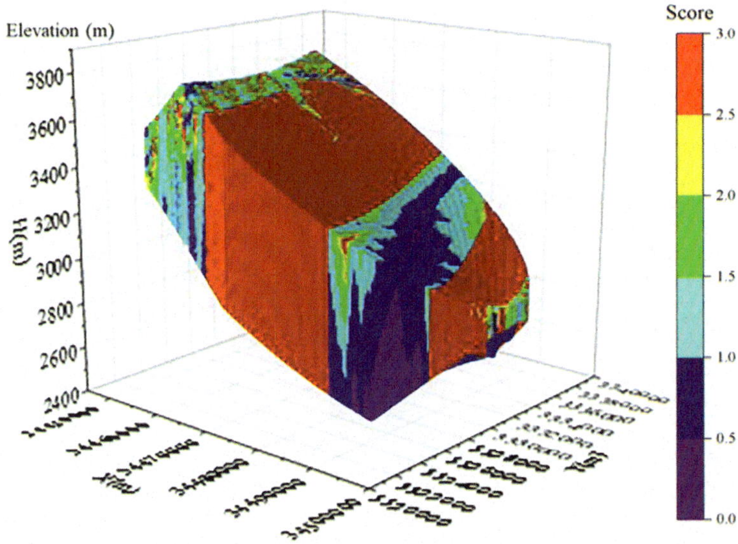

Figure 2.5 Distribution of all voxels' environmental suitability values.

(3) Isolated patterns, where suitable areas are fragmented into isolated "islands" due to constraints such as adverse geological conditions or nature reserves; and

(4) Hybrid patterns, where multiple distribution effects coexist within the same study region.

Each distribution pattern presents distinct challenges for railway alignment, necessitating different alignment strategies:

(1) For connected patterns, a "alignment-first, point-later" sequential search strategy from the start to the end point (or vice versa) is appropriate (see Fig. 2.6a);

(2) For blocked patterns, a "point-first, alignment-later" strategy is more suitable, in which an alignment begins from difficult segments and extends toward easier terrain (see Fig. 2.6b);

(3) For isolated patterns, a three-stage strategy of "island selection → inter-island connection → intra-island alignment" is recommended (see Fig. 2.6c);

(4) For hybrid patterns, a "segment-first, divide-and-conquer" strategy is most effective (see Fig. 2.6d).

To address the varying characteristics of high-suitability environment distribution, corresponding alignment adaptability optimization models have been developed.

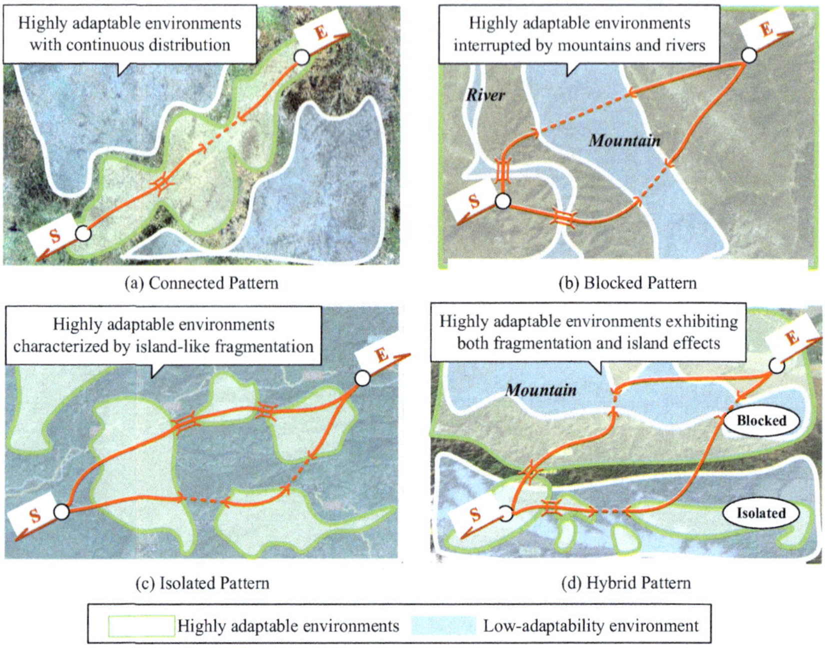

Figure 2.6 Multi-modal distribution patterns of high-suitability environments.

2.4.1 Connected pattern

In the connected mode, high-suitability environments are continuously distributed between the start and end points. In such cases, alignment design typically follows a "alignment-first, point-later" strategy, where the alignment determines the placement of structures [6]. Within high-suitability patches, a point-line collaborative optimization model can be established, allowing sequential search from start to end (or vice versa). The authors have already conducted long-term in-depth research on this pattern, having developed multi-objective optimization models that integrate economic costs, geological hazard risks, and green environmental objectives.

Multiple collaborative search methods have been developed—including distance transform, particle swarm optimization, genetic algorithms, Monte Carlo tree search, and deep reinforcement learning—that are directly applicable to solving this model. These will be elaborated in the following sections.

2.4.2 Blocked pattern

In blocked patterns, alignment optimization must primarily resolve the issues of control structure siting between suitability clusters and intracluster alignment design. A two-level optimization model is proposed with a "point-first, alignment-later" strategy [7]. Design variables include the coordinates of major control structures (e.g., mileage and elevation of bridge/tunnel start and end points) and spatial alignment control points (e.g., coordinates of horizontal intersection points, mileage and elevation of vertical control points). Objective functions involve structure–environment suitability and point-line collaborative environmental suitability, while design constraints include structure siting rules, geometric design standards, and restricted zone boundaries.

The solution method of this problem can be divided into two stages.

2.4.2.1 Structure point selection

For the upper-level model, a multi-directional scanning algorithm is proposed to traverse all voxels in the accessible region. A constraint repair-penalty operator is developed for structure siting, used to build an initial candidate pool of large-scale infrastructure control points. Optimized control point sets are then selected based on the cumulative suitability value along the potential alignments, serving as inputs to the lower-level model.

2.4.2.2 Alignment connection

For the lower-level model, a distance-transform-based point-alignment collaborative search algorithm is developed. Start/end points and control points are treated as target nodes. A generalized distance cost is defined based on the cumulative suitability values between voxels and target points. A spatial neighborhood scanning template dynamically searches voxels to find locally optimized paths with the lowest generalized distance. By propagating these distances and accumulating local optima, global optimized paths are generated. Finally, spline curves are fitted to the paths to obtain smooth, optimized alignment alternatives.

2.4.3 Isolated pattern

For isolated environments, the primary challenges are:

(1) evaluating the comprehensive suitability of high-suitability "island" patches;

(2) selecting optimized inter-island connections;

(3) achieving environment-alignment adaptability within islands.

A three-level optimization model is constructed comprising "island selection → inter-island connection → intra-island alignment." Design variables include the coordinates of island nodes, 3D endpoints of inter-island connections, and intra-island alignment control points [8]. Objective functions include island suitability, inter-island alignment–environment adaptability, and multi-objective suitability (quantified based on topographic, geological, and environmental indicators) of intra-island alignment. Constraint models cover island location, alignment geometry, and spatial zoning requirements, enabling the modeling of control structure siting between islands and intra-island spatial alignment optimization.

Thus, the solution process includes the following three steps:

2.4.3.1 *Island selection*

A suitability evaluation method for high-suitability "islands" is proposed based on the entropy weight method, identifying optimized time-sequenced complex network island nodes, which serve as boundary conditions for the inter-island connection phase.

2.4.3.2 *Inter-island connection*

Island pair connections are modeled as multi-source shortest path problems. A weighted directed graph of the inter-island region is constructed using voxel-level suitability values, and a 3D-Floyd algorithm is developed to search for a non-dominated set of point-alignment collaborative solutions.

2.4.3.3 *Intra-island alignment*

The entrance and exit points of the island nodes serve as start and end points for intra-island alignment. Theta* path planning [9], distance transform, or other methods are used to search for local optimized paths. These are then integrated with inter-island paths to form a globally optimized environment-alignment solution.

2.4.4 Hybrid pattern

In hybrid patterns, alignment optimization begins with segmentation of single-mode regions. Using the spatiotemporal distribution and evolution of strongly connected components, segments are characterized by connected, blocked, or isolated patterns. Based on earlier formulations, each segment is abstracted mathematically according to its dominant challenge, and a "segment-first, divide-and-conquer" intelligent optimization model is established. Each segment employs a tailored hierarchical optimization sub-model, and the results are integrated to form a comprehensive model for the hybrid alignment problem.

For hybrid environments, the Louvain community detection algorithm [10] is introduced based on modularity maximization to distinguish the dominant distribution patterns (e.g., blocked, isolated). A "segment → submodel → integration" strategy is adopted, applying the appropriate optimization model for each segment and integrating results to achieve adaptive global alignment planning.

Due to the inherent complexity and variability of the natural and geological environment, railway alignment optimization cannot rely on a single, fixed algorithm to address all scenarios. Instead, it is necessary to establish a CGIM as the foundational environmental basis. Based on this model, the adaptability of alignments within different environmental regions should be systematically analyzed, taking into account topographical, geological, ecological, and engineering constraints. Notably, the distribution of alignment-favorable regions shows multi-modal characteristics in complex study areas. To adapt different alignment design scenarios, a multi-modal distribution strategy is required to capture the diversity of influencing factors across scales and regions. Finally, by integrating these insights, an adaptation-oriented optimization method is developed, enabling more reliable, efficient, and context-sensitive alignment design.

References

[1] X. Yang, Development of a GIS-based routing model, ProQuest Dissertations & Theses, 1993.

[2] J.C. Jong, M.K. Jha, P. Schonfeld, Preliminary highway design with genetic algorithms and geographic information systems, Comput. Aided Civ. Infrastruct. Eng. 15 (4) (2000) 261–271.

[3] W. Li, H. Pu, P. Schonfeld, et al., Methodology for optimizing constrained 3-dimensional railway alignments in mountainous terrain, Transport. Res. C Emerg. Technol. 68 (2016) 549–565.

[4] H. Pu, T. Song, P. Schonfeld, et al., A three-dimensional distance transform for optimizing constrained mountain railway alignments, Comput. Aided Civ. Infrastruct. Eng. 34 (11) (2019) 972–990.

[5] R. Rostamzadeh, M.K. Ghorabaee, K. Govindan, et al., Evaluation of sustainable supply chain risk management using an integrated fuzzy TOPSIS-CRITIC approach, J. Clean. Prod. 175 (2018) 651–669.

[6] H. Pu, X. Wan, T. Song, et al., A geographic information model for 3-D environmental suitability analysis in railway alignment optimization, Integrated Comput. Aided Eng. 30 (1) (2023) 67–88.

[7] X. Wan, H. Pu, P. Schonfeld, et al., Mountain railway alignment optimization based on landform recognition and presetting of dominating structures, Comput. Aided Civ. Infrastruct. Eng. 39 (2024) 2.

[8] X. Wan, H. Pu, P. Schonfeld, et al., Railway alignment optimization in regions with densely-distributed obstacles based on semantic topological maps, Integrated Comput. Aided Eng. 31 (4) (2024) 421–437.

[9] V.H. Dang, N.D. Thang, H.H. Viet, et al., Batch-theta for path planning to the best goal in a goal set, Adv. Robot. 29 (23) (2015) 1537–1550.

[10] M. Mohammadi, M. Fazlali, M. Hosseinzadeh, Accelerating louvain community detection algorithm on graphic processing unit, J. Supercomput. 77 (6) (2021) 6056–6077.

CHAPTER THREE

Railway alignment design knowledge graph

Abbreviations and corresponding full texts

C-T database Condition-Treatment database
CGIM Comprehensive Geographic Information Model
DT Distance Transform
KG Knowledge Graph
RAD-KG Railway Alignment Design Knowledge Graph

3.1 Overview

Railway alignment design is a process of determining optimized alignment layouts by systematically integrating multi-dimensional professional knowledge (including terrain analysis, geological risk assessment, engineering technical standards, and ecological protection norms) as well as rich practical experience. It embodies the comprehensive application of human knowledge systems—from macro-level strategic judgment to micro-level structural design—where engineers synthesize fragmented information, balance conflicting constraints, and make situation-adaptive decisions. Such a knowledge-intensive and experience-dependent process lays the foundation for the development of alignment optimization, yet it also highlights the need to address knowledge application challenges.

Existing studies on railway alignment optimization have undoubtedly made significant contributions, but a critical limitation persists: the knowledge utilized in this field remains fragmented. This knowledge, often scattered within optimization programs as diverse design parameters, lacks structured and systematic associations. Typically, alignment design principles—defined here as sets of "conditions" (descriptions of the alignment's environment) and their corresponding "treatments" (required disposals)-are stored in fragmented forms. These forms are characterized by a large variety, quantity, and varying value ranges, as well as being multisource and heterogeneous (encompassing text, audio, video, etc.). For example, when addressing macro-level route planning (e.g., aligning with regional transportation networks), meso-level corridor selection (e.g., bypassing

Intelligent Railway Alignment Design
ISBN: 978-0-443-45923-8
https://doi.org/10.1016/B978-0-443-45923-8.00004-2

large-scale ecological zones), or micro-level detail design (e.g., crossing existing traffic facilities), the corresponding treatments (e.g., network connection principles, ecological buffer requirements, intersection control standards) are embedded in programs without clear structural links.

Furthermore, alignment design is an iterative optimization process requiring dynamic adjustments. As the alignment's location changes, the conditions it encounters must be analyzed in real time, demanding updates to corresponding treatments. Treatments directly influence alignment evaluation and modifying design principles (conditions and treatments) are often necessary to generate diverse alternatives - such as either bypassing geo-hazard zones or crossing them with bridges/tunnels. Unfortunately, fragmented knowledge hinders such modifications, extensions, and updates, while human knowledge itself continues to evolve with expanding categories and experience, as seen in Fig. 3.1.

Traditional knowledge engineering has aided in expressing alignment design knowledge but relies on symbolic representations more interpretable

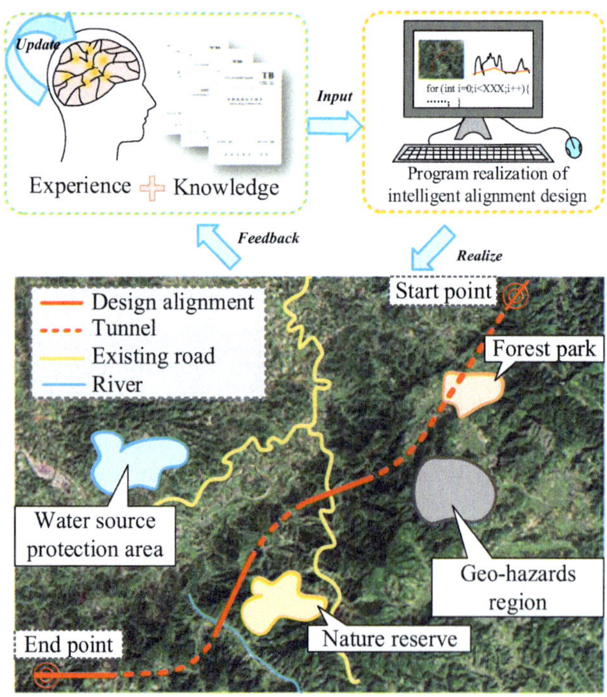

Figure 3.1 Application and updating of knowledge in a railway alignment design.

to humans than computers, limiting progress in automated optimization. Notably, the emergence of Google Knowledge Graph (KG) offers a solution. KG integrates symbolic and vector representations, boasting strong extensibility and compatibility with computer vision and natural language processing. Structured as a graph, it uses nodes for semantic entities and edges for semantic relations, typically expressed as triplets `<head entity, relation, tail entity>` to form a semantic network. KG is widely regarded as key to enabling "knowledgeable" artificial intelligence systems.

Although KG technology has not been previously applied in railway alignment design, many KGs, such as WikiData, Freebase and YAGO, have been constructed in the big data field. KG has been successfully extended in fields such as medical diagnosis, mechanical fault diagnosis, efficient enterprise supply chain management, geoscience studies, surveying and remote-sensing technology by the virtue of relation expression and complex relation reasoning, which provides valuable ideas for the structured knowledge representation and lays the theoretical foundation for introducing KG into railway alignment optimization.

To this end, a two-stage knowledge modeling method for constructing and applying a railway alignment design knowledge graph (as called RAD-KG) with comprehensive knowledge "representation - querying" is proposed. For stage I, a hierarchical and categorized semantic network composed of a set of customized triplets is constructed as the RAD-KG model. Then, for stage II, a rapid knowledge querying method of RAD-KG is proposed to guide an automated alignment optimization process while satisfying the complicated dynamic railway design requirements.

3.2 Stage I: Modeling of RAD-KG

3.2.1 Knowledge representation of railway alignment design

For decades, railway alignment designers have accumulated many design experiences. However, these experiences usually consisted of unstructured information stored in human brains or written specifications. It is challenging to transform them into structured information recognizable by computers. Especially, the conditions encountered by an alignment (such as terrain difficulties and geo-hazards) are generally dissimilar in different study areas. Moreover, as mentioned in Section 3.1, even in the same study area, the con*ditions* encountered by the alignment are dynamically changed in the iterative optimization process. Thus, the design principles must be applied according to specific *conditions*, which also increases the difficulty of modeling knowledge.

To solve this problem, a structural design method for modeling the knowledge involved in railway alignment design is proposed, which includes two aspects: (1) Modeling the network framework of RAD-KG; (2) Detailed property design of every node in RAD-KG.

3.2.1.1 Network framework of RAD-KG

Based on the characteristics of different railway alignment design principles, a customized semantic network framework for modeling RAD-KG is proposed in this section. It may be exemplified as follows: for the principle of "Main railways should prioritize alignment consistency with regional comprehensive transportation hub layouts", the "regional transportation hub planning context" is the "condition", and the "hub-oriented alignment" is the "treatment". Therefore, a railway alignment design principle can be abstracted as a triplet of <*condition* - relation - *treatment*>.

The above principle possesses characteristics of categorization and stratification. On the one hand, in real-world railway design processes, different railway projects may face significantly different actual problems, such as resource limitations, ecological protection, and geological risk. Therefore, according to different *conditions*, different categories of principles should be considered, which demonstrates the "categorization" characteristic of railway design principles. On the other hand, for a specific *condition*, the *treatments* applicable to this situation have the characteristics of stratification. For example, if the alignment is located in a region with various environmentally sensitive areas, a national nature reserve is a sub-factor of the environmentally sensitive area. More specifically, the core area of a national nature reserve is a sub-factor of an entire national nature reserve, which, in general, should be completely bypassed by the alignment, while a "transitional" area of the nature reserve may be crossed by an alignment in some strictly limited ways (e.g., through tunnels or bridges). The above phenomenon of layer-by-layer concretization reflects the stratification characteristics of alignment design principles. Generally, only the specific *conditions* with the lowest level correspond to *treatments*.

Moreover, the categorized characteristics form several semantic relations of "one-to-one","one-to-many" and "many-to-one" between *conditions* and *treatments*. Specifically, a principle such as "When the alignment intersects an existing road, the intersection angle should be at least 30 degrees, and the clearance between road and alignment should be at least 5.5 m"

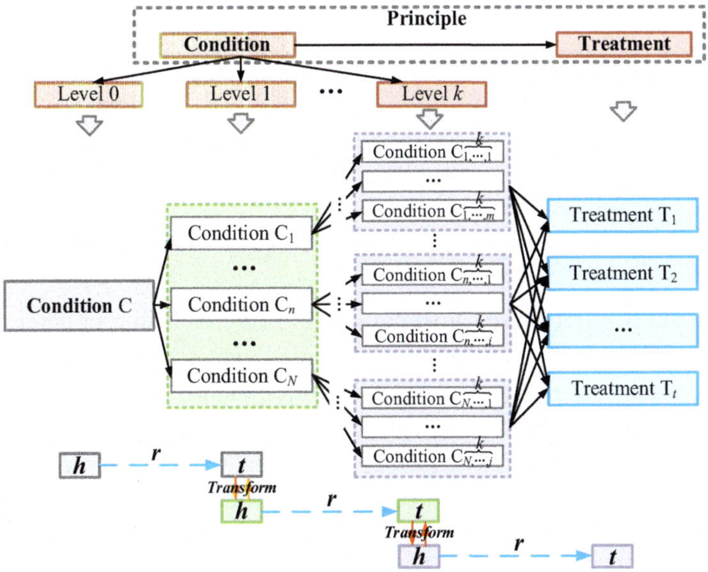

Figure 3.2 Hierarchical and categorized semantic network framework of RAD-KG.

is expressed as a "one-to-many" triplet of <existing road - intersects with alignment - the intersection angle is at least 30 degrees/the clearance is at least 5.5 m >. Similarly, the "one-to-one" and "many-to-one" relations that exist in other principles can be treated similarly to the above example. The above semantic logic pervades in the knowledge system of railway alignment design principles. Hence, the framework of RAD-KG is constructed as shown in Fig. 3.2.

The proposed framework is a multilevel tree structure, in which each node represents an entity of *condition* ($C_{n,\ldots,i}$, If the entity is at level k, the subscript has k entries) or *treatment* (T_t). This design has two merits: (1) Intuitive: the categorization of design principles is gradually expanded from the root node to lower-level nodes; (2) Extendable: newly discovered knowledge can be easily merged into the existing knowledge structure as new tree branches. Meanwhile, it is also convenient to combine the tree-structured RAD-KG with another KG by just merging their respective top-level entity nodes.

However, the forms and sources of information (e.g., images, voices, texts) contained in the RAD-KG are diverse. Moreover, the construction of a complete RAD-KG requires the joint efforts of many stakeholders. Therefore, in the process of constructing subgraphs, some entities may be

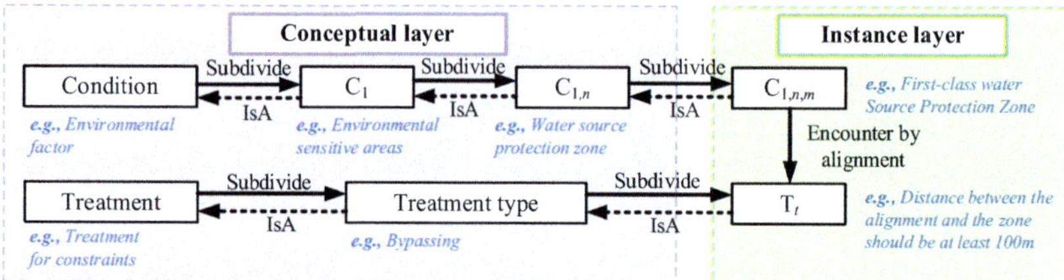

Figure 3.3 Ontology schema of the RAD-KG based on the multilayer chain structure.

given multiple alternative names considering different habits of different researchers, but, sometimes, different entities may adopt the same name, which results in the ambiguity of entities. Thus, for performing the ontology design [1] it is essential to set a unified schema of RAD-KG.

Hence, the design principles of railway alignment optimization are abstracted as a chain structure composed of multilayer triplets, as shown in Fig. 3.3. This chain is composed of multiple entities and relations, including a conceptual layer and an instance layer. The conceptual layer is a collection of entities that represent abstract concepts, such as "environmentally sensitive area" in *condition* entities, and "bypassing" in *treatment* entities. The instance layer is a collection of entities with specific properties.

Taking as an example the principle "The alignment should bypass the First-class water source protection zone, while the separation between this zone and the alignment should be at least 30 m","First-class water source protection zone" is a *condition* instance, which possesses properties such as the name, area and protection level. This instance is a sub-factor of the "Water source protection zone" in the conceptual layer. "At least 30 m" is a *treatment* instance with specific value, which is a sub-*treatment* of the entity "bypassing" in the conceptual layer. Based on this, a standardized schema for the RAD-KG can be accomplished.

3.2.1.2 Property design

The *conditions*, or actual problems faced by railway alignment design are diverse and changeable, which requires the RAD-KG to store a variety of information. Although the network of triplets can express macro-level design principles, each entity contains limited information. Therefore, a data structure for property design of entities in the instance layer is proposed in this section, which is applied to enhance the information storage capacity of entities and to eliminate the possible ambiguities among different entities

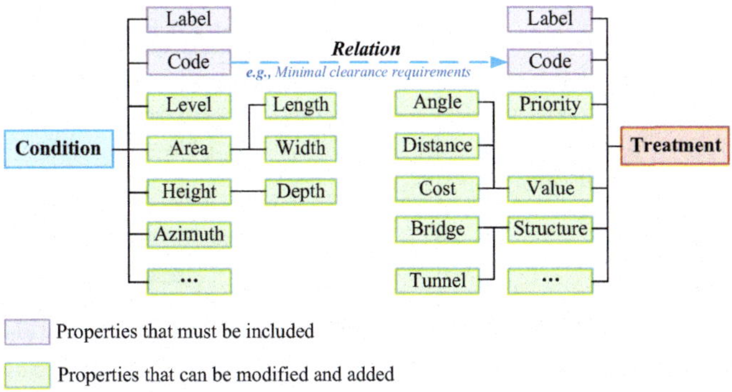

Figure 3.4 Property design of the *condition* and *treatment* instance entities.

in the same *condition* or *treatment* category. The property design for the entities in instance layer (the last level of *conditions* and *treatments*) is shown in Fig. 3.4.

As an instance, each *condition* entity is distinguished according to different properties, such as level, azimuth angle, area, and height. These properties are required to be definite since some *treatments* are related to specific properties of the *condition*. For example, when an alignment crosses a reservoir, the allowable elevation of the alignment must be determined based on the elevation of the reservoir. Similarly, each *treatment* entity stores the property information of a specific *treatment*, including the action, property and limit value. The properties of *condition* and *treatment* entities (see Fig. 3.4) can be modified and added according to the actual situation of the *condition* and *treatment* entities. However, both *condition* and *treatment* entities must contain the properties of "Label" and "Code", where "Label" distinguishes whether the entity is a *condition* or a *treatment*; "Code" is the standardized distinguishing feature among different entities. The encoding rules of the two types of entities will be introduced in detail in Section 3.2.2.

Thus far, a customized structural representation for railway alignment design knowledge has been modeled. This model forms a macroscopically semantic tree network of alignment design knowledge, while also providing microscopically the information storage and demarcation among entities.

3.2.2 Knowledge storage in RAD-KG

Based on the modeling of knowledge in the previous section, a RAD-KG can be constructed. The whole process is shown in Fig. 3.5.

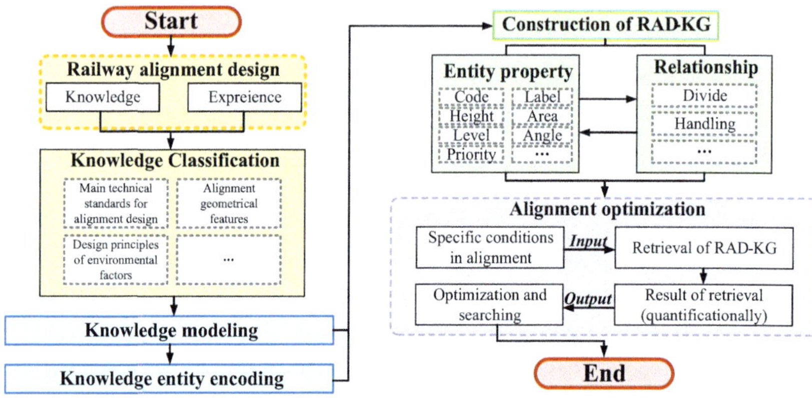

Figure 3.5 Construction process of the RAD-KG.

Currently, a variety of popular graph database platforms are available, including Neo4j, Dgraphs, TigerGraphs and CosmosDB. In this work, considering the convenience of computer operation, the visibility of human-computer interaction [2], and the adaption to railway alignment design knowledge, Neo4j is chosen to construct the RAD-KG. More specifically, Neo4j is a graph database for storing the KG data, which is highly scalable and fully transactional [3]. Based on the Cypher query language, Neo4j can access large and complex graphic data sets to realize automatic query and retrieval of data [4, 5]. The entities and relations are represented in the form of nodes and links, respectively, which form a directed graph network.

The construction of RAD-KG not only provides a systematic and structured knowledge source for railway alignment optimization, but also sets a visual digital carrier for storing human experience. Meanwhile, the independent operation ability provided by Neo4j makes the knowledge in RAD-KG independently-refreshable with the progress of human knowledge.

Based on the constructed RAD-KG, knowledge and experience expressed as design principles in this research can be systematically and structurally modeled. However, railway alignment optimization is a complex process to which many design principles must be applied. Meanwhile, the applied principles are constantly changed with the dynamic adjustment of the alignment location. Hence, it is challenging to flexibly apply the RAD-KG for guiding the alignment optimization process, which is also an important issue to be solved in next section.

3.3 Stage II: RAD-KG-driven alignment optimization

3.3.1 Optimization mainflow

Since massive spatial data (such as data on topography, soil conditions, forbidden zones, and structural unit costs) are important for railway alignment design, a comprehensive geographic information model (CGIM) should be constructed. In the CGIM, the study area is described as a series of regular cells. Moreover, various geographic information data are discretized into cells to simulate the natural environmental characteristics of the real study area. This provides a database for the design principles obtained from the RAD-KG.

The method of RAD-KG-driven alignment optimization based on CGIM is shown in Fig. 3.6.

In the CGIM, *conditions* in cells are given simple codes, which are usually decimal or binary numbers. *Conditions* of the same categories have the same simple codes. When an alignment traverses a CGIM cell, the encountered *conditions* stored in the cell can be obtained. Simple codes of these *conditions* are matched with the standardized codes in the RAD-KG. Thus, the matched *conditions* are retrieved by inputting standardized codes to the RAD-KG to query for the corresponding *treatments*. The encoding rules for standardized codes in RAD-KG will be introduced in Section 3.3.2.2. Then, the *treatments* corresponding to the *conditions* through the standardized codes are transformed into quantitative design information parameters and returned to the cell. During alignment search, the design information in every cell can be invoked to guide the alignment design. For example, in the search process, if the alignment-traversed cell is located

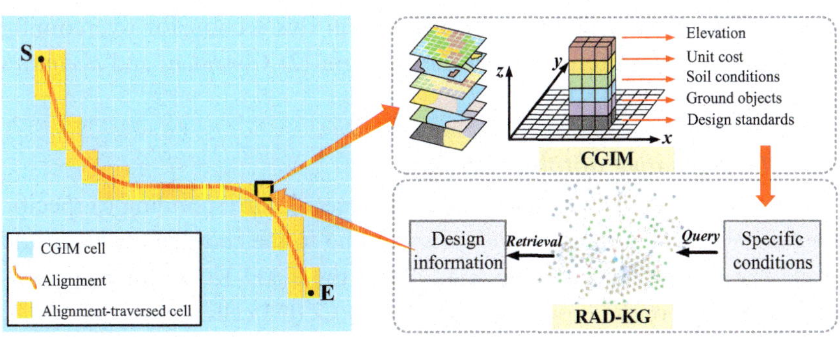

Figure 3.6 Main flow of RAD-KG during alignment optimization based on the CGIM.

in a forbidden zone, the RAD-KG transmits the queried *treatment* "the bypassing separation" and the quantitative separation of "100 m" to the cell. Then, the cells within the bypassing separation (i.e., cells in the whole forbidden zone and the cells within 100 m of the outward extension of the forbidden boundary) are removed from the feasible area in the CGIM, and the alignment is guided to other feasible cells. Other applications of *treatments* in alignment search are similar to this example.

Assisted by the RAD-KG, when an alignment is changed, the cells traversed by the alignment are different, and thus the alignment-surrounding *conditions* stored in CGIM cells are also different. The RAD-KG can provide the design information according to the new *conditions* without manually modifying the design parameters in an optimization program. Moreover, when the design principles are updated, the *treatments* in RAD-KG can be modified independently without influencing the optimization program.

3.3.2 Efficient knowledge-querying approach

As described in Section 3.1, alignment optimization is a dynamic process. The number of cells in the search space depends on the range of study area and the resolution of CGIM. When the cell resolution is high and the study area is large, massive cells in the three-dimensional space should be assessed. Thus, the knowledge-querying process based on the RAD-KG will be quite frequent. In addition, the querying is conducted across various platforms (Microsoft Visual Studio, Python and Neo4j), which increases the difficulty of real-time knowledge-querying when scanning the cells traversed by the alignment.

Therefore, an important problem in applying the RAD-KG-driven railway alignment optimization is to improve the knowledge-querying efficiency. Two critical designs are proposed in this section for overcoming this difficulty: (1) Alternative database design; (2) *Conditions* and *treatments* encoding in the RAD-KG.

3.3.2.1 Alternative database design

Theoretically, the design information (*treatments*) corresponding to specific *conditions* of an alignment is queried according to the codes of *conditions* and *treatments* in RAD-KG (see Fig. 3.4). Then, the RAD-KG transmits parameters which store the design information to CGIM cells.

However, due to the large number of cells in the study area, if the *conditions* obtained from every scanned cell are queried directly from

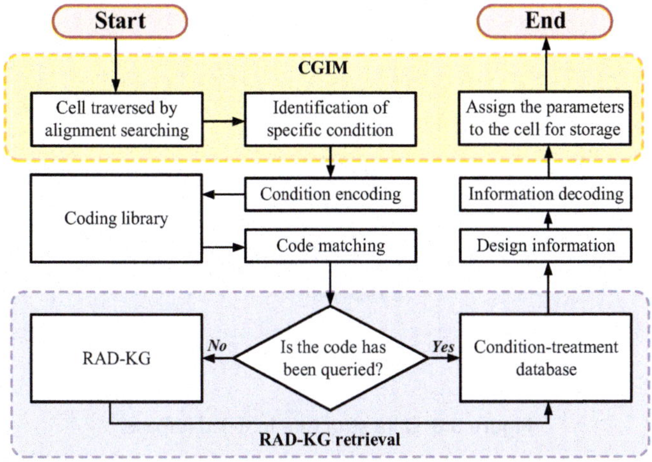

Figure 3.7 Method of rapid querying design information.

RAD-KG, the querying is time-consuming, which seriously affects the efficiency of alignment optimization. Moreover, in the process of alignment search, the *conditions* in cells traversed by the alignment are not completely different from those in the cells traversed by other alignment alternatives. For example, the *condition*s of cells in the same forbidden zone may be exactly the same. It is unnecessary to query the same *condition*s repeatedly from the RAD-KG. Then, a method for rapid knowledge-querying based on an alternative database is proposed and shown in Fig. 3.7.

When a cell traversed by an alignment is scanned, the *condition* stored in the cell can be identified. The *condition*s confronted by the alignment are input into the RAD-KG for querying. Meanwhile, the frequency of querying seriously affects the efficiency of RAD-KG-driven method. As mentioned above, it is unnecessary to query every *condition* stored in the alignment-traversed cell. Therefore, an alternative database is devised to store the intermediate design information between CGIM and RAD-KG, which is defined as a *condition-treatment* database (C-T database). After a cell is scanned, it is judged whether the *condition* has been queried. If yes, the design information corresponding to this *condition* is directly obtained from the C-T database without entering the RAD-KG querying process. Otherwise, the *condition* is input into RAD-KG to obtain the corresponding *treatment*.

The queried *conditions* and retrieved *treatment*s from RAD-KG have one-to-one correspondences. The *treatments* are stored in the C-T database

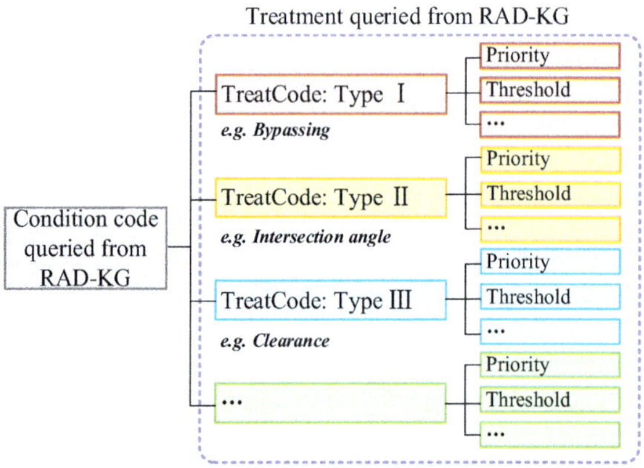

Figure 3.8 Data structure in C-T database.

in the form of design information. The data structure design of the C-T database is shown in Fig. 3.8.

To facilitate the recognition and transmission of data in the C-T database, the properties of the queried *treatment* entity such as the *treatment* type and priority are encoded as enumeration values. Among the *treatment* properties, the priority indicates the execution order of this *treatment* (i.e., a branch of design information in Fig. 3.8 with the same color). In general circumstances, *treatments* with priority 1 should be implemented. In difficult circumstances, *treatments* with lower priority grades should be sequentially executed.

It is worth noting that whenever the *condition* code is queried, the C-T database traverses all the groups of *treatments* corresponding to this *condition* code and returns them to the CGIM cells. The C-T database not only improves efficiency by reducing repetitive knowledge-querying, but also ensures that all *treatments* corresponding to this *condition* can be invoked by the optimization program.

In addition, to improve the retrieval efficiency and avoid omissions, the *conditions* and *treatments* in the C-T database are paired by standardized codes, which are the same as those in the RAD-KG.

3.3.2.2 Encoding rules for knowledge querying and storage in RAD-KG

Although the RAD-KG already has a standardized uniform schema based on the constructed ontology in Section 3.2, different instance entities belonging to the same conceptual layer category may correspond to different *treatments*. For example, the separations between an alignment and two different national forest parks may be different. Hence, it is necessary to uniquely encode each entity (*condition* or *treatment*) for distinguishing it from others, which is also beneficial to the knowledge-query in RAD-KG (see Fig. 3.7).

In this section, two different encoding rules are proposed for *condition* entities and *treatment* entities, respectively. The encoding rules are shown in Fig. 3.9.

For a *condition* entity, the initial code is a 10-character alphanumeric code. The first digit of the code is a letter character, indicating the coding category. The *condition* code is represented as "C". The second and third characters are combined letter characters, which form the name of the subgraph to which the *condition* belongs. For instance, if it is a KG of environmental factors, the second and third characters are "E" and "F", respectively. The fourth digit indicates to which level the entity belongs. If the letter of the fourth digit is "A", it indicates that the entity belongs to the first level (see Fig. 3.9). Similarly, when the letter is "C", it means that the entity is at the third level. Starting from the fifth digit, two adjacent numerical digits are combined to indicate the entity's reference number (RN) at a certain level. That is, the fifth and sixth digits represent the RN at the first level, while the seventh and eighth digits represent the RN at the second level, and the ninth and tenth digits represent the RN at the third level. More specifically, the *condition* code "CEFC010101"

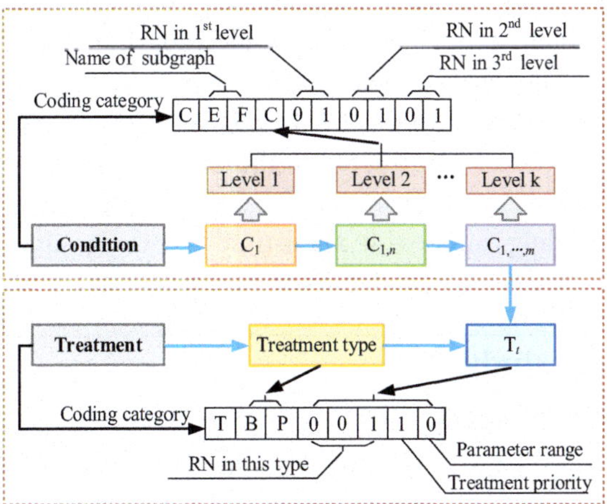

Figure 3.9 Rules of unique encoding for *condition* and *treatment* entities in RAD-KG.

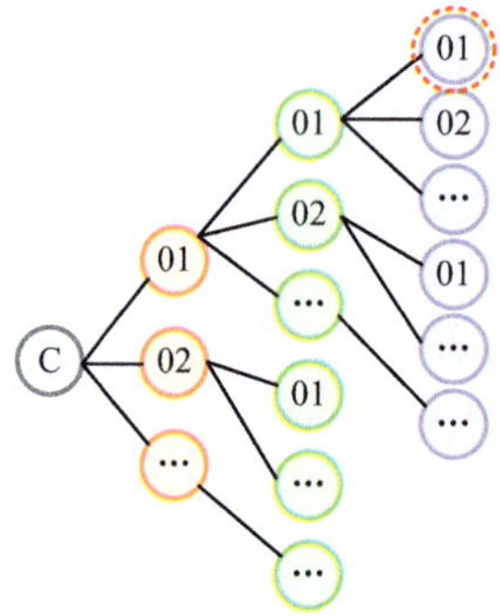

Figure 3.10 Location of entity "CEFC010101" in the RAD-KG.

in Fig. 3.9 means that the node location for this entity should be recognized as shown in Fig. 3.10.

For a *treatment* entity, the code is an 8-character string, where the first digit "T" represents the coding category. The second and third digits are combined letter characters, which represent the type of the *treatment*. For example, "bypassing" is represented by "B" and " P". The following fourth-sixth digits are combined numerical characters, which indicate the order of the *treatment* entity in this *treatment* type, as shown in Fig. 3.9. In addition, the second-to-last digit is numerical, indicating the priority of the *treatment*. The last digit is also numerical, indicating whether the parameter value of the *treatment* is the maximum or minimum limit. Taking as an example the principle that "the intersection angle between the existing road and alignments should exceed 30 degrees","30 degrees" in this principle is the minimum limit of the intersection angle and, thus, the last digit is "0"; otherwise, if "30 degrees" is the maximum limit, the last digit is set as "1". Based on these encoding rules, the unique relations among *condition* and *treatment* entities can be constructed in the RAD-KG.

Thus, the efficient query method based on alternative database and knowledge encoding can provide a real-time knowledge-querying function for RAD-KG-driven railway alignment optimization.

3.4 Case study

3.4.1 Case profile

A real-world railway section located in typical mountainous region is adopted as the case in this paper. This is a complex railway project with densely-distributed environmentally-sensitive factors in the study area.

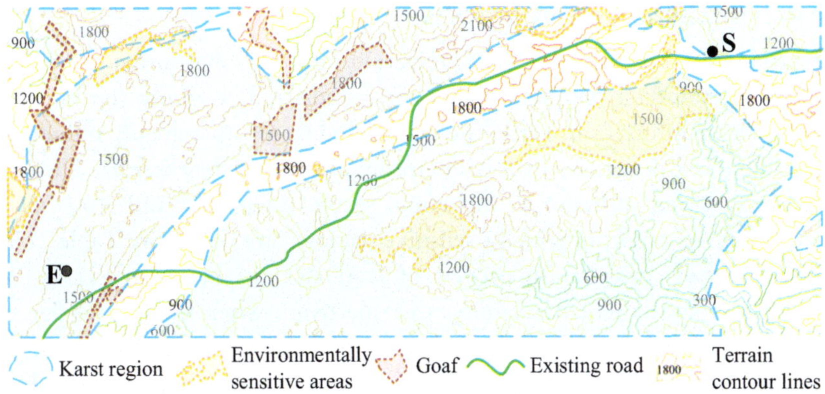

Figure 3.11 Terrain map and environmentally-sensitive areas.

The linear separation and elevation difference between the start and end points are 42.4 km and 205 m, respectively. The study area includes 150 geo-hazards regions (128 goafs and 22 karst regions), one expressway and 87 natural reserves. These environmental factors are fairly complex and can greatly affect the alignment design process. The terrain and environmentally-sensitive factors' distribution are shown in Fig. 3.11. All data required for railway alignment design are provided by the China Railway Siyuan Survey and Design Group Co., Ltd. The major design standards are listed in Table 3.1. The unit costs are shown in Table 3.2.

Table 3.1 Major design standards of the real-world railway case.

Item	Value
Railway grade	High-speed railway
Design speed	350 km/hours
Maximum gradient	20‰
Minimum radius of horizontal curve	7,000 m
Vertical circular curve radius	25,000 m

Table 3.2 Unit costs.

Item	Value	Item	Value
Railway track (10^4¥/km)	460	Right-of-way (¥/m^2)	135
Fill earthwork (¥/m^3)	65	Cut earthwork(¥/m^3)	31
(L < 500 m) bridge (10^4¥/m)	7.50	(L > 500 m) bridge (10^4¥/m)	14.35
(L < 1 km) tunnel (10^4¥/m)	8.50	(1 < L < 10 km) tunnel (10^4¥/m)	7.33
(L > 10 km) tunnel (10^4¥/m)	11.50		

3.4.2 Implementation process

As a practical case, a RAD-KG integrating multiple environmental factors and corresponding design principles is constructed to guide the railway alignment optimization. Here, the controlling factors that greatly influence the alignment design are mainly environmentally-sensitive factors. The integrated design principles, including seven categories and 100 subcategories of environmentally-sensitive factors, are listed in Table 3.3. The stepwise construction process of RAD-KG is shown in Fig. 3.12, which also shows its extendabilty. On one hand, when the knowledge is gradually improved with the growth of human experience, the new principles can be added as the child nodes of existing entity nodes. On the other hand, when a new RAD-KG in another application scenario is constructed, that RAD-KG can be extended by connecting its root node with the root node of a new graph based on the same ontology schema.

To show the effectiveness of the RAD-KG in assisting alignment optimization, a distance transform (DT) algorithm [6], which will be introduced in Section 8, is adopted as the basic search method. However, the application of knowledge to the DT algorithm is still fragmented, while lacking structured and systematic knowledge expression. Thus, it is difficult to update the design principles without modifying the optimization program. In this case, the RAD-KG in Fig. 3.12 is incorporated into the DT algorithm to improve the search process. The RAD-KG-driven DT method is shown in Fig. 3.13. Detailed result analyses will be shown in the following sections.

3.4.3 Results

3.4.3.1 Feasibility verification

The alignment alternative (hereinafter denoted as A_C) generated by the proposed method is compared with the best manually-designed alignment (denoted as A_M), which is produced by experienced engineers in the China Railway Siyuan Survey and Design Group Co., Ltd. The principles applied during the design process are listed in Table 3.3. The horizontal and vertical alignments of A_C and A_M are shown in Figs. 3.14 and 3.15, respectively.

In this case, the design principles about geo-hazard regions (goafs, and karst regions) and existing roads are dominating factors for railway alignment design (see Table 3.3). As shown in Fig. 3.14, A_C completely bypasses the karst regions, while $A_M$crosses some karst regions. These results indicate that A_C reduces the geologic risks for railway construction due to karst

Table 3.3 Factors and corresponding principles in RAD-KG.

Factors	Number of design principles	Example principles in RAD-KG	Modified principles 1	Modified principles 2
Environmentally sensitive areas	30	**Natural reserve:** Core zones should be completely bypassed by alignments. Bridge and tunnel sections are allowed to pass transitional and peripheral zones. **Vibration and noise sensitive areas:** Alignments should completely bypass such areas. Besides, the separation between alignments and these areas should be at least 80 m.		
Rivers and channels	4	**Rivers:** The intersection angle between an alignment and a river should be at least 75 degrees under general circumstances and 45		

(Continued)

Table 3.3 Factors and corresponding principles in RAD-KG.—cont'd

Factors	Number of design principles	Example principles in RAD-KG	Modified principles 1	Modified principles 2
		degrees under difficult circumstances.		
Roads	15	**Expressway:** The intersection angle between an alignment and an expressway should be at least 30 degrees under general circumstances and 15 degrees under difficult circumstances.	**Expressway:** Alignments are allowed to be parallel with existing roads.	**Expressway:** The intersection angle between an alignment and an expressway should be at least 45 degrees under general circumstances and 20 degrees under difficult circumstances.
Railways and rail transits	8	**High-speed railway:**The intersection angle between an alignment and an existing railway should be at least 45 degrees.		
Geo-hazards regions	11	**Karst region:** Karst regions should be bypassed by alignments, but may be crossed in some difficult circumstances.	**Karst region:** Alignments are allowed to cross karst regions.	**Karst region:** Alignments are allowed to cross karst regions.

		Goaf: Goafs should be completely bypassed by the alignment. **Soft soil:** Soft soil areas should be completely bypassed by alignments, or crossed through deep-buried tunnels.	**Goaf:** Alignments should bypass goafs as far as possible.	**Goaf:** Alignments are allowed to cross goafs.
Biochemical hazard zones	16	**Long oil and gas pipelines:** The intersection angle of alignment and pipeline should be at least 30 degrees. Parallel spacing between an alignment and a pipeline should be at least 25 m. **Nuclear power plant:** Nuclear power plants should be completely bypassed by alignments. The separation between an alignment and a plant should be at least 1000 m.		

(Continued)

Table 3.3 Factors and corresponding principles in RAD-KG.—cont'd

Factors	Number of design principles	Example principles in RAD-KG	Modified principles 1	Modified principles 2
Forbidden zones	15	**Dam:**Dams should be completely bypassed by alignments. The separation between an alignment and a dam should be at least 200 m. If the dam is crossed by a bridge, the clearance should be at least 30 m. **Airport:** Airports should be completely bypassed by alignments, and the separation between an alignment and an airport should be at least 200 m.		

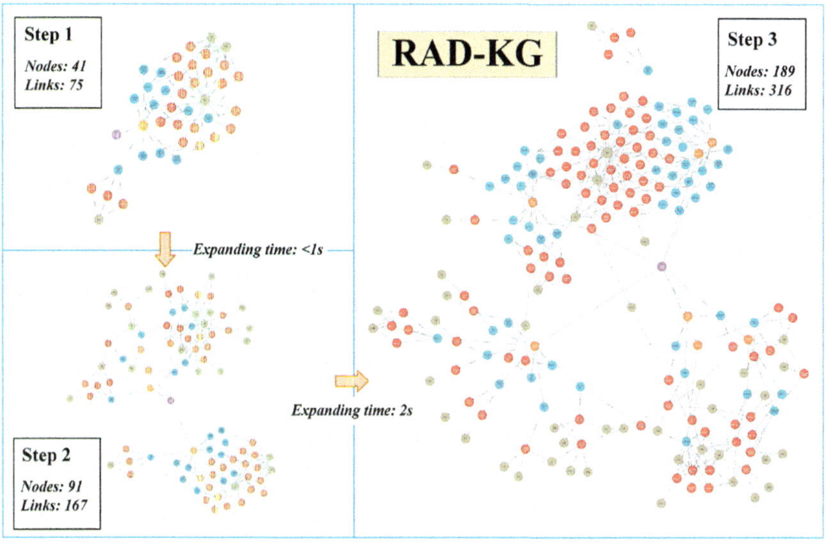

Figure 3.12 Building and expanding processes of the RAD-KG.

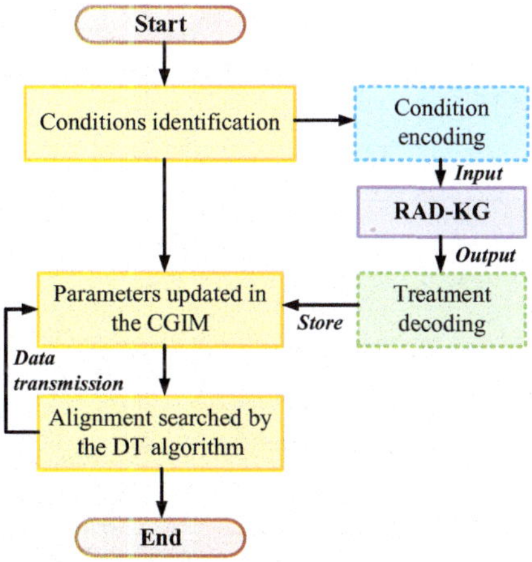

Figure 3.13 RAD-KG-driven alignment optimization process.

caves. Moreover, compared with A_M, the total tunnel length of A_C is reduced by 4,289 m. The maximum tunnel length of A_C is reduced by 3109.5 m. The comparison of construction costs between the A_M and A_C is shown in Table 3.4. It shows that A_C reduces costs by 9.2%, which verifies that the proposed method can generate less costly alternatives than traditional manual work.

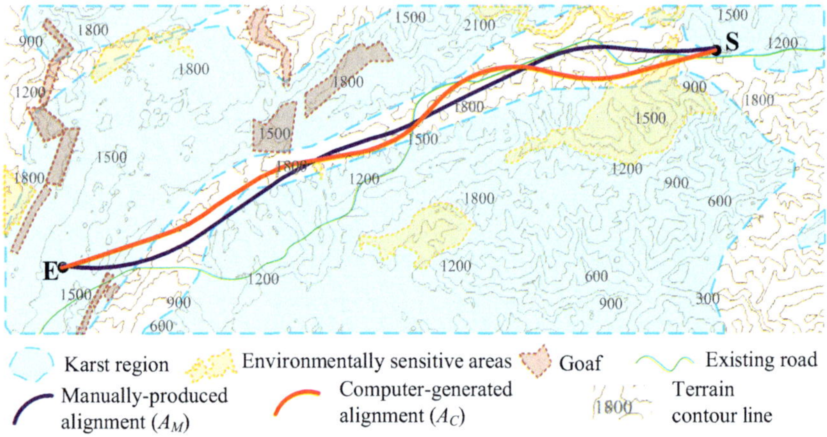

Figure 3.14 Comparison of horizontal alignments A_M and A_C.

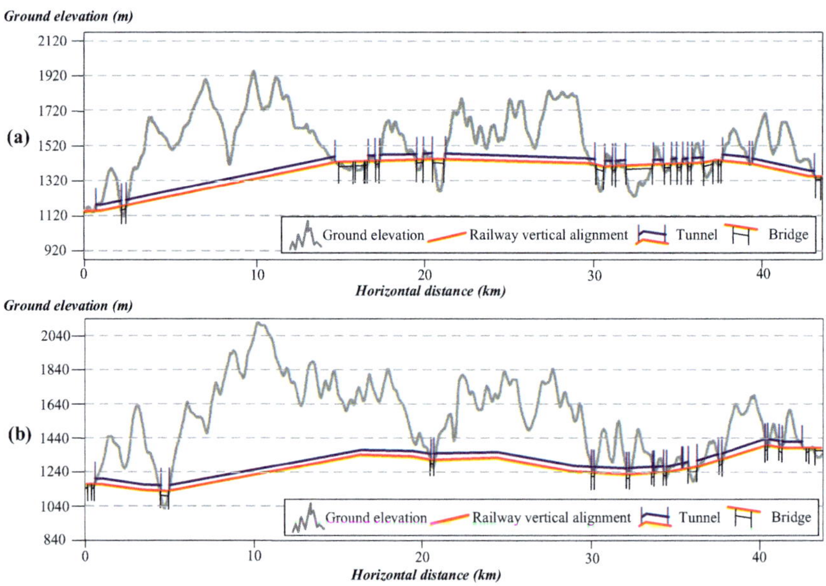

Figure 3.15 Comparison of vertical alignments A_C (a) and A_M (b).

3.4.3.2 Flexibility verification

With the continuous advancement of human knowledge and development of engineering technology, design specifications should be updated frequently. Therefore, the knowledge system in the alignment optimization program should also be updated to reflect current developments. To test the

Table 3.4 Detailed engineering results comparison.

Item	A_M	A_C
Length (m)	43,488	43,671
Right-of-way (m^2)	126,665	280,100
Embankment volume (m^3)	1,165,433	1,317,559
Excavated volume (m^3)	267,493	296,110
Bridge/number-length, m	15–546	14-1946
Tunnel/number- length, m	12–15,471	14-12,362
Total construction cost/million ¥	4608	4075
Total construction cost saving/%	–	9.2%

effectiveness of RAD-KG when the design principles are updated, two modified alignment alternatives (denoted as A_{C1} and A_{C2}) are generated by modifying specific principles in the constructed RAD-KG. The original and modified principles are as follows, and also listed in Table 3.3.

Original principles (for generating A_C): (1) The intersection angle between the alignment and the expressway should be at least 30 degrees under general circumstances and 15 degrees under difficult circumstances. (2) Karst regions should be bypassed by the alignment as far as possible, but are allowed to be traversed at high cost under difficult circumstances. (3) Goafs should be completely bypassed by an alignment.

Modified principle 1 (for generating alignment solution A_{C1}): (1) The alignment is allowed to be parallel with an existing road. (2) The alignment is allowed to traverse karst regions. (3) The alignment should bypass the goaf as far as possible.

Modified principle 2 (for generating alignment solution A_{C2}): (1) The intersection angle between the alignment and the existing road should be at least 45 degrees under general circumstances and at least 20 degrees under difficult circumstances. (2) The alignment is allowed to traverse karst regions. (3) The alignment is allowed to traverse goaf regions.

Based on the above settings, A_C, A_{C1} and A_{C2} are shown in Fig. 3.16. The layouts of A_{C1} and A_{C2} differ significantly from A_C. More specifically, A_{C1} traverses a goaf and a karst region (see Fig. 3.16 (b)). Moreover, a part of A_{C1} almost overlaps with the existing road (see Fig. 3.16 (d)), which demonstrates that the modified principle 1 is activated and works when A_{C1} intersects with the existing road. More than 70% of the total length of A_{C2} is in karst regions (see Fig. 3.16 (a)). Meanwhile, the intersection angle between the existing road and A_{C2} satisfies the requirements of modified principles 1 and 2 (as seen Fig. 3.16 (c)(d)(e)).

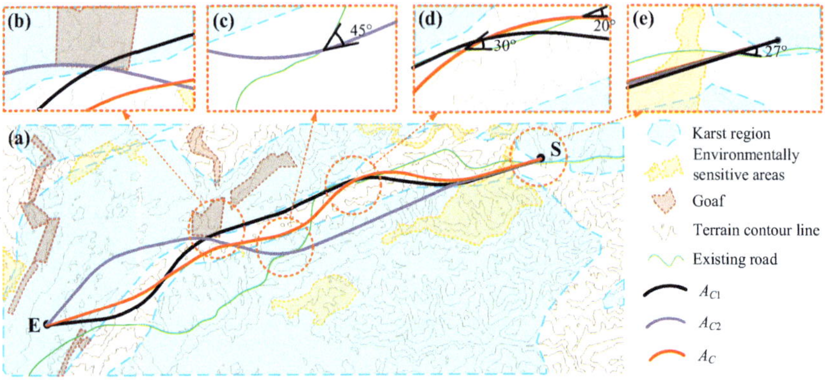

Figure 3.16 Comparison of horizontal alignments A_C, A_{C1} and A_{C2}.

The above results verify that the principles modified in the RAD-KG can significantly affect the layouts of computer-generated alignment alternatives. Moreover, the generated alignments can fully satisfy the required principles set in RAD-KG. This indicates that when the principles are modified, the RAD-KG changes, and further affects the diversity of the alternatives generated by the RAD-KG-driven alignment design method to better support designers. Thus, RAD-KG enables the adjustability of railway alignment optimization according to the designer's preference, which verifies the flexibility of the RAD-KG-driven method.

References

[1] M.A. Osman, S.A.M. Noah, S. Saad, Ontology-based knowledge management tools for kknowledge ssharing in organization - a review, IEEE Access 10 (2022) 43267–43283.
[2] G. Drakopoulos, P. Gourgaris, A. Kanavos, Graph communities in Neo4j, Evol. Syst. 11 (3) (2020) 1868–6478.
[3] M. Saad, Y. Zhang, J. Tian, et al., A graph database for life cycle inventory using Neo4j, J. Clean. Prod. 393 (2023) 136344.
[4] N. Francis, A. Green, P. Guagliardo, et al., Cypher, in: Proceedings of the 2018 International Conference on Management of Data, 2018, pp. 1433–1445.
[5] P. Kotiranta, M. Junkkari, J. Nummenmaa, Performance of graph and relational databases in complex queries, Appl. Sci. 12 (13) (2022) 6490.
[6] H. Pu, T. Song, P. Schonfeld, et al., A three-dimensional distance transform for optimizing constrained mountain railway alignments, Comput. Aided Civil Infrastruct. Eng. 34 (11) (2019) 972–990.

CHAPTER FOUR

Multi-task deep learning methods for determining railway major technical standards

Abbreviations and corresponding full texts

CMTNN Cascaded Multi-Task Neural Network
CNN Convolutional Neural Network
ELRDT Effective Length of Receiving–Departure Track
FC Fully Connected
HMTNN Hybrid Multi-Task Neural Network
MHCCR Minimum Horizontal Circular Curve Radius
MTNN Multi-Task Neural Network
PMTNN Parallel Multi-Task Neural Network
RMTS Railway Major Technical Standard (plural: RMTSs)
STNN Single-Task Neural Network

4.1 Overview

Railway Major Technical Standards (RMTSs) serve as the fundamental criteria for determining the type, capacity, and scale of railway structures and equipment. They exert a crucial influence on railway transport capacity, operational safety, transportation efficiency, investment scale, as well as economic and social benefits. In general, the decision-making process for RMTSs follows a systematic approach: under the premise of meeting transportation demands, all potential RMTS values are proposed based on factors including the role of the designed railways in the railway network, the volume and nature of transportation, natural conditions such as terrain and geology along the alignment, resource distribution, regional economic development, and population density, and the RMTSs of adjacent railway networks. Subsequently, these candidate values are evaluated from multiple perspectives to select the preferred one, including railway network layout, regional economic development, technical and economic rationality, coordination among RMTSs, and environmental protection.

Intelligent Railway Alignment Design
ISBN: 978-0-443-45923-8
https://doi.org/10.1016/B978-0-443-45923-8.00002-9

A primary principle in RMTS decision-making is mutual coordination and compatibility among various RMTSs, which means that the decision-making process must take into account the interrelations among different standards. For instance, design speed is interrelated with the minimum horizontal circular curve radius (MHCCR); meanwhile, parameters such as maximum gradient, locomotive type, and effective length of receiving–departure track (ELRDT) all have direct impacts on railway transport capacity. Collectively, these interrelated standards influence engineering costs and operational expenses. Therefore, a comprehensive optimization approach is essential for RMTS decision-making. The evaluation of RMTS decision quality primarily considers the following aspects:

- ➢ Railway transport capacity and its reserve status, the impact on future capacity enhancement, and the utilization level of existing equipment.
- ➢ The influence on the railway network and regional economic development.
- ➢ The impact on environmental protection.
- ➢ The level of safety, comfort, and travel speed.
- ➢ Construction period and conditions.
- ➢ Engineering indicators and construction costs.
- ➢ Operational indicators and transportation costs.

Conventional comprehensive optimization methods for RMTSs typically select all or part of the technical standards as decision variables. They then construct single-objective or multi-objective optimization functions to evaluate the safety, economy, comfort, and environmental performance of the schemes corresponding to these decision variables. Based on a large number of engineering cases, statistical regression analysis is used to determine the coefficients of the objective functions, thereby establishing an optimization model. Finally, optimization algorithms are applied to solve the model, and a recommended RMTS scheme is derived.

However, the RMTS optimization process is characterized by a large spatiotemporal span, a long decision cycle, and high hysteresis. The multi-dimensional dynamic coupling relations between decision variables and evaluation objective functions are difficult to express using a unified functional formula. As a result, this "model-centric" method struggles to accurately reflect objective laws and lacks universality. In recent years, the rapid development of technologies such as big data and artificial intelligence has provided a "data-centric" new approach to address the RMTS optimization problem. In this chapter, a comprehensive technical framework integrating a voting election strategy with multi-task deep learning methods is proposed, which offers a data-driven approach to enhance RMTS decision-making.

The core technical route begins with a voting election framework designed to balance local and global benefits in the study area, comprising three key stages:

- Stage I: In the constituency partition stage, the railway study area is discretized into uniformly sized constituencies (optimally determined as 333×333 pixels through experiments) to capture local environmental variations; edge areas smaller than the standard size are expanded to maintain consistency, ensuring neither oversimplification nor underrepresentation of terrain features.
- Stage II: The constituency election stage, the framework's core, employs multi-task neural networks (MTNNs) to process environmental factors—characterized as three-channel images integrating elevation, slope, and transport capacity data—and predict RMTSs for each constituency. Here, four MTNN architectures are developed to model interrelations among RMTSs: cascaded MTNN (CMTNN), which determines standards in a hierarchical order mimicking manual design logic; overall parallel MTNN, where all standards share a global feature extractor and are predicted simultaneously; grouped parallel MTNN, which partitions standards into horizontal (design speed, MHCCR) and vertical (maximum gradient, locomotive type, tractive mass, ELRDT) groups with independent shared layers to avoid interference between weakly correlated tasks; and hybrid MTNN, combining parallel and cascaded structures to balance intergroup independence and intragroup hierarchy.
- Stage III: The ballot summary stage aggregates constituency-level predictions using a weighted voting strategy, where weights are determined by each constituency's distance to the straight line connecting the railway's start and end points (prioritizing alignments closer to the optimized path), ensuring the final RMTSs reflect both local terrain specifics and global alignment feasibility.

To optimize these MTNNs, hyperparameter sensitivity analyses are conducted, focusing on convolutional filter size, number of filters, and neurons in fully connected layers. Results indicate that a 5×5 filter size, 8 filters, and 256–512 neurons yield optimized performance, minimizing cross-entropy loss (used to evaluate prediction accuracy against manual decisions). These optimized networks are validated using a representative, high-quality dataset of 338 real-world railway cases (covering 58,351 km, including complex terrains such as mountainous regions), with 20% of cases used for testing. Experimental results show that the grouped parallel MTNN outperforms the other architectures, achieving an average decision accuracy of 86.9% compared to manual decisions—10.7% higher than the previous single-task neural network (STNN) method. This superiority stems from its ability to isolate correlations within horizontal and vertical groups, preserving intra-group dependencies while avoiding interference among weakly related standards.

In summary, this study's technical route—integrating constituency-based spatial partitioning, multi-task deep learning to model RMTS interrelations, and weighted aggregation to balance local and global factors—provides a robust solution to the challenges of RMTS decision-making. By enhancing accuracy and efficiency, it not only addresses the limitations of traditional methods but also strengthens the scientific basis of railway alignment design, making it a valuable approach for optimizing RMTSs in diverse railway projects.

4.2 The voting election framework

Environmental factors within the study area are crucial for RMTS decision-making. Decision-makers must consider both the distribution of global environmental factors within the study area (such as overall terrain distribution and railway transport capacity) and the variations in local environmental factors (such as the changes in local topography). However, study area sizes may vary greatly for different railway cases. To effectively balance the distribution of global environmental factors and the variation of local environmental factors, a voting election strategy for determining RMTSs is proposed, as shown in Fig. 4.1.

First, at the constituency partition stage, the research area is discretized into normally sized constituencies. Then, entering the constituency

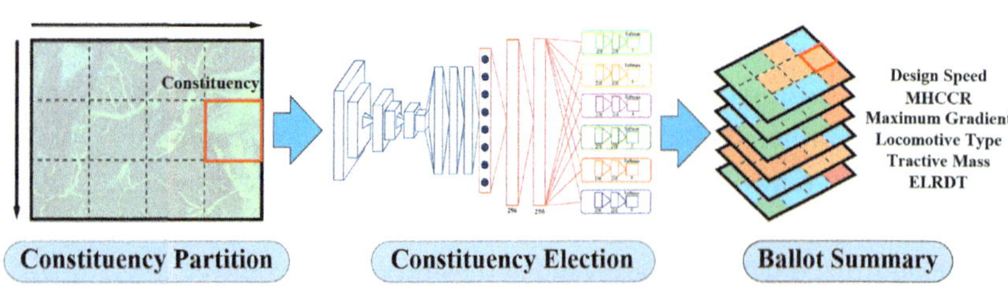

Figure 4.1 The voting election framework.

election stage, several neural networks are designed using multi-task learning methods to complete the extraction of local features of environmental factors within the constituency for RMTS decision-making. Finally, at the ballot summary stage, a ballot summary is conducted among all the constituencies and the final RMTS are determined by considering the overall characteristics within the whole study area. Specifically:

4.2.1 Constituency partition

To account for the benefits within local areas during the voting election process, the study area of every railway case is divided into a number of constituencies with equal size, based on a square lattice template, from the top left to bottom right. Areas at the bottom or right edge that are smaller than a regular constituency are expanded leftward or upward to the normal size.

The size of constituencies during the constituency partition process is a critical factor that affects the effectiveness of the voting election strategy. When predicting RMTSs, adopting an excessively small constituency may neglect variations in local environmental factors. Conversely, when utilizing an overly large constituency, geospatial images may simultaneously involve a broad range of topographical features, such as plains, hills, and mountainous terrain. This can oversimplify and coarsen the features extracted by neural networks, ultimately leading to a decline in prediction performance. Therefore, in this section, experiments with constituencies of different sizes are conducted to make decisions regarding RMTSs. The constituency size with a high decision accuracy is adopted in the subsequent experiments (see Section 4.4 for more details).

4.2.2 Constituency election

Constituency election aims to extract local environmental features from the divided constituencies and make decisions regarding multiple RMTSs within a single constituency. Unlike the study of single RMTS decision-making, the interrelations among RMTSs are the key issue that must be tackled in making multiple RMTS decisions. To achieve this, multi-task learning is a powerful method for recommending a series of mutually matched RMTSs. Therefore, several MTNNs are designed for RMTS decision-making at this stage. The inputs consist of various environmental factors while the outputs include decision results for RMTSs, including design speed, MHCCR, maximum gradient, locomotive type, tractive mass, and ELRDT:

- The design speed is usually determined based on railway transport capacity and railway grade;
- MHCCR ensures safety of train operation and track maintenance;
- The maximum gradient limits the railway alignment gradient within a certain range for the sake of traffic safety and operation cost;
- The locomotive type is directly related to the traction power, which determines the traction features of the locomotive;
- The tractive mass refers to the total mass of a locomotive while it pulls a regular freight train. It is mainly affected by the locomotive type and the maximum gradient;
- ELRDT is the maximum track length in stations that can accommodate the longest arriving and departing train without affecting the operation of the adjacent rolling stock, which is closely related to the tractive mass. Data on the above six kinds of RMTSs are obtained from real-world railway cases.

The detailed structure of MTNNs will be shown in Section 4.3.

4.2.3 Ballot summary

As the optimization of major railway technical standards involves a macroscopic decision-making process, it becomes essential to not only account for the variations in local environmental factors during constituency election but also consider the distribution of global environmental factors. However, it is important to recognize that not all constituencies within the study area carry the same level of significance in RMTS decision-making. A direct computation of votes for all constituencies may result in an underestimation of the impact of valuable constituencies.

During the railway planning and design process, economic factors hold significant weight, with numerous existing studies focusing on reducing the construction costs. Consequently, railways tend to favor relatively straight and direct paths. Environmental factors along these near-straight paths should hold great reference value in the study area. Therefore, this section proposes a weighted ballot summary strategy, as listed in Eq. (4.1). It determines the weight w_i of each constituency by computing the distance from the center of the constituency to a straight line connecting the start and end points of the railway.

$$w_i = \frac{(y_e - y_s)(x_i - x_s) + (x_e - x_s)(y_s - y_i)}{\sqrt{(y_e - y_s)^2 + (x_e - x_s)^2}} \tag{4.1}$$

where $P_i(x_i, y_i)$, $S(x_s, y_s)$, $E(x_e, y_e)$ are coordinates of the constituency center P_i, the start point S, and end point E of the alignment, respectively.

Then, the overall distribution of environmental factors within the study area is taken into account by summing total weighted votes, as seen in Eq. (4.2).

$$V_{j,k} = \sum_{A} w_i v_{i,j,k} \tag{4.2}$$

where A is the set of all constituencies; w_i is the voting weight in Eq. (4.1); $v_{i,j,k} = 0$ or 1 is the vote count of a constituency i with respect to the kth type of the jth kind of RMTS. For each RMTS, its value with the highest number of votes is recommended as the final decision for the whole study area.

4.3 Multi-task deep learning architectures

To find the most suitable MTNN for RMTS decision-making, four kinds of different MTNN architectures are proposed by referring to the experience with the manual decision-making process. These networks are used to offer recommendations of RMTSs (outputs) according to given environment factors (inputs). The general framework contains two parts which are integrated in a fully connected manner: the shared convolutional neural network (CNN) and the MTNN, as shown in Fig. 4.2.

4.3.1 Shared convolutional neural network

The shared CNN is used to extract common image features for RMTS decision-making. A CNN can learn some features of different levels from

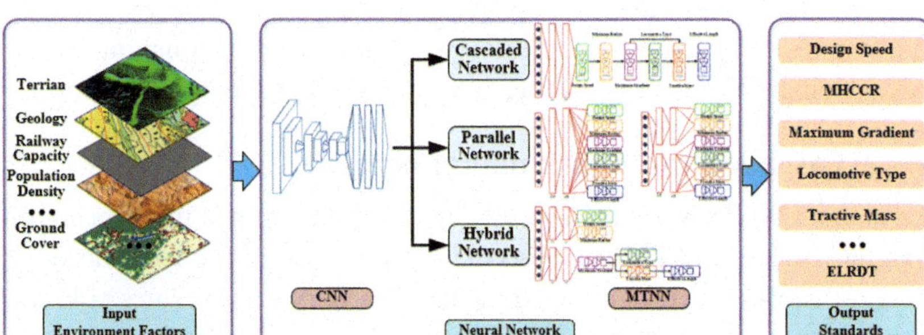

Figure 4.2 The general framework of proposed multi-task neural networks.

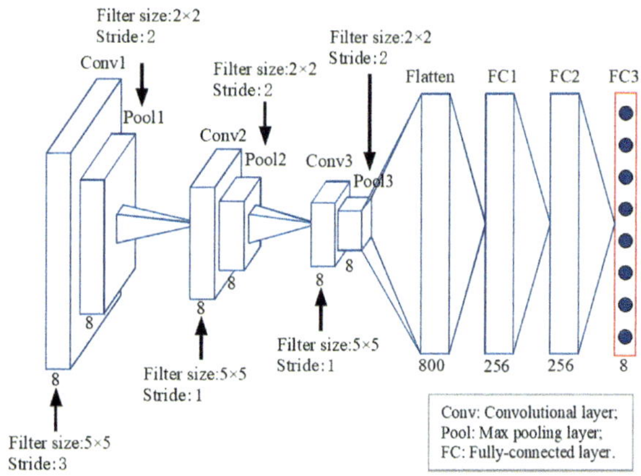

Figure 4.3 Shared convolutional neural network.

feature maps. In its initial layers a CNN captures lower-level features such as edges and textures. As the network deepens, it starts capturing higher-level features such as shapes, object parts, and positions. The detailed network structures of the two parts are described below.

Working from Fig. 4.3, the shared CNN consists of three convolutional as well as pooling layers, a flatten layer, and three fully connected layers:

(1) Convolutional layers are used to extract features of the input image. There are multiple convolutional filters in a convolutional layer. The number of convolutional filters determines the number of features that can be extracted by the layer. The size of each convolutional filter limits the size of the extracted features.

(2) The pooling layers can compress the size of features to save computational resources and improve efficiency. Then the output feature map is converted into a vector by the flatten layer.

(3) The fully connected layers establish a mapping relation between the feature maps and feature vector with an activation function ReLU, which enables the network to have nonlinear mapping capability.

4.3.2 Cascaded multi-task neural network

Tasks in a CMTNN are highly correlated and determined in a specific hierarchical order. CMTNN consists of multiple cascaded subnetworks operating at various levels of abstraction. Each subnetwork executes tasks and shares outputs with the subsequent subnetwork. Learning from shallow-level tasks significantly enhances the acquisition of knowledge required for tackling more complex deep-level tasks. In comparison to STNNs, CMTNN offers various disadvantages and advantages.

On the downside, there are several challenges associated with CMTNN. Firstly, designing and training CMTNN is inherently more complex due to considerations such as architecture, loss functions, and task hierarchy. Moreover, it tends to yield larger models. Another drawback is the potential for negative transfer among tasks, which can adversely affect overall performance. Additionally, CMTNN may require a greater amount of data, impacting both data quality and associated costs. The process of selecting appropriate tasks for designing the task hierarchy is also challenging. Lastly, it is important to note that a CMTNN is inherently less interpretable than single-task networks, which makes it more challenging to understand how the network learns from multiple tasks and makes decisions based on multiple outputs.

On the other hand, a CMTNN offers several key advantages. Firstly, it efficiently shares features and parameters among tasks, which results in reduced computational costs and improved knowledge acquisition for complex tasks. Additionally, it obtains robust representations, reduces overfitting, and enhances the performance on new data. Furthermore, it employs one task's loss as a regularizer, which effectively prevents irrelevant learning. Lastly, CMTNN exhibits the capability of seamlessly incorporating new tasks.

In a CMTNN, RMTSs are determined in a specific order. There can be different decision orders in practice for different railway projects. In most instances, designers make decisions on the design speed and the MHCCR mainly based on the railway transport capacity. With the determined railway grade and design speed, the maximum gradient, tractive mass, and locomotive type of the railway can be decided according to the information about terrain and traffic capacity. Finally, the ELRDT is determined based on the tractive mass. Accordingly, this study proposes a CMTNN for RMTS decision-making following the above task hierarchy. The overall architecture of the proposed CMTNN is illustrated in Fig. 4.4.

According to Fig. 4.4, the MTNN part contains cascaded decisions for six kinds of RMTSs, i.e., design speed, MHCCR, maximum gradient, locomotive type, tractive mass, and ELRDT. It mainly consists of RMTS decision-making modules and the links among them.

Each RMTS decision-making module contains two fully connected layers, both of which have 256 neurons and one output layer. Each RMTS's output layer contains a number of neurons that equals the quantity of categories. This setting is frequently encountered in classification tasks. It

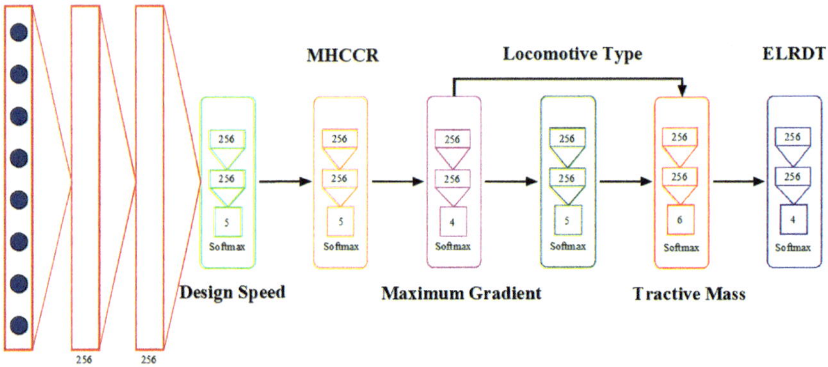

Figure 4.4 The proposed cascaded multi-task neural network.

aligns with the usage of a softmax activation function, which is employed to generate a probability distribution over the available categories. These output layers are not limited to accepting only the RMTS decision result of the adjacent lower level as input but can also use the RMTS decision results at the lower levels as additional information for higher-level standard decisions. For example, in Fig. 4.4, the tractive mass decision considers the maximum gradient at the lower level as an input in addition to its adjacent task's output (i.e., locomotive type).

In this section, the cross-entropy loss function is used as the objective function for the learning process of RMTS decision-making, which can quantitatively evaluate the difference between the manual decision results and the network outputs. It is a common choice for classification tasks, where the model's output represents probabilities between 0 and 1. It quantifies the dissimilarity between the true label and the model's predicted probability. This function is employed in this chapter due to its desirable properties:

- Nonnegativity: The loss is always nonnegative, with lower values indicating better predictions.
- Convexity: It is convex, which ensures a unique global minimum for easy optimization.
- Sensitivity to Probability Divergence: The loss increases as predicted probabilities diverge from true labels, which encourages accurate predictions.

The cross-entropy loss function for multi-classification tasks is shown in Eq. (4.3).

$$L = -\frac{1}{N}\sum_{i=1}^{N}\sum_{c=1}^{M} y_{ic}\log(p_{ic}) \tag{4.3}$$

where N is the number of training samples. M is the number of categories. y_{ic} is a symbolic function, in which 1 represents belonging to category c and 0 represents not belonging to category c. p_{ic} represents the probability that sample i belongs to class c in the neural network. The entire network is trained according to the total loss function, i.e., by summing the loss functions of six kinds of RMTSs to evaluate the network performance.

4.3.3 Parallel multi-task neural network

In contrast to CMTNN, parallel multi-task neural networks (PMTNNs) involve multiple tasks with both similarities and differences. PMTNNs share parameters among similar tasks while maintaining distinct components as parallel branches. This approach has both advantages and disadvantages in managing task interference and complexity.

Task Interference: Sharing features and parameters in PMTNN can lead to task interference. It enhances individual task performance and reduces overfitting by exploiting task correlations. However, competition among subnetworks may also compromise task performance and complicate optimization. To mitigate task interference, task selection can be employed in PMTNN based on relevance or similarity, which enhances generalization while minimizing interference risk.

Complexity: PMTNN's complexity may result in a large model scale, especially with multiple tasks or subnetworks, which increases computational resource consumption. Nonetheless, the complexity of PMTNN comes with the ability to prevent overfitting in individual tasks by sharing network parameters.

Unlike the CMTNN which builds network hierarchies, there is no hierarchy for RMTS decision-making in the proposed PMTNN. Parallel networks are designed in two main ways: overall PMTNN and grouped PMTNN. In the overall PMTNN, all tasks have the same shared layer, while in the grouped PMTNN, tasks in different groups have independent shared layers. The two structural forms of the proposed PMTNNs are illustrated in Fig. 4.5.

Generalization ability is crucial for a neural network, and determines how well it can react to new data input. In the overall PMTNN, although the shared layer can improve the overall generalization ability of the model

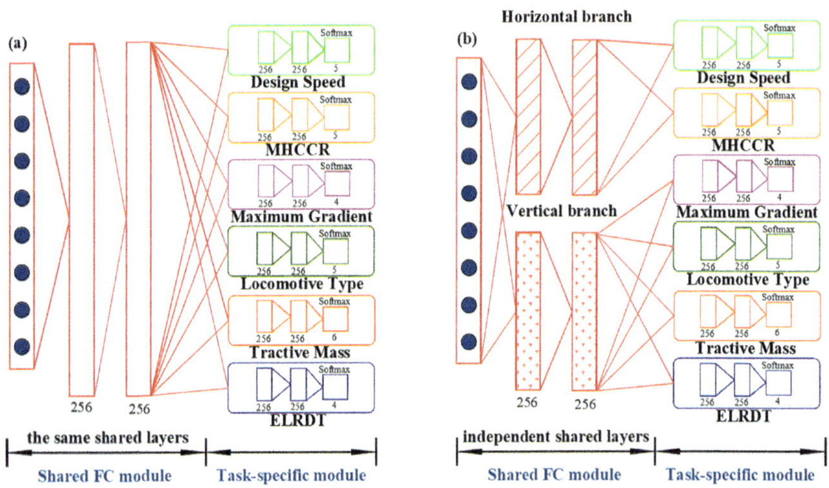

Figure 4.5 The proposed parallel multi-task neural networks: (a) Overall PMTNN; (b) grouped PMTNN.

with "training signals" shared by different RMTS decision tasks, sharing training signals may compromise the model performance if the correlation among RMTSs is not significant, which is known as negative transfer. Therefore, although the overall PMTNN can be considered for RMTS decision-making when the RMTSs are closely correlated, the grouped PMTNN can also be used for RMTS decision-making to avoid the interference of a training signal generated by tasks with relatively low correlations. Thus, two parallel multi-task baseline models are designed and tested in this study.

The main difference between the structure of the overall PMTNN and the grouped PMTNN is the way in which shared layers are set (Fig. 4.5). In the overall parallel network, all subtasks have the same fully connected shared layers (Fig. 4.5a), while in the grouped PMTNN, RMTSs with high correlation have the same shared layer, but RMTSs with low correlation (such as MHCCR and ELRDT, which are RMTSs in horizontal and vertical branch, respectively, in Fig. 4.5b) adopt independent shared layers to avoid signal interferences generated in the training process. In this study, six RMTSs are divided into two branches based on their correlations. These are the horizontal branch and vertical branch that focus on horizontal and vertical railway alignment design standards, respectively.

More specifically, among the six kinds of RMTSs on which this chapter focuses, the design speed and MHCCR mainly affect the horizontal alignment design, while the maximum gradient, locomotive type, tractive mass, and ELRDT are closely related to the vertical alignment design. Therefore, these two groups of RMTSs are determined in two parallel branches (i.e., horizontal branch and vertical branch) of the grouped PMTNN.

Furthermore, the PMTNN can be divided into shared fully connected (FC) modules and task-specific modules. For overall PMTNN, the shared fully connected module contains two fully connected layers. In the task-specific module, each RMTS decision branch contains two fully connected layers and one output layer. The number of neurons in each of the fully connected layers is 256. The only difference between grouped and overall PMTNNs is that the RMTS decision tasks of different groups in a grouped PMTNN have separate shared layers.

4.3.4 Hybrid multi-task neural network

Compared to CMTNN and PMTNN, the hybrid multi-task neural network (HMTNN) stands out by enabling tasks with different cascaded hierarchies to learn in parallel branches. For example, in an HMTNN, one branch can focus on learning the hierarchy of maximum gradient and locomotive type, while another branch can concurrently handle the hierarchy involving maximum gradient, tractive mass, and ELRDT. One parallel branch has a separate subnetwork which can mitigate the task interference. Within parallel branches, highly related tasks can be organized hierarchically, which allows deep-level tasks to benefit from the assistance of shallow-level tasks in sequential decision-making. HMTNN possesses the capability of task-specific learning, which means it can acquire unique features and representations for each task while leveraging shared information among tasks. This capability is rooted in the parallel and hierarchical relations among multiple tasks, which enables HMTNN to capture both dependencies and correlations among tasks.

In order to consider different MTNN architectures more comprehensively, this study also integrates the features of CMTNN and PMTNN to design an HMTNN. That is, some of the subtask branching structures are parallel, while other subtasks present a hierarchical structure in the network.

According to the experience of manual RMTS decision-making, the design speed and MHCCR are closely related in horizontal alignment designs, while the maximum gradient, locomotive type, tractive mass, and the

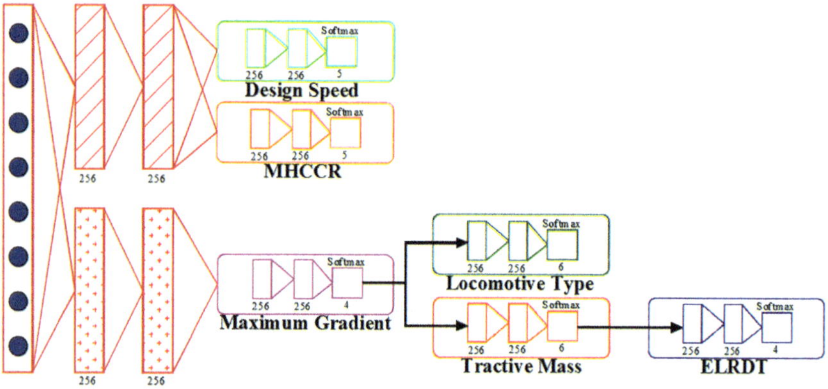

Figure 4.6 The proposed hybrid multi-task neural network.

ELRDT mainly affect vertical alignment design processes. Motivated by the idea of grouped PMTNN, two major parallel branches are designed in the HMTNN, and have their own shared layers, both of which contain two fully connected layers, as shown in Fig. 4.6.

In the horizontal alignment standard branch, the design speed and the MHCCR decision module do not have a strong hierarchical structure. Thus, they are designed as a partially parallel structure. In each decision module, two fully connected layers which contain 256 neurons first process input data and, then, output layers for design speed and MHCCR both adopt a softmax activation function [1] to make decisions regarding these two RMTSs.

In the vertical alignment standard branch, the maximum gradient information facilitates the decision on locomotive type and tractive mass, while the tractive mass has a significant effect on the ELRDT. Therefore, the overall design of the vertical alignment standard branch is a cascaded structure, in which the locomotive type and tractive mass are partially parallelized in view of their weak hierarchical correlation.

4.4 Experiments and analyses

4.4.1 Data preparation and experimental environment

In this section, a set of real-world railway cases are selected to form a sample dataset, which contains 338 railway cases from 68 railway projects.

The dataset is used for testing the effectiveness of the multi-task deep learning method regarding RMTS decision-making. Railway cases in the dataset include various railways in plain and mountainous regions. These railway cases collectively account for a total operational mileage of 58,351 km. The information regarding RMTSs and railway transport capacity is sourced partially from publicly available prefeasibility research reports and partially from China Railway Siyuan Survey and Design Group Co., Ltd., as well as China Railway Eryuan Engineering Group Co., Ltd. The database records the environmental factors and RMTSs of each railway case. For the database in this section, elevation, slope information, and railway transport capacity are considered as major environmental factors, which are characterized as three-channel images. The image characterization of these major environmental factors in the study area is presented in Fig. 4.7. As decision results, six RMTSs for the railway design are recorded in the database in form of text data (i.e., design speed, MHCCR, maximum gradient, locomotive type, tractive mass, and ELRDT).

The environmental factors, including terrain and railway transport capacity information, are input data in RMTS decision-making process. To reflect the terrain conditions in the study area, the digital elevation model data at the 30 m resolution are obtained from the "Geospatial Data Cloud" and generate two grayscale images for elevation and slope maps, as shown in Fig. 4.7. The railway transport capacity of each case is also characterized

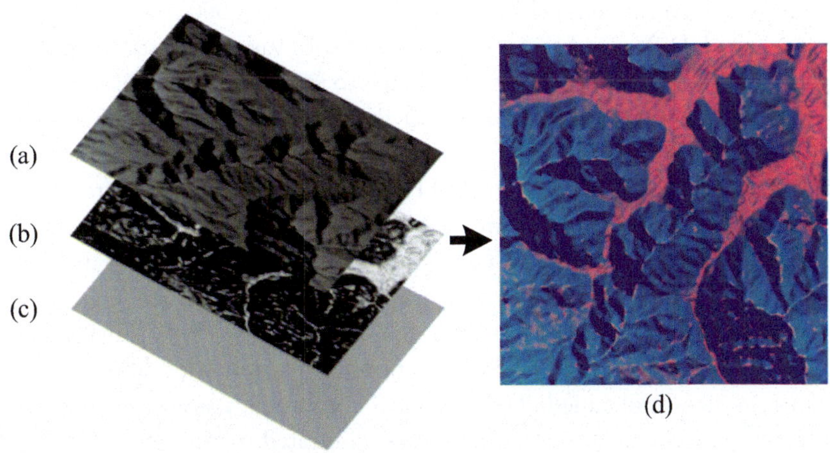

Figure 4.7 Image characterization of environmental factors in the study area: (a) digital elevation map; (b) digital slope information map; (c) transport capacity grayscale map; (d) three-channel image.

with a grayscale image, in which all pixels have the same gray value. Then, the above three grayscale images are merged into a three-channel image.

In the multi-task learning process for RMTS decision-making, the above railway case database is used in the training and testing processes, which therefore includes a training set and a testing set. These are used to train the neural network and evaluate the generalization performance of the neural network, respectively. 20% of the railway cases are included in the testing set.

As explained in the introduction, the balance between overall and local benefit in the study area should be considered in the RMTS decision-making process. Thus, an election voting strategy is proposed (see Section 4.2 for details), which divides the study area into regular constituencies. The dataset is divided into 13,816 constituencies. 20% of the training set is selected as the validation set, which is used to monitor the changes of the loss function during the training process.

The training and testing experiments are all run on a host computer with Intel i7-7700 CPU @3.60GHz, 16.0 GB of RAM, and an NVIDIA GeForce GTX 1050 GPU. They are conducted using PyCharm 2019 and TensorFlow-GPU 2.3.0. The network is trained for 40 epochs with a sample batch size of 16 and a learning rate of 3×10^{-4}.

4.4.2 Influence of constituency size on RMTSs prediction

To determine the appropriate constituency size, an experiment is conducted in which constituency side lengths of 50, 100, 333, 500, and 1000 pixels are selected. Different datasets are generated with varying constituency sizes based on the same railway cases. The aim is to investigate the impact of constituency sizes on the effectiveness of the voting election framework by comparing the performance of different sizes. The proposed voting election framework is employed to offer RMTSs recommendations on the test dataset. These recommendations are then compared to manual decision results to compute their accuracy. When experimenting with datasets of different constituency sizes, the neural networks utilized for constituency elections remain constant, which constitute the baseline PMTNN. Finally, the average accuracy of six RMTSs predictions for datasets with different constituency sizes is obtained. The impact of constituency size on the average prediction accuracy of the six RMTSs is illustrated in Fig. 4.8.

In Fig. 4.8, it is evident that both excessively large (i.e., 1000 pixels) and small (i.e., 50 or 100 pixels) constituency sizes reduce prediction accuracy.

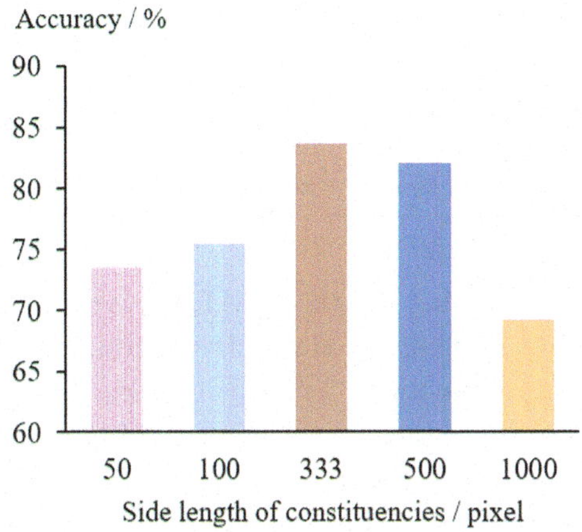

Figure 4.8 Impact of constituency size on average prediction accuracy of the six RMTSs studied.

In contrast, moderate constituency sizes increase RMTSs prediction accuracy. Notably, when the constituency's side length is 333 pixels, the RMTSs prediction accuracy reaches its peak at 83.7%. Based on these findings, a constituency size of 333×333 pixels is selected as the best one. This constituency size is consistently applied to divide study areas of railway cases with varying scales into datasets, which serve as the foundation for the subsequent experiments.

4.4.3 Parameter fine-tuning

Baseline models proposed in Section 4.3 contain many parameters, some of which can be automatically adjusted by the optimization algorithm according to the loss function, while others are hyperparameters that cannot be optimized at the training stage. In this chapter, sensitivity analyses are conducted to determine the appropriate hyperparameters for each MTNN architecture.

The design of hyperparameter sensitivity experiments must specify key hyperparameter fine-tuning items and model performance evaluation metrics. Since the above baseline model mainly consists of convolutional and fully connected layers, their key hyperparameters, including the convolutional filter size, the number of convolutional filters, and the number of neurons of the fully connected layer, are selected as the experimental fine-tuning items. The values of these hyperparameters in the baseline model are shown in Table 4.1.

Table 4.1 Hyperparameters of the PMTNN baseline model.

Hyperparameters	Value
Convolutional filter size	5
Number of convolutional filters	8
Number of neurons of the fully connected layer	256

It should be noted that all the RMTSs have discrete values, and that RMTS decision-making is a multi-classification problem. Therefore, the cross-entropy loss function, which is also known as log loss [2], is selected to evaluate the performance of each subtask. The loss function of each RMTS is summed to evaluate the comprehensive performance of four MTNNs.

4.4.3.1 Cascaded multi-task neural network

In order to find the most suitable hyperparameters for CMTNN, experiments in hyperparameter sensitivity analyses based on the developed baseline model are conducted in this study. The optimized combination of hyperparameters is determined by observing the change of the minimum loss function in the validation set by adjusting the size of the convolutional filter, the number of convolutional filters, and the number of neurons in the fully connected layer separately, by keeping other hyperparameters constant. In order to reduce the influence of the random process in the network training on the experimental results, each experiment is repeated several times for the same hyperparameter combination. The performance of the network is evaluated by averaging the values of repeated experiments. The results are shown in Fig. 4.9.

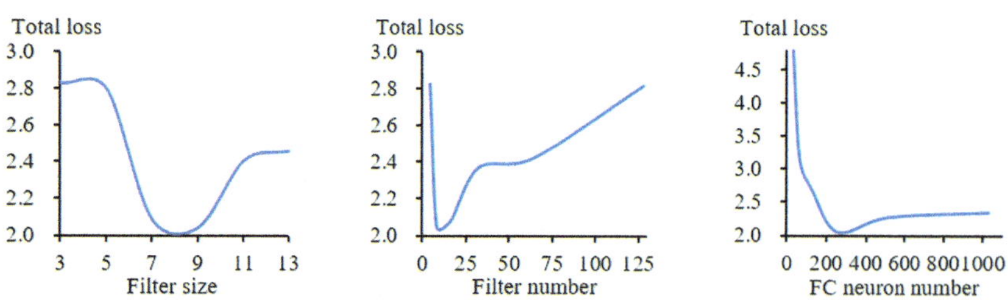

(a) (b) (c)

Figure 4.9 Effect of parameter tuning on the total loss function in the developed CMTNN: (a) effect of convolutional filter size; (b) effect of the number of convolutional filters; (c) effect of the number of neurons in the fully connected layer.

A lower total loss function indicates that the neural network fits the sample pool better. It can be seen from Fig. 4.9 that the network performance does not show a purely positive or negative correlation with the above hyperparameters. The increase of hyperparameters within a certain range improves the model performance. Beyond the reasonable range, it not only degrades the model performance but also impairs the generalization ability of the model.

In Fig. 4.9a, the loss function of CMTNN first decreases and then increases. Feature maps are the products processed by the convolutional layer based on the input image. Under the condition that other hyperparameters of the network remain unchanged, a larger convolutional filter produces a smaller feature map, which facilitates the extraction of feature patterns from the input images by the neural network. However, overly large convolutional filters excessively simplify the feature images, disabling the subsequent fully connected layers from establishing the mapping relation between them and the manual RMTS decisions. Based on the experimental results, a convolutional filter size of 7–11 pixels is recommended for CMTNN used for the RMTS decision-making to obtain concise and easily recognizable feature maps from the input images.

For the number of convolutional filters and the number of neurons of the fully connected layer, their variation patterns of the total loss function are similar, but stem from different causes. In Fig. 4.9b and c, the increases in the number of convolutional filters and the number of neurons of the fully connected layer both strengthen the model's robustness and its fitting ability to complete the learning of the sample pool. However, when the model is too complex with respect to the size of the sample pool, it is very prone to overfitting, i.e., the complex model considers not only the effective features in the training sample pool but also its irrelevant features. Hence, the model shows poor generalization ability in the validation set due to the interference from the irrelevant features. For the proposed CMTNN, this study recommends values of the number of convolutional filters and the number of neurons in the fully connected layer in the range of 4–16 and 128–512, respectively, which can effectively improve the fitting ability of the model under the premise that the complexity of the network model is comparable to the size of training set.

By specifying the baseline model and hyperparameter sensitivity analyses, the structural parameters of cascaded multi-task learning are finally determined. That is, three convolutional layers are set up, which all contain

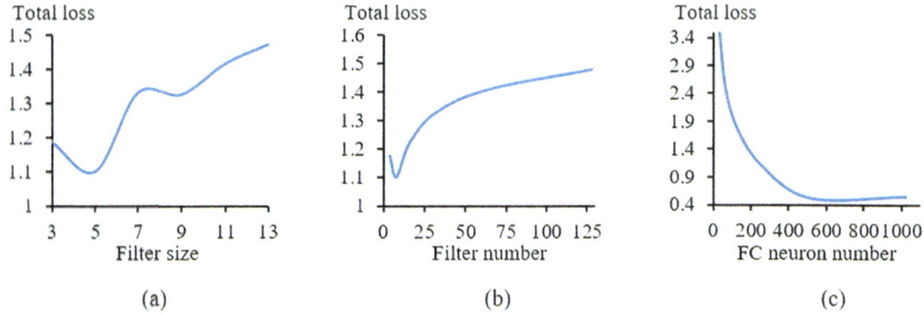

Figure 4.10 Effect of parameter tuning on the total loss function in the developed overall PMTNN: (a) Effect of the convolutional filter size; (b) effect of the number of convolutional filters; (c) effect of the number of neurons in the fully connected layer.

eight convolutional filters of 9×9 pixels size. Besides, each fully connected layer contains 256 neurons.

4.4.3.2 Parallel multi-task neural network

With other hyperparameters fixed, the convolutional filter size, the number of convolutional filters, and the number of neurons of the fully connected layer of the baseline model of the PMTNN are adjusted individually to determine the best hyperparameter combination by observing the change in the minimum loss function of the validation set. The experimental results are shown in Figs. 4.10 and 4.11.

As shown in Figs. 4.10 and 4.11, the total loss function of the network first decreases and then increases with the increase of tested parameters. In the overall and grouped PMTNNs used for RMTS decision-making, it is recommended that the suitable convolutional filter size be from 3 to 7,

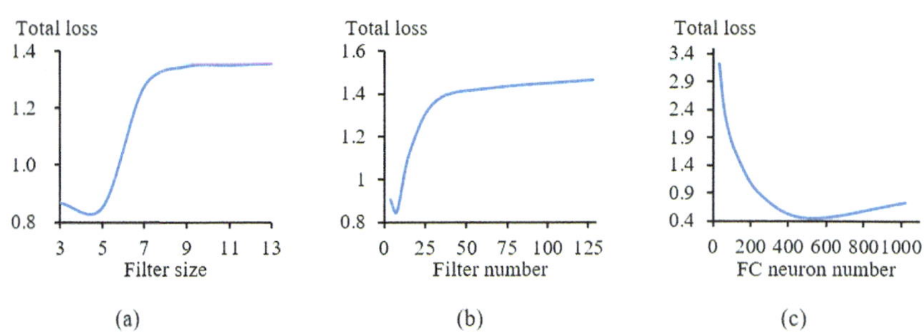

Figure 4.11 Effect of parameter tuning on the total loss function in the developed grouped PMTNN: (a) Effect of the convolutional filter size; (b) effect of the number of convolutional filters; (c) effect of the number of neurons in the fully connected layer.

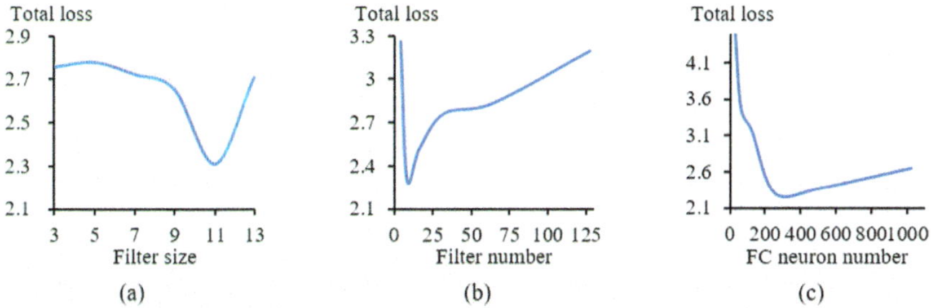

Figure 4.12 Effect of parameter tuning on the total loss function in the developed HMTNN: (a) effect of the convolution filter size; (b) effect of the number of convolution filters; (c) effect of the number of neurons in the fully connected layer.

the number of convolutional filters be between 4 and 16, and the number of neurons of the fully connected layer be from 256 to 1024. Thus, overall and grouped parallel multi-task learning networks are set up with three convolutional layers containing eight convolutional filters of 5×5 pixels size. In addition, each fully connected layer contains 512 neurons.

4.4.3.3 Hybrid multi-task neural network

With other hyperparameters fixed, the convolutional filter size, the number of convolutional filters and the number of neurons of the fully connected layer of the baseline model of the HMTNN are adjusted separately to determine better hyperparameter combinations by observing the change of the minimum loss function of the validation set. The experimental results are shown in Fig. 4.12.

Therefore, in the HMTNN, this study recommends a convolutional filter size from 9 to 13, 4 to 6 convolutional filters, and 128 to 512 neurons of the fully connected layer. Moreover, the hybrid multi-task learning network is set up with 3 convolutional layers containing 8 convolutional filters of 11×11 size. Each fully connected layer contains 256 neurons.

4.4.4 Comparison among multi-task neural networks

After determining the hyperparameter combination, a testing case set is used to study the generalization capability of the four MTNNs (see Section 4.3). To evaluate the generalization ability of different MTNNs, the manual RMTS decisions collected in the database are used as benchmarks. The average accuracy of the six kinds of RMTSs recommended by different MTNNs, compared to the manual decisions, is also computed as the evaluation index of the model performance. At the model testing stage, different

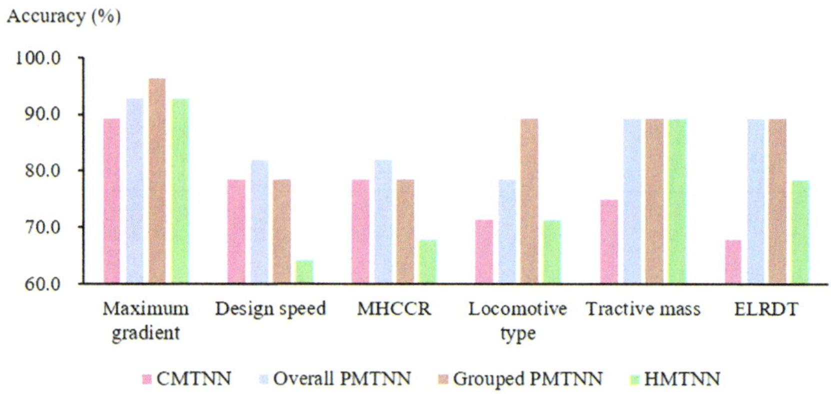

Figure 4.13 Accuracy comparison of MTNNs.

Table 4.2 Performance comparison of MTNNs.

Networks	Average accuracy rate (%)	Total network loss
CMTNN	76.8	1.66
Overall PMTNN	85.7	0.45
Grouped PMTNN	86.9	0.29
HMTNN	77.4	1.57

MTNNs are tested under the same hardware conditions with the same testing set. The results are shown in Fig. 4.13.

The following conclusions can be drawn from the experimental results.

From Fig. 4.13, it can be seen that among all MTNNs, the accuracy of the maximum gradient exceeds that of the other five RMTSs, indicating that the maximum gradient learning is the most effective in the network training process. The low accuracy of other five RMTS decisions is probably caused by the complex mapping and coupling relations among environmental factors and decision results.

As shown in Table 4.2, the accuracy of grouped PMTNN is improved compared to that of overall PMTNN, indicating that the low correlation of design speed and MHCCR with the other four RMTSs leads to performance losses during the sharing of training signals in overall PMTNN.

Among the four types of networks, the grouped PMTNN performs the best with an average accuracy of 86.9%. Details of comparisons between recommended RMTSs by grouped PMTNN and manual decision results are listed in Table 4.3 and differences are marked as bold. The results

Table 4.3 Comparison of RMTSs recommended by grouped PMTNN and the results of manual decision-making.

RMTS item	Design speed (km/h)		MHCCR (m)		Maximum gradient (‰)		Locomotive type		Tractive mass (kg)		ELRDT (m)	
Grouped PMTNN recommended (R) or manual decision (M) result	**R**	**M**	**R**	**M**	**R**	**M**	**R**	**M**	**R**	**M**	**R**	**M**
Line 1	100	100	800	800	6	6	SS4	SS4	5000	5000	1050	1050
Line 2	100	100	800	800	6	6	SS4	SS4	5000	5000	1050	1050
Line 3	100	100	800	800	6	6	SS4	SS4	5000	5000	1050	1050
Line 4	120	120	1200	1200	6	6	SS1	SS1	2500	2500	**750**	**850**
Line 5	160	160	2000	2000	6	6	**SS4**	**SS6**	4000	4000	850	850
Line 6	**100**	**200**	**800**	**3500**	6	6	SS4	SS4	5000	5000	1050	1050
Line 7	80	80	600	600	12	12	SS3	SS3	**2500**	**2000**	850	850
Line 8	80	80	600	600	12	12	**SS3**	**SS1**	2500	2500	750	750
Line 9	80	80	600	600	12	12	SS1	SS1	2500	2500	750	750
Line 10	80	80	600	600	12	12	SS1	SS1	2500	2500	750	750
Line 11	**80**	**160**	**600**	**2000**	12	12	SS3	SS3	2500	2500	850	850
Line 12	**80**	**160**	**600**	**2000**	12	12	SS3	SS3	2500	2500	**750**	**850**
Line 13	**120**	**160**	**1200**	**2000**	12	12	SS4	SS4	5000	5000	1050	1050
Line 14	120	120	1200	1200	18	18	SS4	SS4	**3500**	**3000**	850	850
Line 15	**160**	**120**	**2000**	**1200**	18	18	**HXD**	**SS4**	3500	3500	850	850
Line 16	120	120	1200	1200	**30**	**18**	SS4	SS4	**2000**	**4000**	850	850
Line 17	120	120	1200	1200	18	18	SS4	SS4	4000	4000	850	850
Line 18	160	160	2000	2000	18	18	HXD	HXD	3500	3500	850	850
Line 19	160	160	2000	2000	18	18	HXD	HXD	3500	3500	850	850
Line 20	**120**	**160**	**1200**	**2000**	18	18	HXD	HXD	3500	3500	850	850
Line 21	200	200	3500	3500	30	30	HXD	HXD	2000	2000	650	650
Line 22	200	200	3500	3500	30	30	HXD	HXD	2000	2000	650	650
Line 23	80	80	600	600	30	30	SS1	SS1	2000	2000	**650**	**750**
Line 24	120	120	1200	1200	30	30	HXD	HXD	3000	3000	650	650
Line 25	120	120	1200	1200	30	30	HXD	HXD	3000	3000	650	650
Line 26	120	120	1200	1200	30	30	HXD	HXD	3000	3000	650	650
Line 27	200	200	3500	3500	30	30	HXD	HXD	2000	2000	650	650
Line 28	200	200	3500	3500	30	30	HXD	HXD	2000	2000	650	650

Table 4.4 The performance of single-task neural networks.

STNN performance	Accuracy (%)	Loss function
Design speed	71.4	0.41
MHCCR	71.4	0.40
Maximum gradient	89.3	0.08
Locomotive type	75.0	0.23
Tractive mass	78.6	0.14
ELRDT	71.4	0.48

indicate that the grouped parallel architecture is the most suitable for decision-making regarding multiple RMTSs.

4.4.5 Comparison between multi-task and single-task networks

To illustrate the effectiveness of multi-task learning, the model performance between MTNNs and STNNs is also compared. An STNN for maximum gradient decision-making was proposed by Pu et al. [3]. In this chapter, STNNs were used for six kinds of RMTS decision-making. The STNNs for six kinds of RMTSs are the same as corresponding branches in the overall PMTNN. Experiment combinations and data inputs remain constant for STNNs and MTNNs. The decision-making performance of STNNs is shown in Table 4.4.

The experiment data in Table 4.4 yield an average decision accuracy of 76.2% for all RMTSs and a total loss function of 1.74. The following findings should be noted with the above data: Compared with the testing results of MTNNs in Table 4.2, all six STNNs perform worse than the four proposed MTNNs. Specifically, the best MTNN, i.e., the grouped PMTNN, has a significant enhancement of 10.7% in average decision accuracy compared to these STNNs. Thus, these outcomes reveal the improvement of the proposed multi-task mechanism in comprehensive RMTS decision-making.

References

[1] A. Krizhevsky, I. Sutskever, G.E. Hinton, ImageNet classification with deep convolutional neural networks, Commun. ACM 60 (6) (2017) 84–90.
[2] A. Painsky, G.W. Wornell, On the universality of the logistic loss function, in: IEEE International Symposium on Information Theory (ISIT), 2018, pp. 936–940.
[3] H. Pu, H. Zhang, P. Schonfeld, et al., Maximum gradient decision-making for railways based on convolutional neural network, J. Transp. Eng. A Syst. 145 (11) (2019), 04019047.

CHAPTER FIVE

Construction and operation cost optimization model for railway alignment

Abbreviations and corresponding full texts

CAET Annual Train Running Energy and Time Costs
CB Cost of Bridge
CC Construction Cost of the Project
CE Cost of Earthwork
CET Life-Cycle Train Running Energy and Time Costs
CL Length-dependent Construction Costs
CM Maintenance Cost of Fixed Equipment
CO Operating Costs
CR Right of Way Costs
CT Cost of Tunnel
DEM Digital Elevation Model
HAO Highway Alignment Optimization
HIP Horizontal Intersection Points
VIP Vertical Intersection Points

5.1 Introduction to the optimization model

In the field of intelligent optimization, the optimization model is usually based on reasonable simplifications and assumptions. It describes actual engineering problems using functional expressions and other methods to enable the parametric characterization and intelligent evaluation of real-world solutions. Generally, an optimization model consists of three parts: design variables, constraints, and the objective function. Its mathematical expression is as follows Eq. (5.1):

$$\begin{aligned} &\min f(\mathbf{x}) \\ &\text{s.t. } g_i(\mathbf{x}) \leq 0, i = 1, 2, \ldots, M \\ &\quad\; h_i(\mathbf{x}) = 0, i = 1, 2, \ldots, N \end{aligned} \tag{5.1}$$

where $\mathbf{x}$ is a column vector consisting of the decision variables, $f(\mathbf{x})$ is the objective function, *s.t.* (subject to) denotes constrained to, $g_i(\mathbf{x})$ and $h_i(\mathbf{x})$ are the constraints.

Intelligent Railway Alignment Design
ISBN: 978-0-443-45923-8
https://doi.org/10.1016/B978-0-443-45923-8.00009-1

Based on the comprehensive geographic information model in the previous chapter, this chapter describes the basic alignment optimization model, which mainly includes the decision variables describing the spatial position of the alignment, the constraints restricting the spatial shape of the alignment, and the objective function for evaluating the alignment alternative. The fundamental optimization model described in this chapter primarily considers an economic evaluation objective function, while subsequent chapters will extend it.

For the railway alignment optimization problem, decision variables serve as the fundamental parameters for determining the spatial position of the alignment. From a geometric perspective, five key sets of decision variables are used to characterize the three-dimensional alignment: the coordinates of intersection points (X_k, Y_k) and horizontal curve radii R_k define the horizontal alignment, while the mileage points of grade changes K_i, vertical curve radius R_v and the corresponding design elevations H_i determine the vertical profile. Since railway alignment design codes provide recommended transition-curve lengths for different railway classes and curve radii, and given their relatively minor impact on the objective function, this parameter can be fixed according to the code rather than treated as a decision variable in order to reduce model complexity. The objective function defines the relation between the decision variables and the optimization goal, serving both to evaluate alignment alternatives and to guide the optimization process. Since engineering and operational costs are significantly influenced by the spatial position of the alignment, they play a critical role in alignment optimization. Existing studies on alignment optimization models primarily focus on engineering costs, which comprehensively account for various construction cost components—such as earthwork, bridges, and tunnels—by establishing quantitative models linking spatial alignment parameters to cost estimates. For operational costs, standard cost computation methods are commonly used, estimating costs-based alignment length, curvature, and gradient [1]. However, the energy consumption and travel time are not only affected by the spatial characteristics of the alignment but also closely tied to train speed. Substantial differences in energy use and travel time can occur at different speeds along the same alignment, which cannot be accurately captured by quota-based methods. To address this limitation, this section introduces an efficient optimization approach for generating velocity-distance curve that considers both energy consumption and travel time. The method enables real-time generation of optimized velocity-distance curve, offering a reasonably accurate estimation of energy and time-related costs. When combined with fixed equipment maintenance expenses, these components constitute the operation and maintenance costs. Further details are provided in Section 5.4.2.

Constraints are the limitations on decision-making variables, which are the key to ensure that the alignment program can be practically applied. In addition to considering the minimum curve radius, maximum gradient, and other geometric constraints of the alignment, it is also necessary to account for complex and hazardous areas with sharp topographic relief.

In these areas, numerous bridges and tunnels are set. The maximum bridge span is limited by the design codes for train loading. If the required bridge length exceeds this span, the bridge type must be changed or additional piers must be set. In regions with significant terrain differences, the bridge height can reach several hundred meters. However, due to constraints such as engineering feasibility and economic viability, excessively high piers are unacceptable, which may yield an impractical alignment plan. Therefore, the mountain area alignment design needs to meet the maximum bridge height constraints. In addition, the maximum tunnel length, elevation trend and other constraints must be considered. This chapter systematically reviews and analyzes the constraints regarding alignment, structures and environment, and classifies them into geometric constraints, structural constraints and regional constraints according to different constraint characteristics and mathematical portrayal, as detailed in Section 5.3.

Below, three aspects of the constructed intelligent alignment basis optimization model will be described: decision variables, objective function, and constraints.

5.2 Decision variables

The spatial representation of the alignment serves as the foundation for constructing alignment optimization models. Decision variables are the fundamental parameters used to describe the spatial position of the alignment. Different alignment optimization models adopt different methods of spatial representation, resulting in variations in the corresponding decision variables. Early alignment optimization models typically use a polyline [2,3] or cubic spline curve [4] to express the spatial location of the alignment, and the decision variables are the coordinates of the endpoints of the polyline or the parameters of the spline curve. However, these methods express the spatial position of the alignment too roughly and have large deviations from the actual alignment positions. To address this problem, Jong developed a highway alignment optimization model (Highway Alignment Optimization, HAO) model based on the intersection point method [5], which can more accurately represent the actual spatial position of the alignment. For railways, to facilitate track maintenance and repair, vertical curves and transition curves cannot overlap, and the vertical intersection points in the vertical profile should not coincide with horizontal curves. However, in the HAO model, the number of horizontal intersection points is equal to the number of vertical intersection points in the vertical profile, and the vertical intersection points are set at the same mileage as the midpoints of horizontal curves. This inevitably causes overlaps between horizontal and vertical curves, which violates railway design codes. Therefore, considering the characteristics of railway alignments, the following decision variables for horizontal and vertical alignment should be selected:

(1) Horizontal alignment decision variables

The horizontal alignment consists of a series of basic elements, including tangent (straight) segments, circular curves, and transition curves. Designing the horizontal alignment involves the combined design and computation of these fundamental components to provide a safe and smooth-running trajectory for trains. Specifically, the positions of tangent and circular curve segments can be determined by the coordinates of horizontal intersection points and the radii of circular curves, which serve to control the horizontal alignment. Transition curves function as connectors between tangent and circular curve segments. Their positions can be derived based on the lengths of the transition curves after the locations of tangent and circular curve segments are fixed. For railway alignments, the length of transition curves can be selected according to the radius of circular curves. The Code for Design of Railway Lines (TB10098-2017) [6] specifies the lengths of transition curves associated with different circular curve radii. Therefore, to simplify the number of decision variables and reduce model complexity, this study selects the coordinates of horizontal intersection points (HIP) (X_k, Y_k) and the radii of circular curves R_k as the horizontal decision variables for the alignment, where k denotes the index of the intersection point. The method for determining the spatial position of the alignment based on the coordinates (X_k, Y_k), curve radii R_k, and corresponding transition curve lengths l_0 can be found in Ref. [7].

(2) Vertical alignment decision variables

The vertical alignment of a railway consists of grade segments with varying lengths and gradients, connected by vertical curves between adjacent segments. According to common design practices and standards, the radius of the vertical curve is typically set as a fixed value based on the railway classification. The characteristics of each grade segment can be described by its length L_i and gradient g_i, as illustrated in Fig. 5.1, where Vertical Intersection Points (VIP_i) denote the ith grade break point.

The slope section length L_i in Fig. 5.1 can be calculated based on the mileage (K_{i-1}, K_i) of adjacent grade break slope points (VIP_{i-1}, VIP_i), as in Eq. (5.2); and the slope gradient g_i can be calculated based on the difference in elevation between the adjacent grade break points in vertical alignment, ΔH_i, and the length of the slope section L_i, as in Eq. (5.3).

$$L_i = K_i - K_{i-1} \tag{5.2}$$

$$g_i = \frac{\Delta H_i}{L_i} = \frac{H_i - H_{i-1}}{L_i} \tag{5.3}$$

Therefore, the spatial position of the vertical alignment can be determined by a set of grade break points containing mileage and elevation information. After obtaining the position along the vertical alignment, the start and end positions of bridges and tunnels can be further determined based on the ground elevation along the alignment, as well as the fill height at the subgrade-bridge transition and the excavation depth at the subgrade-tunnel transition, as shown in Fig. 5.2.

Therefore, the station Ki and elevation Hi of grade break points are adopted as the decision variables for the vertical alignment of the railway.

In summary, the three-dimensional spatial position of the alignment can be represented by the following five column vectors:

Column Vectors of Horizontal Intersection Points:

X-Coordinates: $\mathbf{X} = [X_1, X_2, \ldots, X_M]^T$ Y- Coordinates: $\mathbf{Y} = [Y_1, Y_2, \ldots, Y_M]^T$

Horizontal Circular Curve Radii: $\mathbf{R} = [R_1, R_2, \ldots, R_M]^T$

Column Vectors of Vertical Intersection Points:

Mileage: $\mathbf{K} = [K_1, K_2, \ldots, K_N]^T$

Design Elevation: $\mathbf{H} = [H_1, H_2, \ldots, H_N]^T$

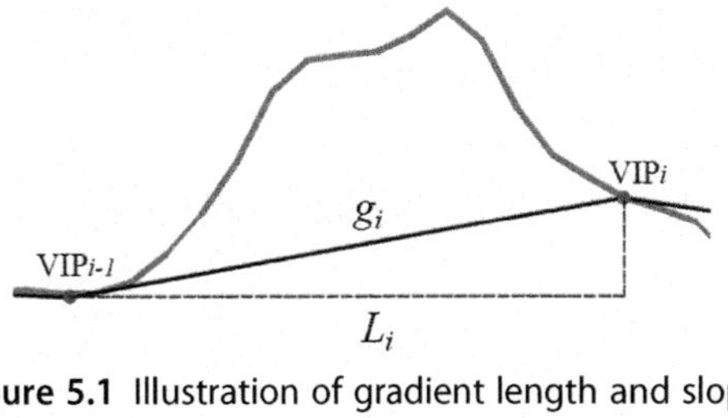

Figure 5.1 Illustration of gradient length and slope.

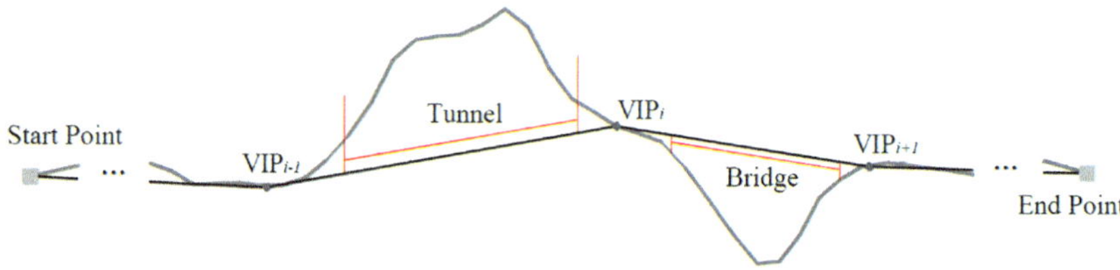

Figure 5.2 Determination of bridge and tunnel locations.

where M and N denote the number of horizontal intersection points and grade break points in the vertical alignment, respectively.

5.3 Constraints

5.3.1 Geometric standard constraints

According to the China's Code for Design of Railway Lines [6], the alignment shall satisfy the following geometric constraints: minimum curve radius, minimum curve length, minimum tangent length between two adjacent horizontal curves, minimum gradient section length, maximum gradient, algebraic difference between slope sections, and other constraints.

5.3.1.1 Horizontal alignment constraints

(1) Minimum curve radius is calculated as follows Eq. (5.4):

$$R_{\min} - R_i \leq 0 \quad i = 1, 2, \cdots, M \tag{5.4}$$

where $R_{\min}$ is the minimum radius of the specified circular curve; R_i is the curve radius of the *i*th circular curve; M is the number of horizontal intersection points.

(2) Minimum circular curve length is calculated as follows Eq. (5.5):

$$K_{\min} - \alpha_i \cdot R_i \leq 0 \quad i = 1, 2, \cdots, M \tag{5.5}$$

where $K_{\min}$ is the prescribed minimum circular curve length; α_i is the turning angle of the *i*th curve; R_i is the curve radius of the *i*th circular curve; M is the number of horizontal intersection point.

(3) Minimum tangent length is calculated as follows Eq. (5.6):

$$D_{\min} - \left[\sqrt{(x_i - x_{i-1})^2 + (y_i - y_{i-1})^2} - R_{i-1}\left|\tan\frac{\alpha_{i-1}}{2}\right| - R_i\left|\tan\frac{\alpha_i}{2}\right|\right] \leq 0, \quad i = 1, 2, \cdots, M \tag{5.6}$$

where $D_{\min}$ is the length of the shortest tangent insert length between neighboring curves; x_i, y_i are the coordinates of the ith horizontal intersection point; x_{i-1}, y_{i-1} are the coordinates of the i-1st horizontal intersection point; R_{i-1}, R_i are the radii of the i-1st and ith circular curves; $\alpha_{i\text{-}1}$, α_i are the angle of the i-1st and ith curves; M is the number of horizontal intersection points.

(4) Transition curve length
For railway alignments, the length of a transition curve is determined by the radius of the corresponding circular curve. Standard lengths for different curve radii can be obtained from the Code for Design of Railway Lines (TB10098-2017) [6] and used directly in alignment design.

5.3.1.2 Vertical alignment constraints

(1) Minimum slope segment length is calculated as follows Eq. (5.7):

$$K_{i+1} - K_i \geq d_{\min} \quad i = 1, 2, \cdots, N \tag{5.7}$$

where K_{i+1}, K_i are the mileage of the i+1st and ith vertical intersection points, respectively; $d_{\min}$ is the length of the minimum slope section; N is the number of vertical intersection points.

(2) Maximum gradient is calculated as shown in Eq. (5.8):

$$\frac{|H_i - H_{i-1}|}{K_i - K_{i-1}} - g_{\max} \leq 0 \quad i = 1, 2, \cdots, N \tag{5.8}$$

where H_i, $H_{i\text{-}1}$ are the design elevation of the ith and i-1th vertical intersection point; K_i, $K_{i\text{-}1}$ are the mileage of the ith and i-1th vertical intersection points, respectively; $g_{\max}$ is the maximum gradient value; N is the number of vertical intersection points.

(3) Gradient difference between adjacent slope sections
In order to ensure the safety of trains, the difference in gradient between two adjacent slope sections (as shown in Fig. 5.3) must not exceed the maximum specified value, as listed in Eq. (5.9):

$$\left| \frac{H_{i+1} - H_i}{K_{i+1} - K_i} - \frac{H_i - H_{i-1}}{K_i - K_{i-1}} \right| - \Delta g_{\max} \leq 0 \quad i = 1, 2, \cdots, N \tag{5.9}$$

where H_{i+1}, H_i, $H_{i\text{-}1}$ are the design elevations of the i+1st, ith and i-1st vertical intersection points; K_{i+1}, K_i, $K_{i\text{-}1}$ are the mileage of the i+1st, ith and i-1st vertical intersection points, respectively; $\Delta g_{\max}$ denotes the maximum gradient difference between two adjacent slope sections; N is the number of vertical intersection points.

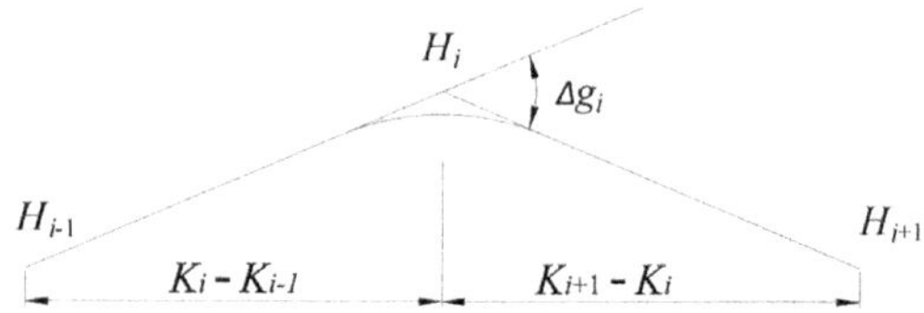

Figure 5.3 Slope algebraic difference constraints.

5.3.2 Structural constraints

Bridges, tunnels, subgrade and other structures must satisfy their corresponding constraints, such as the following: Bridges must satisfy the maximum bridge height constraints and the prohibition of vertical curves on open-decks. Tunnel design must comply with the constraints on the maximum allowable length and the requirements for either a unidirectional or an inverted-V longitudinal gradient. Subgrades must satisfy the ground gradient constraints.

5.3.2.1 Bridge constraints

(1) Maximum bridge height, calculated as follows Eq. (5.10):

$$BH_i - BH_{\max} \leq 0 \quad i = 1, 2, \cdots, N_B \tag{5.10}$$

where $BH_{\max}$ is the maximum allowable pier height; BH_i is the maximum pier height of the ith bridge; N_B is the number of bridges.

(2) No vertical curves on open-deck is calculated as follows Eq. (5.11): According to the China's Rules for the Repair of Railway Bridge and Tunnel Buildings [8], railway alignments with open-deck should be designed without vertical curve:

$$P_i \notin B, \forall P_i \in P \quad i = 1, 2, \cdots, N_P \tag{5.11}$$

where P_i is the ith vertical intersection point requiring a vertical curve; B is the set of vertical intersection points on a bridge with open-deck; P is the set of vertical intersection points requiring vertical curves along the whole alignment; N_P is the number of vertical intersection points where vertical curves are needed.

5.3.2.2 Tunnel constraints

(1) Maximum tunnel length is calculated as follows Eq. (5.12):

$$TL_i - TL_{\max} \leq 0 \quad i = 1, 2, \cdots, N_T \tag{5.12}$$

where TL_{max} is the maximum allowable tunnel length; TL_i is the length of the ith tunnel; N_T is the number of tunnels.

(2) Unidirectional or inverted-longitudinal gradient in tunnels

The vertical alignment of a railway tunnel can be designed with either a unidirectional or an inverted-V longitudinal gradient. Let the set of all vertical intersection points along the full alignment be denoted as: $VIP = \{VIP_0, VIP_1, ..., VIP_n\}$. The subset of vertical intersection points within the tunnel is defined as: $VIP_{SE} = \{VIP_S, VIP_{S+1}, ..., VIP_E\}$, where S and E, represent the grade change points before the tunnel entrance and after the tunnel exit, respectively. Let M denote the index of the highest vertical intersection point within VIP_{SE}. The VIP_{SE} is then divided into two subgroups: $VIP_{SM} = \{VIP_S, VIP_{S+1}, VIP_M\}$ and $VIP_{ME} = \{VIP_M, VIP_{M+1}, VIP_E\}$. If g_i denotes the gradient between the ith and $i+1$st grade change points, then the vertical slopes within the tunnel must satisfy the following conditions, as shown in Eq. (5.13):

$$\begin{cases} g_i \geq 0 \ \forall i \in [S, M-1] \\ g_i \leq 0 \ \forall i \in [M, E-1] \end{cases} \tag{5.13}$$

5.3.2.3 Other structural restraints

(1) Constraints on the subgrade transverse gradient

To ensure subgrade stability, the alignment must pass through an area in a way that complies with the constraints on the subgrade transverse gradient, as listed in Eq. (5.14).

$$g_c \leq g_{c\,max} \tag{5.14}$$

where g_c is the fitted gradient at the ground point of the subgrade cross-section; g_{cmax} represents the maximum allowable transverse slope of the ground for the subgrade cross-section.

5.3.3 Positional constraints

5.3.3.1 Restraints in the restricted/compulsory zone

(1) Exclusion zone constraints

In order to protect the natural environment, while reducing engineering and operation costs, the alignment should avoid wetlands and environmentally protected areas as much as possible. Here, the given wetlands and environmental protection areas are set as forbidden areas, and the intersection of all cells U_F in the forbidden areas and cells U_A in the area where the alignment passes through should be an empty set, as listed in Eq. (5.15).

$$U_F \cap U_A = \emptyset \tag{5.15}$$

(2) Must-pass area constraints
Due to economic, military, strategic, and other needs, the alignment may need to pass through certain cities, factories or mining areas. Usually, the designers will set up the must-pass area before the alignment design. To satisfy requirements, the must-pass area U_M should be included in the line traveling area U_A, as listed in Eq. (5.16):

$$U_M \subseteq U_A \tag{5.16}$$

5.3.3.2 Crossing constraints

The alignment may cross rivers, existing railways (highways) and other existing features. When the alignment crosses these features, it must satisfy the clearance constraints.

The crossing points can be found precisely where the horizontal alignment (which depends on **X, Y, R**) intersects with the center polyline of existing rivers, roads, or railways. The vertical profile of the new alignment is determined by the station **K** and elevation **H** of the vertical intersection points. Consequently, the elevation at the intersection point between the new and existing alignments can be expressed as a function $h(\mathbf{X}, \mathbf{Y}, \mathbf{R}, \mathbf{K}, \mathbf{H})$. The crossing constraints can thus be abstracted into Eq. (5.17):

$$\Delta h_{\min} - |h(\mathbf{X}, \mathbf{Y}, \mathbf{R}, \mathbf{K}, \mathbf{H}) - h_E| \leq 0 \tag{5.17}$$

where $\Delta h_{\min}$ is the minimum clearance requirement; h_E is the elevation of the existing railway, highway or river at the crossing point.

5.3.3.3 Elevation trend constraints

In sections where the natural ground gradient significantly exceeds the maximum allowable gradient, hereafter referred to as steep gradient sections, designers must continuously increase or decrease the alignment elevation until reaching the target point, without introducing reverse gradients so as to avoid losing elevation. This feature is referred to as the elevation trend, which frequently occurs in alignment selection within complex mountainous areas. Therefore, when a substantial elevation difference exists between the current point and the target point—i.e., during alignment design in steep sections—the alignment search must satisfy the elevation trend constraint.

When the target point elevation is below the current point elevation, the alignment elevation in the search region H_i should satisfy Eq. (5.18):

$$H_{i+1}\text{-}H_i < 0 \tag{5.18}$$

When the target point elevation is above the current point elevation, the alignment elevation in the search region H_i should satisfy Eq. (5.19):

$$H_{i+1}\text{-}H_i > 0 \tag{5.19}$$

5.4 Economic evaluation objective function

The objective function of the economic evaluation of the alignment mainly includes the construction, maintenance and operation costs, and the computation method for each objective function is described in Sections 5.4.1 and 5.4.2.

5.4.1 Construction cost of the main track works

The construction cost is closely related to the spatial position of the alignment, which is the main objective function of the alignment optimization model. For the alignment planning and design stage, the main consideration is the cost of earthwork (C_E), bridge (C_B), tunnel (C_T), land acquisition (C_R), and other engineering costs related to the length of the alignment (C_L). Each of these costs can be computed using the following alignment design process:

Step 1: Determine the spatial coordinates of the alignment in the horizontal plane based on the design variables **X**, **Y**, **R**;

Step 2: A series of stake points along the railway centerline is generated at specified intervals, with additional stakes introduced at critical locations. The ground elevation at each stake point is subsequently obtained through interpolation from the Digital Elevation Model (DEM);

Step 3: Determine the spatial positions of the vertical alignment based on the vertical profile design variables **K** and **H**;

Step 4: Determine the elevation at each stake point according to the vertical alignment.

Step 5: Compute the difference between ground elevation and alignment elevation at each stake point Δh.

Step 6: Based on Δh, together with the preset filling height at the subgrade–bridge transition and the preset excavation depth at the subgrade–tunnel transition, the number and start–end points of bridges and tunnels are automatically determined. This procedure completes the structural design of the subgrade, bridges, and tunnels, resulting in an integrated alignment alternative that includes the subgrades and all associated bridge and tunnel structures.

After obtaining the alignment alternative, the computation methods for the construction costs of the various engineering components are as follows:

5.4.1.1 Cost of earthwork C_E

The cost of earthworks can be computed by performing double integration to obtain the volumes of cuts and fills, which is given by Eqs. (5.20) and (5.21), after which the final cost is determined based on their respective unit prices, as shown in Eq. (5.22).

$$V_F = \int_0^L \int_{-W_l}^{W_r} \max\{h(s,x) - H(s,x), 0\} dxds \tag{5.20}$$

$$V_C = \int_0^L \int_{-W_l}^{W_r} \max\{H(s,x) - h(s,x), 0\} dxds \tag{5.21}$$

$$C_E = \sum_{i=1}^{n_f - 1} uf_i \times V_F + \sum_{i=1}^{n_c - 1} uc_i \times V_C \tag{5.22}$$

where V_F, V_C are the volume of filling and excavation, respectively; L is the total length of the alignment; W_l, W_r are the width of the left side and the right side of the subgrade, respectively; x is the lateral distance from the center pile of the alignment, s is the distance along the alignment, and $s \in [0,L]$; H is the ground elevation; h is the design elevation; uf_i, uc_i are respectively the unit prices of filling and excavation for the ith section.

5.4.1.2 Cost of bridge C_B

The bridge construction cost can be computed by multiplying the bridge length by the unit cost per linear meter. The unit cost varies with the bridge height and length; higher and longer bridges are generally more expensive. In addition, the construction costs of abutments at both ends of the bridge should also be considered, as shown in Eq. (5.23):

$$C_B = \sum_{i=1}^{N_B} u_{bi} \times l_{bi} + 2c_a \times N_B \tag{5.23}$$

where: N_B is the number of bridges; l_{bi} is the length of the ith bridge; c_a is the cost of a single abutment; u_{bi} denotes the unit cost per meter of the ith bridge, which can be expressed as a piecewise function of the pier height and span length.

5.4.1.3 Cost of tunnel C_T

Tunneling costs can be computed based on the tunnel length and the unit cost per linear meter of tunnel. The unit cost per linear meter of tunnel varies according to the tunnel length. In addition, the cost of constructing the tunnel portal at each end of the tunnel should be considered, as shown in Eq. (5.24),

$$C_T = \sum_{i=1}^{N_T} u_{ti} \times l_{ti} + 2c_w \times N_T \tag{5.24}$$

where: N_T is the number of tunnels; l_{ti} is the length of the ith tunnel; c_w is the cost of a single portal; u_{ti} is the cost of the ith tunnel per meter, which can be represented as a piecewise function of burial depth and total tunnel length.

5.4.1.4 Right of way costs C_R

After the cross-section design is finalized, a land widening allowance is added at the toe of the slope and at the crest of the valley for each cross-section, defining the effective land width of the cross-section. Considering that the land unit price may vary across different regions along the route, the corresponding unit price should be extracted according to the land-use type at each station number, and the land acquisition cost should be computed and summed for each stake point, as shown in Eq. (5.25),

$$C_R = \sum_{i=1}^{N_K-1} u_{ri} \times \left[(k_{i+1} - k_i) \times \frac{w_i + w_{i+1}}{2} \right] \tag{5.25}$$

where: N_K denotes the total number of stakes; u_{ri} represents the unit price of land acquisition for the ith cross section, which can be expressed as a function of the land type; k_i denotes the mileage of the ith station; w_i is the land width of the ith station.

5.4.1.5 Length-dependent construction costs C_L

Length-dependent construction costs, including track-laying costs and electric facility costs, can be computed by multiplying the unit length costs with the alignment length, as listed in Eq. (5.26):

$$C_L = u_l \times L \tag{5.26}$$

where: u_l is the cost of the project per linear meter associated with the length and L is the length of the alignment.

In summary, the construction cost of the project, C_C, is:

$$C_C = C_E + C_B + C_T + C_R + C_L \tag{5.27}$$

The quantitative model of the construction cost of the alignment can be determined with Eq. (5.27), thus determining the value of the total cost objective function.

5.4.2 Maintenance and operating costs of the mainline

The major components of operating costs—namely, train operation energy consumption, passenger time cost, and fixed facility maintenance cost—are strongly influenced by the spatial characteristics of the alignment. Therefore, the alignment optimization model in this chapter focuses on these three operating costs.

5.4.2.1 Train operation energy cost and passenger travel time cost

The train operation energy and passenger time costs constitute an important component of the mainline maintenance and operation expenses. These costs can be computed based on train energy consumption, travel time, and the corresponding unit prices.

Train energy consumption is primarily determined by the operating distance and the magnitude of the resistance encountered, which includes basic resistance and additional resistance. The additional resistance, which depends on the alignment, comprises gradient resistance, curve resistance, and tunnel resistance. In contrast, the basic resistance is independent of the alignment and is mainly influenced by the train operating speed, as expressed in Eq. (5.28):

$$w_0 = a + bv + cv^2 \tag{5.28}$$

where: w_0 is the unit basic resistance; v is the train operating speed; and a, b and c are empirical constants. Specifically, a primarily reflects resistance related to train weight, including friction between axle journals and bearings, as well as wheel–rail rolling friction. *b* mainly accounts for resistance caused by wheel slip on the rails, impacts and vibrations due to track irregularities, and wheel tread wear. *c* primarily represents aerodynamic resistance.

Train energy consumption exhibits significant variation when operating at different speeds along the same alignment, with lower speeds generally resulting in reduced energy expenditure. To minimize operational energy consumption, trains should ideally operate at relatively low speeds. However, excessively slow operation increases travel time, thereby raising passenger travel time costs. Consequently, the trade-off between train operation energy and travel time costs for a given alignment can be evaluated based on an optimized velocity-distance curve.

For a given alignment, numerous optimization methods for velocity-distance curve have been proposed, such as ant colony optimization and genetic algorithms. These methods typically employ iterative evolutionary search strategies, generating a large number of candidate velocity-distance curves during the optimization process. As a result, optimizing a velocity-distance curve for a single alignment may require several minutes to several hours. Moreover, in the context of alignment optimization, the velocity-distance curve optimization problem occurs in a dynamically changing environment, where new alignment schemes continuously emerge. Applying the aforementioned methods to optimize velocity-distance curve for all generated alignments would substantially increase computational time, rendering the approach impractical.

To ensure computational efficiency, existing alignment optimization approaches often rely on quota-based methods, in which train operation is computed according to alignment length, curvature, and gradient, or assume a constant target speed such as a balanced speed or maximum allowed speed, without explicitly optimizing the velocity-distance curve. Both approaches neglect the effect of operational speed on energy consumption, and quota-based methods fail to account for passenger travel time costs, resulting in inaccurate estimates of operating expenses.

To address these limitations, an efficient velocity-distance curve optimization method is proposed, which simultaneously considers train energy consumption and travel time, enabling a scientifically accurate evaluation of train operation energy and passenger time costs. The method is detailed as follows:

(1) An optimization method for velocity-distance curves considering both energy consumption and travel time

In the process of alignment optimization, accurately evaluating the operating cost of candidate alignments requires real-time acquisition of the velocity-distance curve along each alignment. Based on this curve, a train's energy consumption and time-related costs can be computed to guide the optimization direction. To address the requirements of the alignment optimization stage, this section introduces an efficient method for optimizing velocity-distance curves that accounts for both energy consumption and time costs. By formulating functional relations among train speed, energy consumption, and time cost, and then applying differentiation, the optimal speed that minimizes the combined cost is derived as follows:

Train running resistance constitutes the primary source of traction energy consumption. The relation between unit-distance energy consumption and train speed can be approximated by the relation between running resistance and speed. It can be expressed as a function of train speed. Additional resistance is essentially speed-independent and primarily determined by track conditions—such as gradients, curves, and tunnels—and can be expressed as a function of the equivalent gradient. When the equivalent gradient varies, the relation between running resistance and speed is affected accordingly, resulting in different unit-distance energy consumption–speed relationships for different track segments. Therefore, the alignment must be divided into segments according to the equivalent gradient, and for each segment, a functional relation between unit-distance energy consumption and train speed is formulated. By differentiating these functions, the optimal target speed for each segment can be obtained, which then guides train traction control to produce an optimized velocity-distance curve.

① Alignment Sector Calculation Unit Classification

The preceding analysis indicates that the relation between train running resistance and speed is influenced by the alignment's equivalent gradient. Accordingly, the alignment can be partitioned into computational segments based on this gradient. To ensure precise stopping at the alignment terminus, the braking initiation point must be calculated prior to segmenting the alignment.

In order to compute the position of the braking point, the following two assumptions are made in this paper:

1) The common braking deceleration during actual train operation is 50%–80% of the emergency braking [9].

2) The running speed of the train during braking is controlled as the speed at the position of the transition point between constant moment and constant power in the traction characteristic curve of the train V_n.
Based on the above two assumptions, the train stopping brake position S_b can be computed with Eqs. (5.29) and (5.30) according to the alignment length L:

$$L_b = \frac{V_n^2}{2 \times \lambda \times a_{\max}} \tag{5.29}$$

$$S_b = L - L_b \tag{5.30}$$

where: L_b is the running distance required for the train to decelerate from V_n down to 0; λ for the common braking acceleration coefficient, take 0.5–0.8; $a_{\max}$ for emergency braking acceleration.

After obtaining the location of the train braking point, the alignment can be divided according to the calculated gradient and the location of the braking point, each slope section with the same calculated gradient before the braking point is a subsection, and after the braking point (including the slope section where the braking point is located) is a subsection, as shown in Fig. 5.4.

② The target velocity solution for each subsector
The energy consumption of train operation is closely related to its running resistance. For e.g., when a train operates at constant speed on a line with a constant equivalent gradient, its energy consumption can be approximated by the running resistance and operating distance. Therefore, for a given unit distance of alignment, the energy consumption of train operation (E(v)) can be computed as expressed in Eqs. (5.31) and (5.32).

$$E(v) = \frac{[w_0(v) + w_a] \times M \times 1000}{3.6 \times 10^6} \quad (kwh) \tag{5.31}$$

$$w_0(v) = a + bv + cv^2 \quad (N/t) \tag{5.32}$$

where $w_0(v)$ is the basic resistance (N/t), $w_a(v)$ is the additional resistance (N/t), M is the mass of the train (t), v is the train speed (km/h), a, b, c are the empirical constants. a is mainly affected by the weight of the train related to the impact of the resistance, including axle necks and bearings between the

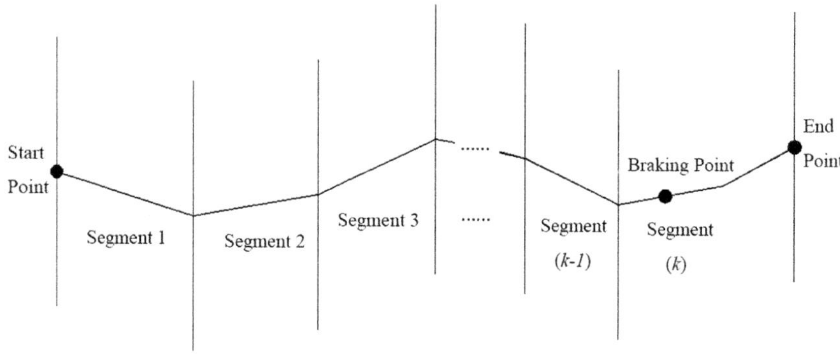

Figure 5.4 Alignment segmentation.

friction resistance, the wheels and the rail rolling friction resistance; b is mainly affected by the wheel on the rail sliding friction resistance, track unevenness and wheel tread scuffing caused by impact and vibration resistance; c is mainly affected by air resistance.

Then, the train operating energy cost $C_e(v)$ can be computed based on the unit energy cost c_e and the train operating energy consumption $E(v)$, as shown in Eq. (5.33):

$$C_e(v) = E(v) \times c_e \quad (rmb) \tag{5.33}$$

Since each sub-section is divided based on the calculated gradient of the alignment, the additional train resistance w_a remains constant in each sub-section, and the energy cost is only affected by the basic train resistance $w_0(v)$. The relation between the energy cost of train operation and its operating speed is shown in Fig. 5.5.

For a given unit distance of alignment, the train running time is also closely related to its running speed, as given in Eq. (5.34):

$$T(v) = \frac{1}{v} \quad (h) \tag{5.34}$$

Then, the train running time cost C_t can be computed from the unit time cost c_t and the train running time $T(v)$, as expressed in Eq. (5.35):

$$C_t(v) = T(v) \times c_t \quad (rmb) \tag{5.35}$$

Based on the analysis above, the relation between train operating energy and time costs and the operating speed for a given unit length of alignment (e.g., 1 km) can be expressed as in Eq. (5.36):

$$C_{E\&T}(v) = \frac{\left[\left(a + bv + cv^2\right) + w_a\right] \times M \times 1000}{3.6 \times 10^6} \times c_e + \frac{1}{v} \times c_t \qquad (5.36)$$

By taking the derivative of v, Eq. (5.37) is obtained.

$$C'_{E\&T}(v) = \frac{(b + 2cv) \times M \times 1000}{3.6 \times 10^6} \times c_e - \frac{1}{v^2} \times c_t \qquad (5.37)$$

This chapter assumes that only one type of train will operate on that alignment., the parameters b, c, and M in Eq. (5.37) are known constants. Therefore, the operating speed v_o, which minimizes the energy and time costs of train operation, can be computed from the given unit energy and time costs.

In addition to the optimal speed, the target speed of each sub-sector is also constrained by the equalization speed and limiting speed of the alignment. Therefore, the target speed v_t of each sub-section is the minimum of the optimal speed v_o, the equilibrium speed v_a and the limiting speed v_l, as shown in Eq. (5.38).

$$v_t^i = \min\left\{v_o^i, v_a^i, v_l^i\right\} \qquad (5.38)$$

The speed limit v_l is the maximum speed permitted to ensure the safety of train operation on a downhill section, which can be computed based on the alignment's gradient.

The equilibrium speed v_a is primarily influenced by the locomotive's traction power and the train's running resistance. It is defined

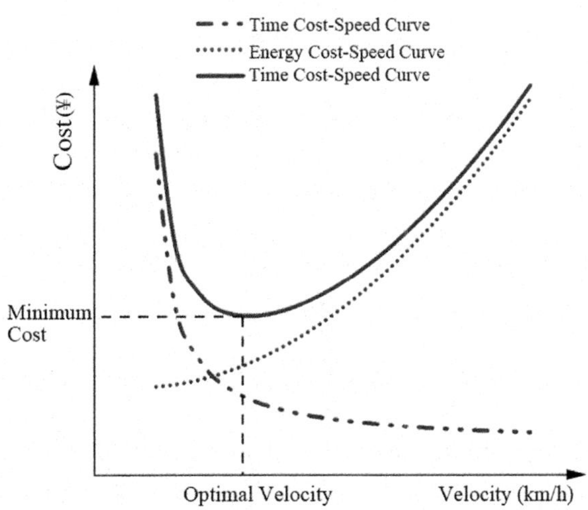

Figure 5.5 Plot of cost versus speed.

as the train's running speed when the net force acting on it is zer-o—that is, when the traction force is equal to the running resistance. Since the train is subjected to different additional resistances in various sub-sections, its equilibrium speed also varies accordingly.

③ Obtain the optimized train operating speed curve according to the target speed of each sub-section

Train operational modes include acceleration, constant-speed running, coasting, and braking. As a train enters each sub-section, its operational mode is automatically switched based on a comparison between its current running speed $V_c(S_i)$ and the target speed $V_t(S_i)$ for that sub-section. The specific conditions for mode switching are as follows:

1) When $V_c(S_i) < V_t(S_i)$, the train first accelerates, and when the target speed $V_t(S_i)$ is reached for the subsector, it stays constant until the end of the subsector.
2) When $V_c(S_i) = V_t(S_i)$, the train operates at a constant speed until the end of the sub-section.
3) When $V_c(S_i) > V_t(S_i)$, the train initially decelerates Once the target speed $V_t(S_i)$ for the sub-section is reached, it stays constant until the end of the sub-section.
4) Braking is applied under two conditions: when the train reaches a designated braking point to stop, or when its running speed reaches the speed limit and the running resistance is in the same direction as its movement.

(2) The method of computing the energy consumption and time of train operation

Under different operation modes, there are significant differences in the energy consumption and time required for a train to travel the same distance. In the acceleration mode, the magnitude of the train's traction acceleration can be calculated based on the resultant force acting on the train. Moreover, to ensure passenger comfort and safety, the acceleration should be computed using Eq. (5.39):

$$a(v) = \frac{F(v) - w(v) \times M}{M \times 1000} \quad \left(m / s^2\right) \tag{5.39}$$

where $a(v)$ is the train's acceleration at speed v (m/s^2); $F(v)$ is the train's traction force at speed $v(N)$; $w(v)$ is the train's running resistance at speed v (N/t); M is the train's mass (t); the train's running speed v is in units m/s.

The energy consumption $E_t(v)$ and time $T_t(v)$ required to increase the train speed by one speed interval can be computed after obtaining the train operating acceleration $a(v)$ by using Eqs. (5.40)–(5.42):

$$E_t(v) = \frac{1}{2} M \times 10^3 \times \left[(v + \Delta v)^2 - v^2 \right] + w(v) \times M \times L_t \quad (J) \tag{5.40}$$

$$L_t(v) = \frac{(v + \Delta v)^2 - v^2}{2 \times a(v)} \quad (m) \tag{5.41}$$

$$T_t(v) = \frac{\Delta v}{a(v)} \quad (s) \tag{5.42}$$

When the operating speed reaches the target speed, switch the operation mode to constant speed. At this time, a traction force with the same magnitude as the train's running resistance but in the opposite direction is applied. Therefore, in the constant - speed operation mode, the train's energy consumption $E_u(v)$ depends on the train's running resistance and the distance traveled, and its calculation formula is as follows in Eq. (5.43):

$$E_u(v) = w(v) \times M \times L_u \quad (J) \tag{5.43}$$

The running time $T_u(v)$ in the uniform running mode can be calculated based on running distance and operating speed, as shown in Eq. (5.44).

$$T_u(v) = \frac{L_u}{v} \quad (s) \tag{5.44}$$

When the train's operating speed exceeds the target speed (where the target speed is lower than the speed limit), a coasting mode is adopted. In this mode, the train is subjected only to running resistance, and the traction force is zero. As a result, the energy consumption during coasting is zero. The train's travel time is influenced by its operating speed and the running resistance it encounters. If the running resistance is positive (i.e., acting opposite to the direction of motion), the train decelerates. Conversely, if the running resistance is negative (i.e., acting in the same direction as the train's motion), the train accelerates. The acceleration or deceleration of the train is computed using Eq. (5.45):

$$a_c(v) = \frac{-w(v)}{1000} \quad (m/s^2) \tag{5.45}$$

Based on $a_c(v)$, the running time of the train at each speed interval can be computed for the coasting mode of operation using Eq. (5.46):

$$T_{coa}(v) = \begin{cases} \dfrac{-\Delta v}{a_c(v)} & a_c(v) < 0 \\ \dfrac{\Delta v}{a_c(v)} & a_c(v) > 0 \end{cases} \quad (s) \tag{5.46}$$

When the train's operating speed exceeds the speed limit and the running resistance is in the same direction as its movement, a braking mode of operation is required. To maintain a constant speed, a braking force equal in magnitude and opposite in direction to the running resistance is applied. Assuming the train is equipped with a regenerative braking system, a portion of the energy generated during braking can be recovered. For the purposes of this study, it is assumed that this recovered energy is stored on-board the train, and is computed by Eq. (5.47):

$$RE_1(v) = \eta \times w(v) \times M \times L_{b1} \quad (J) \tag{5.47}$$

where, η is the regenerative energy conversion rate; L_{b1} is the train braking distance (m).

The train running time in braked operation mode can be determined by Eq. (5.48):

$$T_{bra1}(v) = \frac{L_{b1}}{v} \quad (s) \tag{5.48}$$

When the train reaches the braking point, it is also necessary to adopt the braking operation mode to stop the train at the end of the alignment. The train recovers the energy generated by braking and, at the same time, overcomes resistance to produce operating energy consumption. The energy calculation as listed in Eq. (5.49):

$$RE_2(v) = \eta \times \left[\frac{1}{2} M \times 10^3 \times \left[(v - \Delta v)^2 - v^2\right] + w(v) \times M \times L_{b2}(v)\right] \quad (J) \tag{5.49}$$

where L_{b2} is the braking distance (m) under the maximum braking acceleration when the running speed of the train is reduced by 1 m/s, which is given by Eq. (5.50).

$$L_{b2}(v) = \frac{(v - \Delta v)^2 - v^2}{-2 \times a_{\max}} \quad (m) \tag{5.50}$$

The running time of the braking phase is expressed as Eq. (5.51):

$$T_{b2}(v) = \frac{v}{a_{\max}} \quad (s) \tag{5.51}$$

Based on the above formulas, the train running energy and time for different operational modes can be obtained. By accumulating the train running simulations for a given speed interval, the energy consumption and running time for a single one-way trip along the alignment can then be determined.

(3) Life-Cycle train running energy and time costs
Based on the methods described above, the one-way trip energy consumption E_r and time T_r for a train can be determined. Using the unit prices for energy consumption c_e and time c_t, the total energy and time costs for a single round trip can then be calculated. Note that energy consumption and travel times may vary between the two directions. By combining the annual transportation demand V_a with the transport capacity per round trip V_r, the annual number of train round trips can be found, which in turn gives the annual train running energy and time costs C_{AET}, the formula is shown in Eq. (5.52):

$$C_{AET} = \frac{V_a}{V_r} \times (E_r \times c_e + T_r \times c_t) \tag{5.52}$$

Based on the annual running energy consumption and time costs of trains C_{AET}, the total life-cycle running energy, annual maintenance costs and time costs C_{ET} can be obtained using the compound interest present value formula. The computation is listed in Eq. (5.53):

$$C_{ET} = C_{AET} \times \frac{(1+i)^n - 1}{i(1+i)^n} \tag{5.53}$$

where: i is the interest rate and n is the design life.

5.4.2.2 Maintenance of fixed equipment

Based on the train operating cost calculation method in the Railway Engineering Design Manual (Alignment) [10], the formula for fixed equipment maintenance costs can be expressed as Eq. (5.54):

$$C_M = L \times c_{xz} + \sum N_c \times c_{zz} \tag{5.54}$$

where: C_M is the maintenance cost of fixed equipment; L is the Mainline length; N_c is the number of stations (including intermediate stations and section stations); c_{xz} refers to the comprehensive maintenance standard cost associated with the mainline length, including mainline maintenance costs, maintenance costs for block section signaling and communication equipment, as well as for traction substations and overhead contact line systems; c_{zz} refers to the comprehensive maintenance standard cost associated with railway stations.

The operating costs C_O can be obtained as shown in Eq. (5.55):

$$C_O = C_{ET} + C_M \tag{5.55}$$

In summary, a fundamental optimization model for railway alignment is established, with engineering construction, maintenance and operation costs as the objective functions, and geometric, structural, and regional constraints as the constraint conditions, as shown in Eq. (5.56).

$$\begin{aligned} &\min f_e(\mathbf{X}, \mathbf{Y}, \mathbf{R}, \mathbf{K}, \mathbf{H}) \\ &s.t.\ c_g\,(\mathbf{X}, \mathbf{Y}, \mathbf{R}, \mathbf{K}, \mathbf{H}) \le 0 \\ &\qquad c_s\,(\mathbf{X}, \mathbf{Y}, \mathbf{R}, \mathbf{K}, \mathbf{H}) \le 0 \\ &\qquad c_a\,(\mathbf{X}, \mathbf{Y}, \mathbf{R}, \mathbf{K}, \mathbf{H}) \le 0 \end{aligned} \tag{5.56}$$

where: $f_e(\mathbf{X}, \mathbf{Y}, \mathbf{R}, \mathbf{K}, \mathbf{H}) = \Delta\cdot(C_C + C_O)$, representing the railway construction and the maintenance and operation costs; $c_g(\mathbf{X}, \mathbf{Y}, \mathbf{R}, \mathbf{K}, \mathbf{H})$, $c_s(\mathbf{X}, \mathbf{Y}, \mathbf{R}, \mathbf{K}, \mathbf{H})$, and $c_a(\mathbf{X}, \mathbf{Y}, \mathbf{R}, \mathbf{K}, \mathbf{H})$, respectively, representing the geometrical constraints, structural constraints, and regional constraints, respectively.

5.5 Additional considerations

5.5.1 Two-directional operations

Most operations on designed alignments, especially those of passenger trains, consist of train round trips. The performance and operating costs are unlikely to be symmetrical in the two directions, especially in hilly and mountainous regions. Thus, the train operations and costs should be evaluated separately in the two directions, even if the tracks in the two directions are parallel and adjacent along the entire alignment. The constraints on the candidate alignments should be applied from whichever of the two directions imposes the more stringent constraints. Then, the costs of train round trips can be determined by adding the costs obtained separately for the two directions.

5.5.2 Multiple train types

An alignment may be used by different train types (e.g., passenger and freight, express and slower local) in the near and distant futures. Tradeoffs among the various expected train types should affect the alignment optimization. These tradeoffs can be optimized by adding the costs of various train types with appropriate weighting factors for such trains (e.g., based on their relative numbers, values or importance) in a total cost objective function. The constraints on the alignment should be those due whichever train type imposes the more stringent constraints.

5.5.3 Effects of traffic and user characteristics

The amounts and values of the expected traffic on an alignment can significantly affect its optimized design. Thus, an optimized alignment in a hilly or mountainous region should tend to become straighter, more reliant on bridges and tunnels, and more expensive, as the expected number and size of trains using it or the values of passenger and freight times increase. An appropriate objective function will automatically guide an intelligent alignment search toward such characteristics. Uncertainties about future traffic, user characteristics (including values of time), operating costs, and other factors can be explored with sensitivity analyses and incorporated in the alignment design process through robust optimization methods [11–13].

References

[1] W. Li, H. Pu, P. Schonfeld, et al., Methodology for optimizing constrained 3-dimensional railway alignments in mountainous terrain, Transport. Res. C Emerg. Technol. 68 (2016) 549–565.
[2] N.A. Parker, Rural highway route corridor selection, Transport. Plann. Technol. 3 (1977) 247–256.
[3] A.K. Turner, A decade of experience in computer aid route selection, Photogramm. Eng. Remote Sens. 44 (1978) 1561–1576.
[4] E.P. Chew, C.J. Goh, T.F. Fwa, Simultaneous optimization of horizontal and vertical alignments for highways, Transp. Res. Part B Methodol. 23 (5) (1989) 315–329.
[5] J.C. Jong, P. Schonfeld, An evolutionary model for simultaneously optimizing three-dimensional highway alignments, Transp. Res. Part B Methodol. 37B (2) (2003) 107–128.
[6] National Railway Administration, Code for Design of Railway Lines (TB10098-2017), China Railway Publishing House, 2017.
[7] D. Hirpa, W. Hare, Y. Lucet, et al., A bi-objective optimization framework for three-dimensional road alignment design, Transport. Res. C Emerg. Technol. 65 (2016) 61–78.

[8] National Railway Administration, Rules for Repair of Railway Bridge and Tunnel Structures, China Railway Publishing House, 2010.

[9] S. Yi, Principles of Railway Location and Design, Academic Press, 2017.

[10] Ministry of Railways First Survey and Design Institute, Railway Engineering Design Manual: Alignment, China Railway Publishing House, 1994.

[11] T. Song, H. Pu, P. Schonfeld, et al., Robust optimization method for Mountain railway alignments considering preference uncertainty for costs and seismic risks, ASCE-ASME J. Risk Uncertain. Eng. Syst., Part A Civil Eng. 8 (1) (2022) 1–18.

[12] T. Song, H. Pu, T.Y. Yang, et al., Optimizing net present values of risk avoidance for Mountain railway alignments with seismic performance evaluation, Comput. Aided Civ. Infrastruct. Eng. 39 (7) (2024) 944–962.

[13] T. Song, H. Pu, T.Y. Yang, et al., Data-driven distributionally robust optimization of railway alignments in earthquake-prone regions considering active fault zone risks, Comput. Aided Civ. Infrastruct. Eng. 40 (2025) 4321–4341.

CHAPTER SIX

Alignment optimization model considering geological disaster risk assessment

Abbreviations and corresponding full texts

B-regions Buffer Regions
CGIM Comprehensive Geographic Information Model
F-regions Fuzzy Regions
IV Information Value
IVM Information Value Model
KaRAD-KG Karst Railway Alignment Design Knowledge Graph
KG Knowledge Graph
NDVI Normalized Difference Vegetation Index
O-regions Geological Outbreak Regions
R-region Reachable Region
RAD-KG Railway Alignment Design Knowledge Graph

In complex and rugged mountainous regions, the terrain is highly undulating, and gravity-driven slope hazards are prevalent. Geological disasters such as rockfalls, landslides, and debris flows occur frequently, making geological disaster risk a critical factor that threatens operational safety and significantly influences the decision-making process in alignment optimization. Therefore, in addition to the economic evaluation of alignment alternatives, it is a practical necessity to carry out quantitative assessments of geological disaster risk when optimizing railway alignments in such environments.

The geological disaster risk along a railway alignment is jointly determined by two components: the susceptibility of geological hazards and the vulnerability of alignment structures. Geological hazard susceptibility refers to the likelihood or probability of a hazard occurring within a certain period, which is typically evaluated through a comprehensive analysis of multiple natural factors, including topography, geological conditions, as well as meteorological and hydrological parameters. Existing methods for evaluating susceptibility include multi-factor comprehensive evaluation, information entropy models, and artificial neural networks [1]. The multi-factor comprehensive evaluation method involves selecting indicators, assigning weights, and integrating the indicators to obtain a susceptibility value. While this approach allows flexibility for decision-makers to adjust

Intelligent Railway Alignment Design
ISBN: 978-0-443-45923-8
https://doi.org/10.1016/B978-0-443-45923-8.00014-5

weights based on practical experience, it relies heavily on subjective judgment and does not consider interdependencies among indicators. In contrast, the information entropy model provides an objective means to quantify uncertainty by measuring the amount of information each indicator contributes to hazard susceptibility. This method establishes a mapping between influencing factors and susceptibility from an objective standpoint but does not account for the correlation or complex dependencies between indicators. Artificial neural networks, on the other hand, are capable of capturing complex nonlinear relations between influencing factors and susceptibility. They can leverage data correlations to improve predictive accuracy and have shown strong performance in predicting hazard susceptibility around existing railway infrastructure. However, their generalizability to alignment planning scenarios in the design phase still requires further investigation.

The vulnerability of alignment structures refers to the degree and likelihood of damage that such structures may sustain under the influence of geological hazards. Common assessment methods include limit equilibrium analysis [2], numerical simulation [3], reliability theory [4], and field investigation [5]. The limit equilibrium method evaluates structural vulnerability by analyzing equilibrium conditions under critical states, but its reliance on simplified assumptions may introduce uncertainties in the results. Numerical simulation methods build mathematical models of structures and simulate their behavior under various load conditions, allowing sensitivity analyses of key parameters. Reliability theory further incorporates uncertainties in materials, loads, and other parameters to quantitatively assess structural reliability; however, its applicability is limited in complex real-world scenarios due to the assumption of specific probability distributions. Field investigations rely on direct observation and data collection to assess the current state, strength, and stability of structures, but they can only capture observable information and must be complemented by theoretical analyses to identify key contributing factors to vulnerability.

Overall, current research on geological disaster risk along railway alignments is primarily focused on post-evaluation for existing lines. Risk preassessment during the design stage remains largely qualitative, relying on empirical guidelines rather than quantitative analysis, which limits its application in intelligent alignment optimization. Therefore, this chapter first introduces spatial avoidance and mitigation strategies for alignments passing through areas prone to rockfalls, landslides, and debris flows. To quantitatively assess geo-hazard risk, the study area is first discretized into a set of

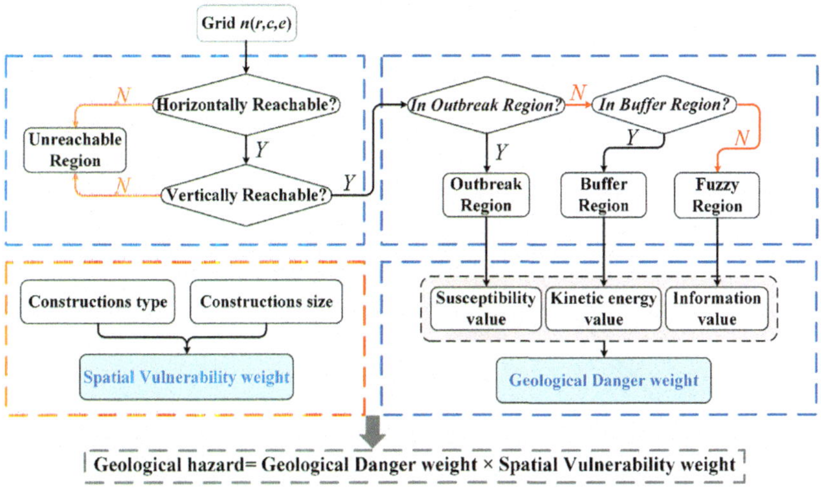

Figure 6.1 Main process of geological hazard assessment.

voxels. When hazards are unavoidable, the affected area is further divided into an outbreak zone, buffer zone, and uncertain zone. For each zone, different evaluation criteria are applied to assess geological hazard susceptibility. Meanwhile, a weighted integral method is used to evaluate the vulnerability of alignment structures. Finally, the geological hazard susceptibility and structural vulnerability are combined to derive a risk value for each voxel within the accessible area. The overall process of quantitative geological disaster risk evaluation for railway alignments is illustrated in Fig. 6.1.

6.1 Geological hazard zone classification

Digitization of the design information is the first step in railway alignment optimization. In our work, all kinds of geographical information required for railway alignment optimization are stored discretely into a regular lattice, which consists of a set of grids, to form a comprehensive geographic information model (*CGIM*, Fig. 6.2).

6.1.1 Determine the railway reachable region

For railway alignment optimization, the two endpoints should usually be pre-determined in the landscape. Then, due to the constraints of the maximum allowable circuity coefficient ($\gamma_{\max}$) and maximum gradient

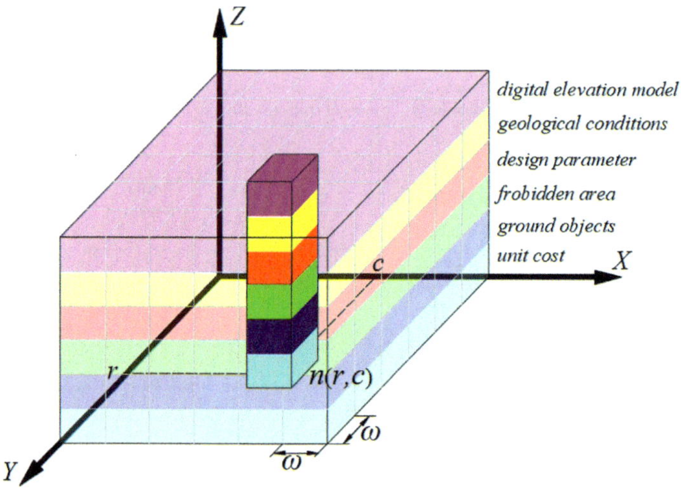

Figure 6.2 Comprehensive geographic information model.

(G_{max}), the reachable region (*R-region*) of a railway alignment is limited. More specifically, the *R-region* of a railway can be determined by analyzing the horizontally and vertically reachable range. Detailed explanations are provided below.

Below, $S\ (r_s, c_s, H_S)$ and $E\ (r_E, c_E, H_E)$, respectively, denote the grids that contain the start and end points, where r_s and r_e are, respectively, the row of the start and end points; c_s and c_e are, respectively, the column of the start and end points; H_S and H_E are the start and end points' design elevations, respectively; L_S and L_E are the linear distances from a random grid $n(r,c)$ to the start and end grid, respectively.

(1) Determine the horizontally reachable range

In order to satisfy the requirement of γ_{max}, $n(r,c)$ must satisfy the following conditions (since otherwise $n(r,c)$ is unreachable) Eqs. (6.1)–(6.3):

$$L_S = \sqrt{(r - r_s)^2 + (c - c_s)^2} \times \omega \tag{6.1}$$

$$L_E = \sqrt{(r - r_E)^2 + (c - c_E)^2} \times \omega \tag{6.2}$$

$$\frac{L_S + L_E}{L_{SE}} < \gamma_{max} \tag{6.3}$$

where ω is the grid width. The horizontal reachable region is shown in Fig. 6.3a.

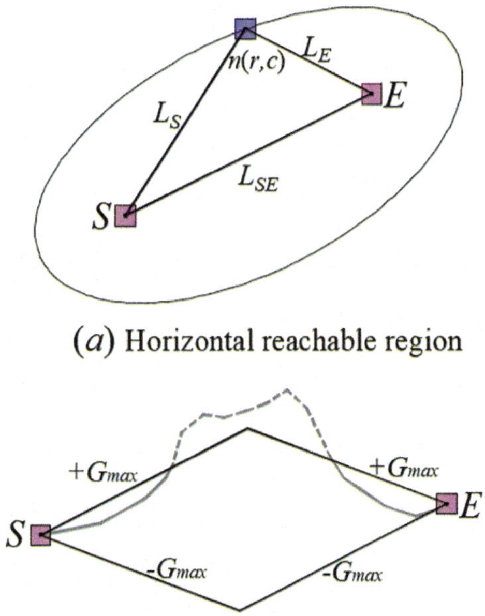

(b) Vertical reachable region

Figure 6.3 Railway reachable region.

(2) Determine vertically reachable range

For each grid within the horizontal reachable region, we first generate a so-called spatial voxel set, which is obtained by extending the grid's ground elevation upward and downward to determine several feasible $\omega \times \omega \times \varepsilon$ voxels, where ε is the height of each voxel.

The upper boundary ($H_{\max}$) is determined by considering the constraints of $i_{\max}$, $\gamma_{\max}$ and the maximum allowable height of bridges ($H_{\max b}$), as shown in Eq. (6.4). In addition, given that the geological hazards studied here are surface geological hazards, if the railway is buried in a tunnel and the overlying soil layer reaches a certain thickness ($H_{\max t}$, assumed to be 25m in this study), the influences of geohazards are deemed to be negligible. Therefore, the lower boundary ($H_{\min}$) is determined according to Eqs. (6.5) and (6.6). The vertical reachable regions are shown in Fig. 6.3b.

$$H_{\max} = \min(H_S + i_{\max} \times L_S \times \gamma_{\max}, H_E + i_{\max} \times L_E \times \gamma_{\max}, H_{\max\,b} + H_n) \tag{6.4}$$

$$H_{\min} = \max(H_S - i_{\max} \times L_S \times \gamma_{\max}, H_E - i_{\max} \times L_E \times \gamma_{\max}, H_n - H_{\max\,t}) \tag{6.5}$$

$$H_{max} > H_n > H_{min} \tag{6.6}$$

where H_{min}, H_{max} denote the minimum reachable depth and height at grid $n(r, c)$, respectively; H_n is the ground elevation at the grid.

6.1.2 Geological hazard zone classification

After obtaining the railway *R-region*, it is divided into three kinds of subregions, i.e., geological outbreak regions, buffer regions and fuzzy regions.

Here, the regions where existing rockfalls (U_R), landslides (U_L) and debris flows (U_D) are located are called the geohazard outbreak regions (*O-regions*). The grids beside the *O-region* are threatened by the existing geohazards. The rock and soil masses may expand outward and destroy the railway constructions when the geological hazard breaks out. These grids affected by the *O-region* constitute the buffer region (*B-region*), whose main hazard source is the moving masses. Remaining grids outside the *B-region* are basically not affected by the *O-region*. However, these grids may be impacted by potential geohazards because the information on potential geohazards is generally difficult to obtain, especially for mountainous regions. Therefore, we cannot directly determine such geological hazards and the corresponding regions are defined as fuzzy regions (*F-regions*). A zone classification diagram of a study area is shown in Fig. 6.4.

At the alignment design stage, *O-regions* are explicitly bounded and can be obtained from geological survey and historical data. Thus, the main aim of the geological hazard zone classification is to determine the *B-region* which reflects the transition between the *O-region* and *F-region*. To achieve this, a geological hazard energy-slope model is applied to generate the *B-region*. Finally, except for the *O-regions* and *B-regions*, other areas in the *R-region* are *F-regions*.

(1) Energy-slope model

Generally, the soil movement process is regarded as the process of energy conversion [6]. During this process, part of the gravitational potential energy is converted into kinetic energy while the rest is dissipated. The energy dissipation of the mass is mainly caused by the internal and external frictions and the damping motion. This process is expressed in Eq. (6.7):

$$E_n = F(h, \beta, m) = \iiint \left(W_0 + W_{\Delta H} - A_k - W_f\right) dh d\beta dm \tag{6.7}$$

where h, β and m are the elevation, sliding direction and weight of the soil mass, respectively; W_0 and $W_{\Delta H}$ are the initial energy and gravity potential

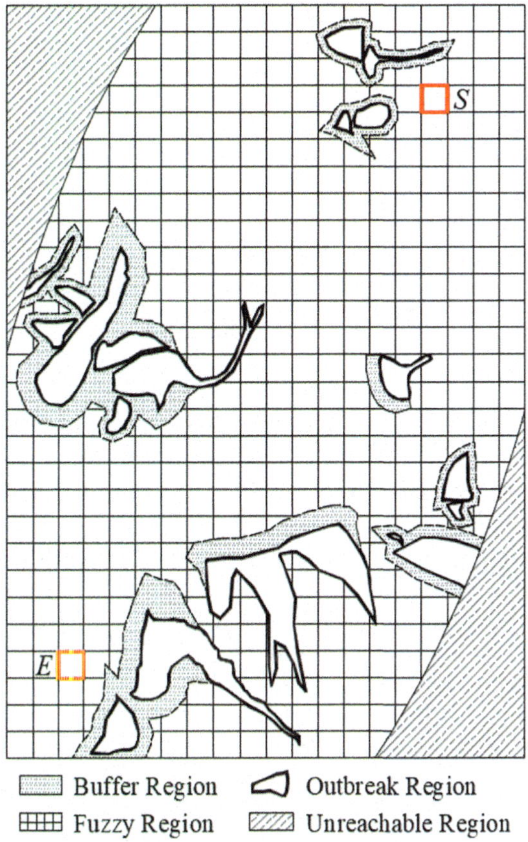

Figure 6.4 Geological hazard zone classification.

energy conversion, respectively; W_f and A_k are the energy spent on overcoming the total friction and damping motion, respectively. The method for computing Eq. (6.7) will be provided below.

(2) Geological hazard energy-slope model

The run-out distance and impact force of mountain geological hazards can be determined by using an energy-slope model, which is a generic model explaining the movement in specific geomorphologic processes [6]. The soil and rock masses of rockfalls, landslides, and debris flows slide from the top to the bottom along multiple single-slope sections. Besides, in the alignment optimization process, the computation accuracy is determined by the *CGIM* grid width [7]. Thus, based on the energy-slope model, each single-slope section is evaluated by analyzing the energy conversion of all the grids that the slope section traverses.

The single-slope motion process can be abstracted as the horizontal motion (Fig. 6.5a and b) and vertical slide (Fig. 6.5c). In the horizontal motion, the mass moves from the nth grid to the (n+1) grid along the horizontal direction (β_n) and the maximum vertical slope (α_n). The vertical slope (α_n) between the nth and (n+1) grids can be computed with Eqs. (6.8) and (6.9).

$$\alpha_n = \frac{H_n - H_{n+1}}{\omega} \tag{6.8}$$

$$\alpha_n = \frac{H_n - H_{n+1}}{\sqrt{2}\omega} \tag{6.9}$$

After determining the horizontal motion direction, the vertical sliding process of the soil mass can be simulated (Fig. 6.5c). Taking Fig. 6.6a as an example, this process can be specified as in Eqs. (6.10–6.13) according to the law of energy conservation which is expressed in Eq. (6.7). The case shown in Fig. 6.5b can be similarly obtained.

From the nth grid to the (n+1) grid, Eqs. (6.10)–(6.13) are expressed as:

$$\frac{1}{2}m\left|\overrightarrow{v_{n+1}}\right|^2 = \frac{1}{2}m\left|\overrightarrow{v_n}\right|^2 + W_{\Delta H} - A_k - W_f \tag{6.10}$$

Among them,

$$W_f = F_f \cdot L_n = \left[mg\sin\alpha_n \cdot (\mu_i + \mu_e)\right] \cdot \sqrt{(H_n - H_{n+1})^2 + \omega^2} \tag{6.11}$$

$$A_k = \frac{1}{2}m\left[\left|\overrightarrow{v_n}\right| \cdot \tan(\beta_n - \beta_{n-1})\right]^2 \tag{6.12}$$

$$W_{\Delta H} = mg(H_{n+1} - H_n) \tag{6.13}$$

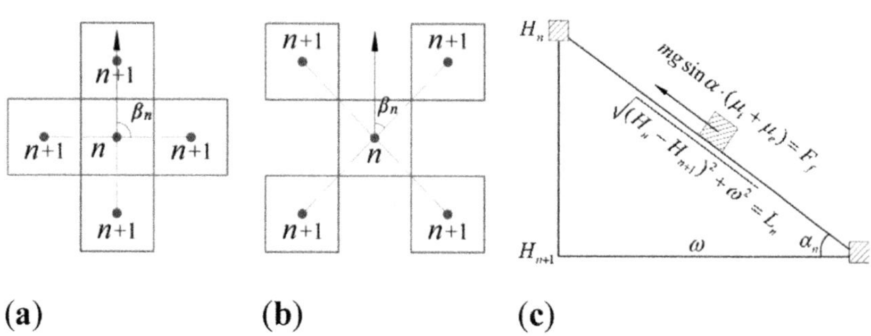

Figure 6.5 Single slope sliding section: (a) Directly adjacent girds; (b) Diagonally adjacent grids; (c) Vertical sliding path for a single slope.

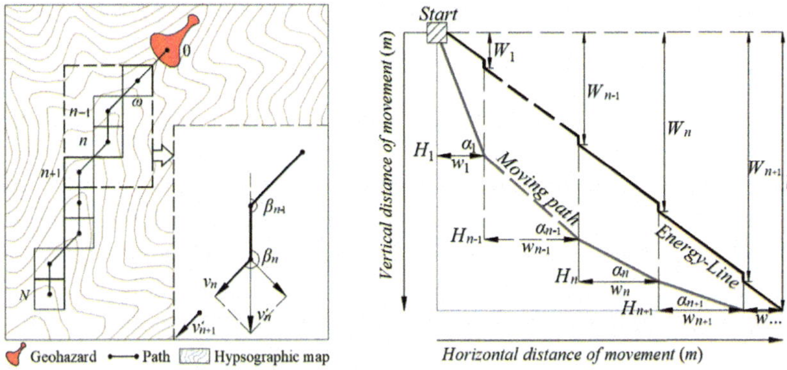

(a) (b)

Figure 6.6 Diagram of the slope-energy model: (a) Horizontal motion path of a geological hazard; (b) vertical sliding path of a geological hazard.

where F_f is the friction, L_n is the single slope sliding distance, μ_i is the internal friction coefficient computed as $\mu_i = \tan\varphi$ while μ_e is the external friction coefficient computed as $\mu_e = \mathrm{e}^{-7.13+9.33\mathrm{NDVI}}$, where φ is the angle of internal friction and the *NDVI* (Normalized Difference Vegetation Index) reflects the vegetation growth status and vegetation coverage of a region.

When the soil mass moves from the starting position and reaches the *nth* grid, the energy conversion can be computed with Eq. (6.14), as shown in Fig. 6.6.

$$\frac{1}{2}m\left|\overrightarrow{v_0}\right|^2 + mg(H_0 - H_n) = \sum_{i=1}^{n}\frac{1}{2}m\left|\overrightarrow{v_i}\right|^2 \cdot \tan^2(\beta_i - \beta_{i-1}) + \sum_{i=1}^{n} mg\sin\alpha_{i-1}\cdot(\mu_i + \mu_e) \cdot\sqrt{(H_{i-1} - H_i)^2 + \omega^2} + \frac{1}{2}m\left|\overrightarrow{v_n}\right|^2 \quad, n \in N \tag{6.14}$$

Denoting that Eq. (6.15):

$$W_n = \sum_{i=1}^{n}\left[\frac{1}{2}m\left|\overrightarrow{v_i}\right|^2 \cdot \tan^2(\beta_i - \beta_{i-1}) + mg\sin\alpha_{i-1}\cdot(\mu_i + \mu_e) \cdot\sqrt{(H_{i-1} - H_i)^2 + \omega^2}\right] \tag{6.15}$$

the kinetic energy of the mass can be determined when it reaches the *nth* grid Eq. (6.16):

$$E_n(r,c) = \frac{1}{2}m\left|\overrightarrow{v_n}\right|^2 = \frac{1}{2}m\left|\overrightarrow{v_0}\right|^2 + mg(H_0 - H_n) - W_n \tag{6.16}$$

The process of energy conversion is slightly different for different types of geological hazards: (1) For rockfalls and landslides, the very slow start velocity of the soil mass can be neglected. (2) For rockfalls, the air friction of the falling rock is negligible. Therefore, the boundary conditions of different geological hazards are set as follows (Eqs. (6.17)–(6.19)):

$$W_0^{\text{landslide}} = F_{W_0}^{\text{landslide}}(v, m) = 0 \tag{6.17}$$

$$W_0^{\text{rockfall}} = F_{W_0}^{\text{rockfall}}(v, m) = 0 \tag{6.18}$$

$$W_f^{\text{rockfall}} = F_{W_f}^{\text{rockfall}}(\mu_i, \mu_e, h, m) = 0 \tag{6.19}$$

The above procedure will be used in processing all the geological hazard regions in the study area to generate their corresponding *B-regions*. Therefore, the *F-regions* can also be determined and the geological hazard zone classification of the study area can be completed, as shown in Eq. (6.20):

$$\begin{cases} n(r,c) \in U_D \cup U_L \cup U_R, n(r,c) \in O\text{-}region; \\ E_n(r,c) > 0, n(r,c) \in B\text{-}region; \\ E_n(r,c) = 0, n(r,c) \in F\text{-}region; \end{cases} \tag{6.20}$$

6.2 Geological danger evaluation

For different geological hazard subregions, i.e., *O-regions, B-regions* and *F-regions*, different methods are used to estimate the degree of threats by using geological danger values.

6.2.1 Geological danger of O-regions

Referring to our previous study, in an *O-region*, a susceptibility value S can be obtained by analyzing the environmental factors such as slope angle, slope aspect, *NDVI* value, area and Gravelius index. Thus, in this work, we use S to quantify the geological danger in an *O-Region*. Then, S is normalized as shown in Eq. (6.21):

$$D_O(r,c) = \frac{S(r,c) - S_{\min}}{S_{\max} - S_{\min}}, n(r,c) \in O\text{-}region \tag{6.21}$$

where $D_O(r,c)$ is the geological danger value of the grid *(r,c)* which is in *O-regions*; $S(r,c)$ is the susceptibility value of the grid *(r,c)*; $S_{\max}$ and $S_{\min}$ are the maximum and minimum susceptibility values in *O-regions*.

6.2.2 Geological danger of B-regions

When a railway passes through a *B-region*, it may be threatened by geohazards due to the potential expansion of the *O-region*. As explained in Section 6.1, we can quantify such danger with an energy-slope. As the energy of the rock and soil masses dissipates, their kinetic energy also decreases and, hence, in a *B-region*, the destructive power of a specific kind of geohazard is reflected by the kinetic energy [6]. Therefore, the kinetic energy obtained in Section 6.1 can be normalized with Eq. (6.22) and used as the danger value in *B-regions*.

$$D_B(r,c) = \frac{E_n(r,c) - E_{n\,\min}}{E_{n\,\max} - E_{n\,\min}}, n(r,c) \in B\text{-}region \tag{6.22}$$

where $D_B(r,c)$ is the geological danger value of the grid *(r,c)* which is in *B-regions*; $E_n(r,c)$ is the kinetic energy value of the grid *(r,c)*; $E_{\max}$ and $E_{\min}$ are the maximum and minimum kinetic energy values in *B-regions*.

In addition, if a grid is located in the intersection regions of several *B-regions*, its danger value can be determined by accumulating the kinetic energy of every traversed *B-region*.

6.2.3 Geological danger of F-regions

The grids in an *F-region* may be far from *O-regions* and are basically not affected by existing geohazards. However, there may be potential and unknown geohazards in *F-regions*. In such conditions, we propose to use a susceptibility value to measure the geological danger in the F-region. In this analysis, rockfalls, landslides and debris flows have similar hazard-generating factors and, hence, can be treated as generalized landslides for simplicity.

In this field, many researchers have used different methods to assess the susceptibility of geohazards on different scales according to different factors, such as an artificial neural network model [8], a stacking ensemble machine learning method [9], and an information value model (IVM) [10,11]. For large-scale railway design problems, we adopt the IVM because it can assess the geo-susceptibility of a broad area very quickly.

The main idea in the IVM is that geological hazards are likely to occur in regions where environmental conditions are similar to regions which already experienced such hazards [12]. The occurrence of landslides is influenced by various factors (x_i) [13], such as altitude, slope angle, slope aspect, *NDVI* value, distance to rivers, distance to faults and lithology [12,14]. Moreover, the contributions of different factors to landslide occurrences are different, and can be measured using the Information Value (*IV*): The greater the *IV*, the greater the contribution of this factor to the occurrence of specific geohazards.

Here, we use the area of the landslide as a parameter in computing the information value. The information value of x_{ij} (the *jth* subclass of the factor x_i) is determined (as shown in Fig. 6.8) as follows:

$$IV\left(x_{ij}\right) = \ln \frac{N_{ij}/N}{S_i/S} \tag{6.23}$$

where N_{ij} is the area of landslide in subclass x_{ij} (km^2); N is the total area of a landslide in the *R-region* (km^2); S_i is the total area of subclass i (km^2); S is the total area of *R-region* (km^2). The area data can be obtained from a geohazard region map.

As shown in Fig. 6.7, according to the value of the x_i factor in the grid n (r,c), the *IV* of the x_i factor can be determined with Eq. (6.23). Then, the total *IV* for each grid (r,c) can be computed by summing up all the *IV* of each factor layer. The susceptibility value can thus be obtained by normalizing the *IV* of each grid in the *F-region*. As shown in Fig. 6.8, after the above process, different geological danger values are set for different geohazard regions. These geological danger values $D(r,c)$ are then recorded in the *CGIM*, as shown in Fig. 6.9.

6.3 Spatial geohazard assessment

6.3.1 Spatial construction vulnerability

The construction vulnerability reflects the different resistances of different railway constructions with respect to specific geohazards. To estimate the construction vulnerability, a simple but effective score ranking method is proposed. More specifically, the vulnerability is determined by considering the type and the elevation difference from the ground of different alignment constructions, including bridges, tunnels, fills and cut

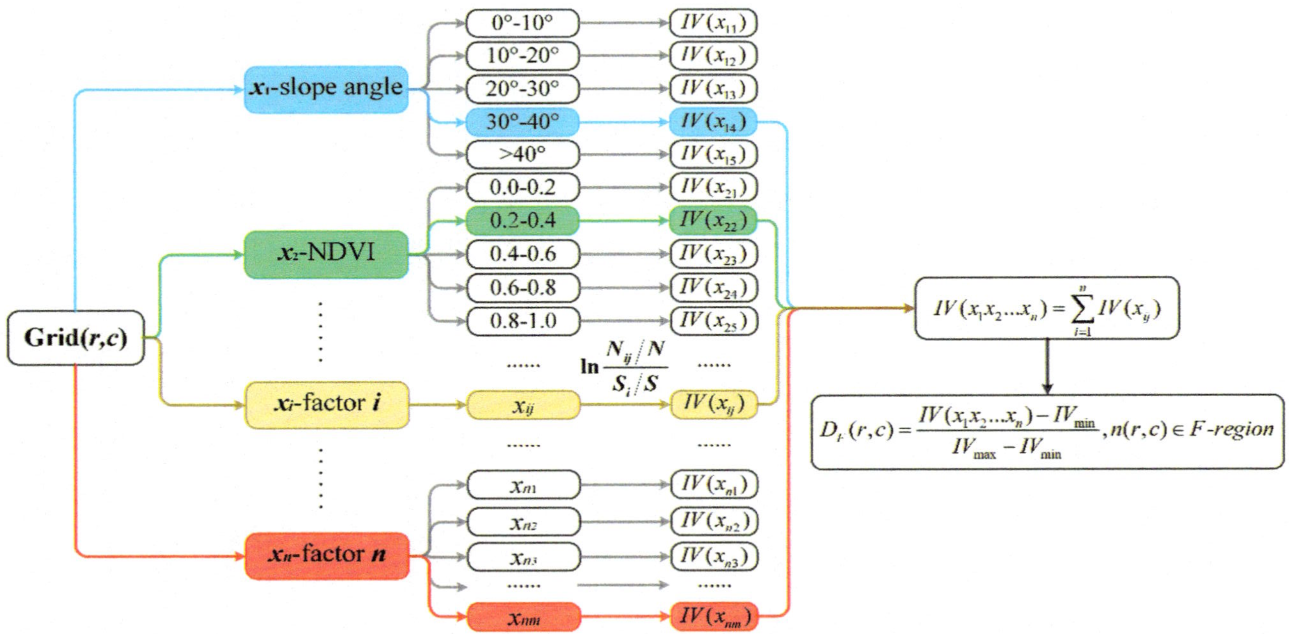

Figure 6.7 Flowchart for determining the hazard value of the *F-region*, where $D_F(r,c)$ is the geological danger value of the grid (r,c) which is in *F-regions*; $IV(r,c)$ is the *IV* of the grid (r,c); IV_{max} and IV_{min} are the maximum and minimum *IVs* in *F-regions*.

sections. In this process, the type of a construction is determined by checking the fill height or cut depth (Δh), which equals the difference between the railway design elevation (e_d) and the ground elevation (e_g), as specified in Eqs.(6.24) and (6.25):

$$F_{BTS}(\Delta h) = \begin{cases} \text{Bridge}, & \Delta H_{B-\max} > \Delta h > \Delta H_B \\ \text{Fill sections}, & \Delta H_B > \Delta h > 0 \\ \text{Cut sections}, & 0 > \Delta h > \Delta H_T \\ \text{Tunnel}, & \Delta H_T > \Delta h \end{cases} \tag{6.24}$$

$$\Delta h = e_d - e_g \tag{6.25}$$

where ΔH_B denotes the fill-bridge threshold height; ΔH_T denotes the cut-tunnel threshold depth; $\Delta H_{B\text{-max}}$ is the maximum allowable bridge height. This process is illustrated in Fig. 6.10.

Generally, bridges are safer than cut and fill sections and hence the preliminary vulnerability weight (V_P) for bridges is set at 2.0, while that for cut and fill sections is set at 3.0. In addition, tunnel sections can well avoid the

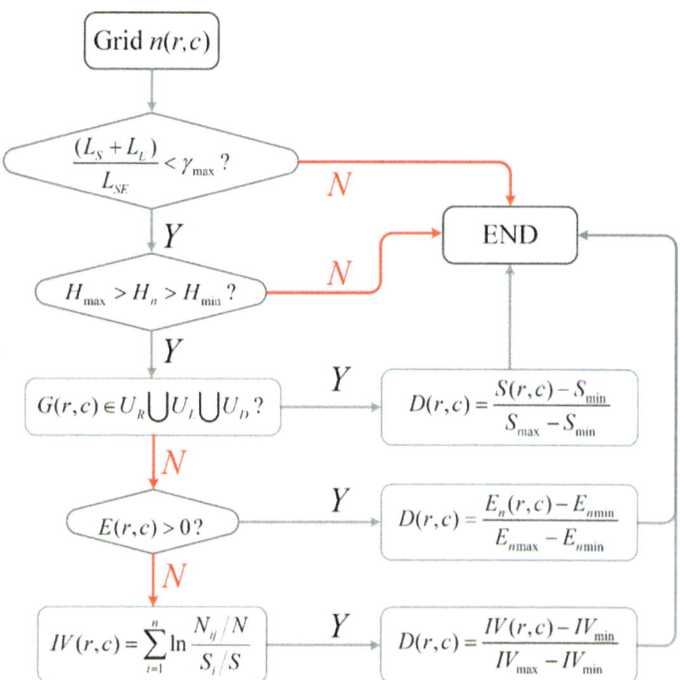

Figure 6.8 Flowchart for computing the geological danger value.

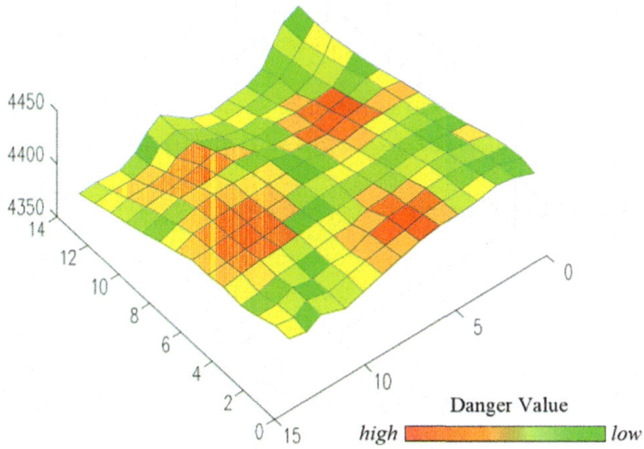

Figure 6.9 Geological danger map.

adverse effects of ground hazards and hence their vulnerability weights are 1.0. Then, each type of construction can be further classified according to its elevation difference from the ground surface (Δh).

As shown in Table 6.1: Bridges, can be classified as general bridges, high bridges, and super-high bridges based on two threshold-heights of $\Delta H_{B\text{-}1}$ and $\Delta H_{B\text{-}2}$. The buried depth of shallow tunnels should be less than ΔH_{min}. When the buried depth of a tunnel section exceeds ΔH_{min}, it is deemed a deep tunnel section which will not be affected by surface geological hazards. Lastly, the earthwork sections whose fill height or cut depth are within the range of $[\Delta H_C, \Delta H_F]$ are deemed general earthwork sections. Otherwise, a cut section whose excavation height exceeds ΔH_F is called a deep cut and, similarly, a fill section whose height exceeds ΔH_C m is deemed a high embankment. Constructions of the same type but with different elevation differences from the ground also have different vulnerabilities. Each construction with a different elevation difference is assigned a correction factor c (1, 2 and 3) according to its resistance abilities to geological hazards. After determining V_P and c, the final vulnerability is determined as Eq. (6.26):

$$V(\Delta h) = V_P \times c \tag{6.26}$$

Afterward, the ground elevation of each grid cell in the *CGIM* can be extended upward and downward to generate the longitudinal space which can be divided into a series of stacked voxels ($\omega \times \omega \times \varepsilon$), thereby extending the geological hazard attributes of the ground surface to 3D spaces. The geological hazard value of each grid can be determined with Eq. (6.27):

$$\text{Hazard}(r,c,e_d) = D(r,c) \times V\left[e_d - e_g\right] \tag{6.27}$$

where r and c are the row and column of each spatial voxel, respectively.

The spatial geological hazard map is shown in Fig. 6.11.

6.3.2 Railway geohazard determination

To determine the geohazard value of a railway, we first convert the alignment into a set of station points, as shown in Fig. 6.12, based on the *CGIM* grids that the railway traverses. Then, the total hazard value of a railway can be obtained by accumulating the geological hazard values of all station points, as expressed in Eqs. (6.28) and (6.29):

$$F_{\text{Hazard}}\left(X_i, Y_i, R_i, K_j, H_j\right) = \sum_{S}^{E} \text{Hazard}\left(x_k, y_k, z_k\right) = \sum_{k=0}^{L/\omega} \text{Hazard}(r_k, c_k, z_k) \tag{6.28}$$

$$r_k = \frac{x_k}{\omega}; \quad c_k = \frac{y_k}{\omega}; \tag{6.29}$$

where $\left(x_k, y_k, z_k\right)^{\mathrm{T}}$ denotes the coordinates of the kth station point on the alignment. r_k, c_k denote the row and column of the kth grid.

The steps in computing the total geological hazard value are.

STEP 1: Convert the railway alignment into a set of station points according to the ω of the *CGIM* grids that the alignment traverses.

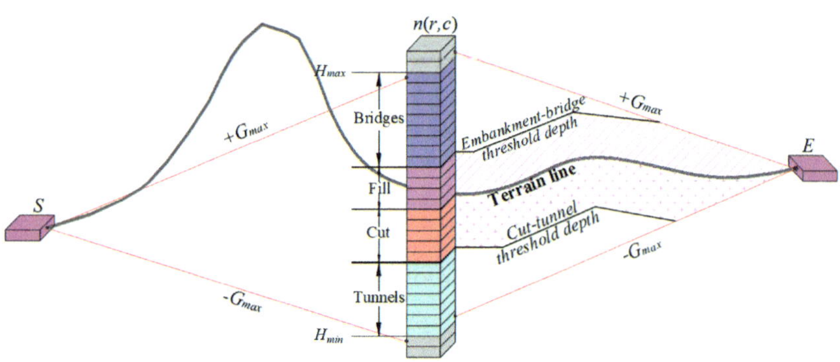

Figure 6.10 Construction configuration.

Table 6.1 Construction vulnerability.

Construction type	Construction subclass	Δh	V_P	c	V
Bridge	Super-high bridge	$[\Delta H_{B\text{-}2}, \Delta H_{max}]$	2.0	3	6.0
	High bridge	$[\Delta H_{B\text{-}1}, \Delta H_{B\text{-}2}]$		2	4.0
	General bridge	$[\Delta H_B, \Delta H_{B\text{-}1}]$		1	2.0
Embankment	High embankment	$[\Delta H_F, \Delta H_B]$	3.0	2	6.0
	Low embankment	$[0, \Delta H_F]$		1	3.0
Cut	Cut	$[\Delta H_C, 0]$		1	3.0
	Deep cut	$[\Delta H_T, \Delta H_C]$		2	6.0
Tunnel	Shallow tunnel	$[\Delta H_{min}, \Delta H_T]$	1.0	2	2.0
	Deep tunnel	$(-\infty, \Delta H_{min})$		0	0.0

STEP 2: Interpolate the ground elevation of the stations based on a digital elevation model.

STEP 3: Compute Δh for all stations.

STEP 4: Determine the type and locations of railway constructions according to Δh of each grid, i.e., Eq. (6.24).

STEP 5: Accumulate the geological hazard values of all the station points using Eq. (6.28) to obtain the total hazard value of an alignment.

6.4 Knowledge-driven karst hazard evaluation

The representative studies mentioned above have provided outstanding contributions for the research and development of railway alignment optimization. Unfortunately, the knowledge currently used in alignment optimization is fragmented, i.e., it is usually scattered in an optimization program in the form of various design parameters. No structured and systematic data associations among these parameters are specified. More specifically, a principle of alignment design is in general a manifestation of knowledge, which is defined here as a set of conditions with their corresponding treatments. That is, a condition is the specific description of the environment in which the alignment is located, and a treatment is a possible solution that should be applied when the alignment traverses the environmental region described by that condition. For instance, when an alignment encounters specific conditions (such as existing railways and roads, or nature reserves), the corresponding treatments (such as separation distances and allowed intersection angles among alignment and the conditions) are usually stored in the optimization programs as design parameters, in ways that are fragmented (i.e., with numerous kinds, quantities, as well as different value ranges), multisource and heterogeneous (e.g., texts, audios and videos).

Furthermore, alignment design is an iterative optimization process in which many alternatives are evaluated. The location of an alignment should be dynamically adjusted during the iterative design process. Thus, the specific conditions encountered by an alignment should also be dynamically analyzed with the change of railway location, which requires real-time updates of the corresponding treatments. Moreover, the treatments corresponding to conditions significantly affect the evaluation of the alignment. Sometimes, for a specific alignment, the applied principles, consisting of conditions and corresponding treatments, must also be changed to produce a diversity of railway alignment alternatives. For example, in a study area with dense geo-hazards regions, two sets of different design principles are applied to guide the alignment for "completely bypassing geo-hazards regions" or "allowing the alignment to cross these regions with bridges or tunnels". These requirements place high demands on the modifiability of design principles. Moreover, human knowledge is constantly extended or updated with the enrichment of knowledge categories and changes in experience, as shown in Fig. 6.13. However, previous alignment optimization methods are usually based on fragmented knowledge, which limits the modification, extension and updating of knowledge.

In the realm of alignment design, although traditional knowledge engineering has contributed to the feasible expression of alignment design knowledge, the symbolic expression of knowledge engineering is more easily understood by humans rather than computers, which limits the development of traditional knowledge engineering in the field of automated alignment optimization. In recent years, the advent of Google Knowledge Graph (KG) provided an innovative method for solving the above problem. KG technology includes both symbolic and vector representations with strong extendability. Furthermore, it can also be integrated with computer

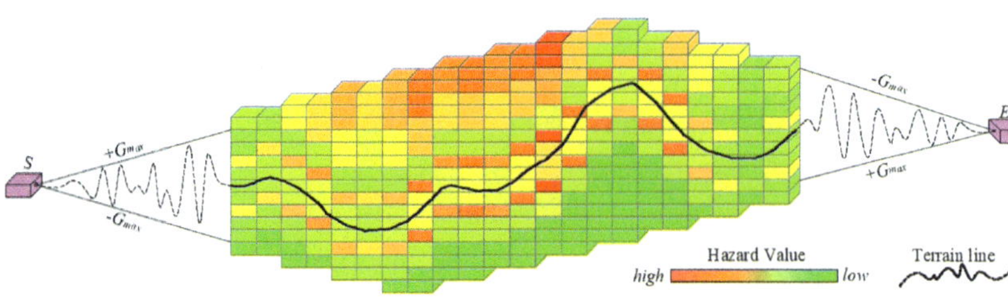

Figure 6.11 Diagram of spatial geological hazard map.

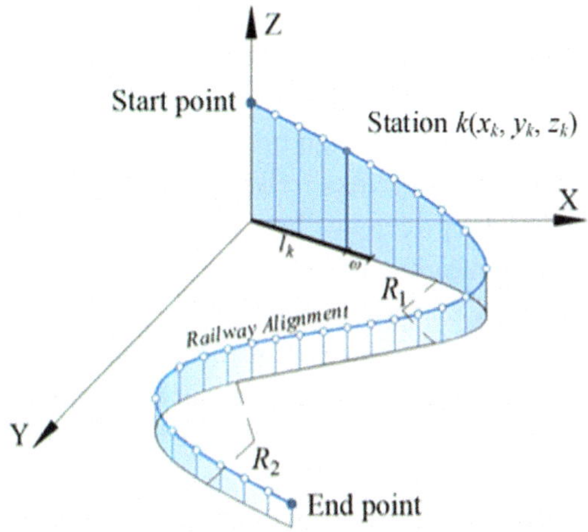

Figure 6.12 Demonstration of station point along an alignment.

vision and natural language processing. KG is a knowledge database with graph structure, in which nodes represent semantic entities and links represent semantic relations. Generally, the entities and relations are represented in the form of triplets (<head entity h, relation r, tail entity t>). A set of end-to-end triplets can compose a huge semantic network. As discussed in several publications, KG is considered the key for making artificial intelligence (AI) system "knowledgeable". Hence, constructing a knowledge graph for railway alignment design (RAD-KG) is a promising way to realize a systematic and structured knowledge expression for railway alignment optimization.

Currently, many KGs, such as WikiData, Freebase and YAGO [15], have been constructed in open fields, which act as the data warehouse of the open source community. In addition, the KG has been successfully extended in other fields including medical diagnosis [16], enterprise efficient supply chain management [17], as well as surveying and remote-sensing technology [18]. In our recent research, the first-known railway alignment design knowledge graph (RAD-KG) that can be applied in railway alignment optimization was constructed. The RAD-KG functioned as a complex representation of relations between alignments and environment to guide the search program according to specific engineering conditions.

However, the triple-based representation often oversimplifies the complexity of the data stored in the KGs, especially for the hyperrelations connecting two or more entities, which may lead to loss of higher-order information. Especially for railway alignment design in karst mountainous regions with highly interacting effects of environmental factors, the expression ability of a general knowledge graph is limited.

In the literature, a hypergraph can be represented as a multidimensional matrix of nodes (entities) and links (relations) [19,20]. Each link connects two or more nodes. The preference between pairs of nodes is reflected by the connecting link's weight. This characteristic is highly similar to the knowledge structure of railway alignment design in karst regions, where coupling environmental impacting factors form high-order relations among *conditions* and *treatments*. Meanwhile, the preferences among different *treatments* determine the weights of the relations. Furthermore, multiple relations are pervasive in open knowledge graphs. In Freebase, more than 1/3 of relations are multiple. More than 1/6 of relations in Wikipedia are multiple [21]. These laid the theoretical foundations for the construction and application of a RAD-KG with hypergraph semantics in karst-dense regions.

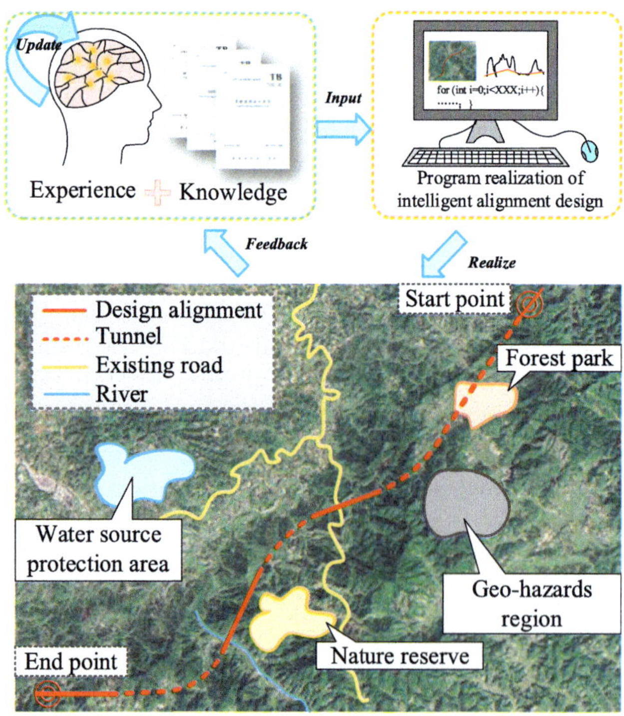

Figure 6.13 Application and updating of knowledge in a railway alignment design.

Therefore, a knowledge graph is developed with weighted hyperrelations incorporating various railway alignment design principles in karst mountainous regions, which is called KaRAD-KG.

6.4.1 Semantic network framework of RAD-KG

In a recent study, a modeling method for Railway Alignment Design Knowledge Graph (RAD-KG) was developed for railway alignment optimization. The semantic network structure of RAD-KG is shown in Fig. 6.14.

In Fig. 6.14, a design principle, which represents a manifestation of knowledge, was abstracted into the *condition* encountered by an alignment and the *treatments* that should be executed according to this specific *condition*. Therefore, the RAD-KG was constructed as a tree structure composed of a series of <*condition* (***h***) - relation (***r***) – *treatment* (***t***) > triplets, in which each node represented an entity of *condition* ($C_{n,\ ...,i}$) or *treatment* (T_t). Additionally, a property design, as detailed in Chapter 3.2.1, was conducted to enhance the information storage capacity of entities and eliminate the possible ambiguities among different entities in the same *condition* or

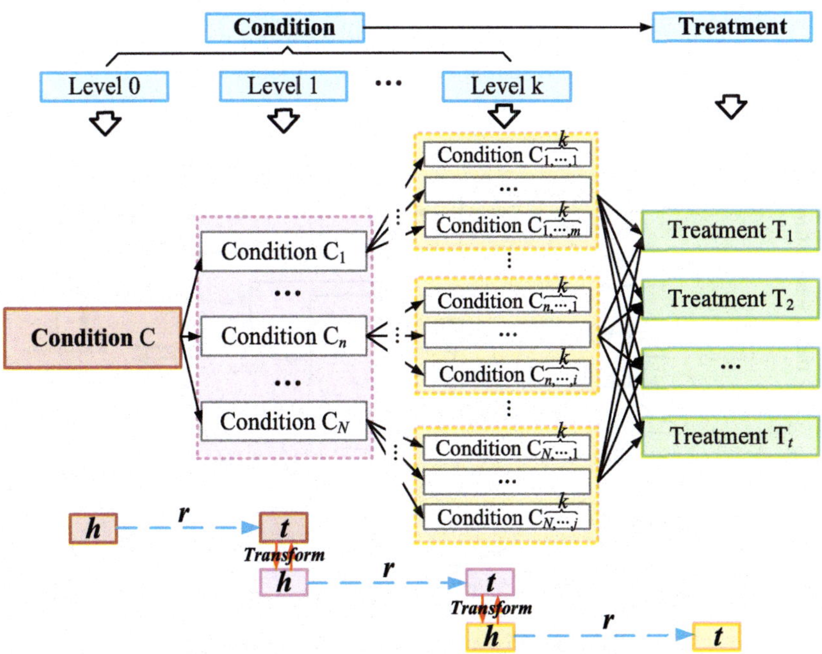

Figure 6.14 Network framework of the RAD-KG.

treatment category. The modeling of RAD-KG provided the foundation of basic semantic logic and property design for this research.

6.4.2 Major criteria for alignment design in karst-dense mountainous regions

(1) Karst hazard inducing factors

Karst is a geologic phenomenon in which water chemically dissolves soluble rocks with mechanical action such as surface subsidence and water erosion. The dissolution expands the void in soluble rock into channels, leading to increased groundwater flow and the occurrence of karst subsidence. The inducing factors of karst regions are summarized in Fig. 6.15. Moreover, the karst develops gradually due to water erosion.

(1) Climate: Karst regions are commonly located in areas with pleasant climate and abundant rainfall, which is the main source of a karst water system.

(2) Topography: The topography of karst mountainous regions is usually conducive for water collection and infiltration.

(3) Geological conditions, such as the distribution of soluble and insoluble rock, folding of strata, as well as breaking of rocks, determine the underground spatial distribution of karst water.

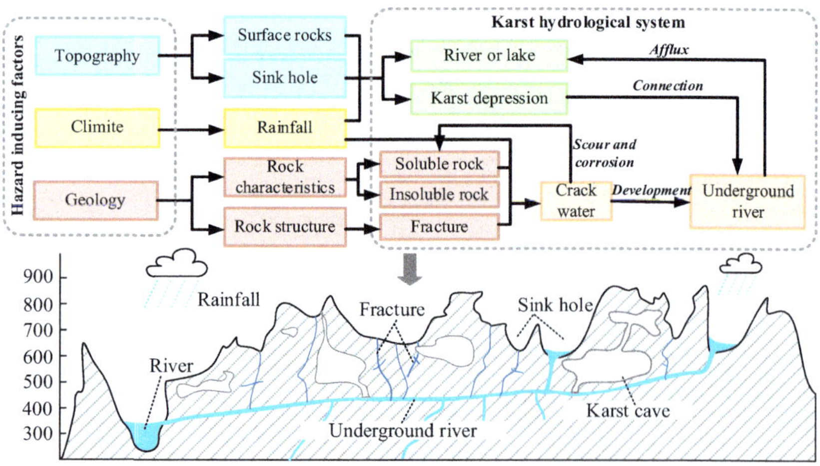

Figure 6.15 Factors inducing karst hazard.

(2) Karst hazard representation

Karst hazard representation is closely related to hazard inducing factors and strongly affects the life-cycle safety of a railway. The direct hazards caused by karst regions and other surface hazards are considered and analyzed.

1. Direct hazard from water

With significant spatial stratification characteristics, the distribution of karst water along the depth direction is shown in Fig. 6.16. Different aquifers pose differentiated impacts to railway alignments: (1) The epikarst zone is an important water source for humans, livestock, and farmland irrigation, with generally 5–30m depth. Railway tunnels in such zones not only pollute the ecology, but are also threatened by ground subsidence. (2) The water flow in an aeration zone is spatially discontinuous while its temporal distribution depends on the amount and duration of rainfall. Moreover, there is generally no hydrostatic pressure in this zone. Thus, the threat is controllable when a railway tunnel is located in an aeration zone. (3) Tunnels in transitional zones are threatened by seasonal rainfall, which may lead to water and mud outbursts during the rainy season. (4) A karst usually develops in shallow and pressure saturation zones with abundant underground water, which pose a long-term risk to tunnels. (5) Although karst development in the deep slow-flow zone is generally stable, tunnel construction is not recommended in such a zone.

2. Direct hazard from rock

Soluble rock is regional on the horizontal plane and stratigraphic on the vertical profile (as shown in Fig. 6.17). The lithology difference

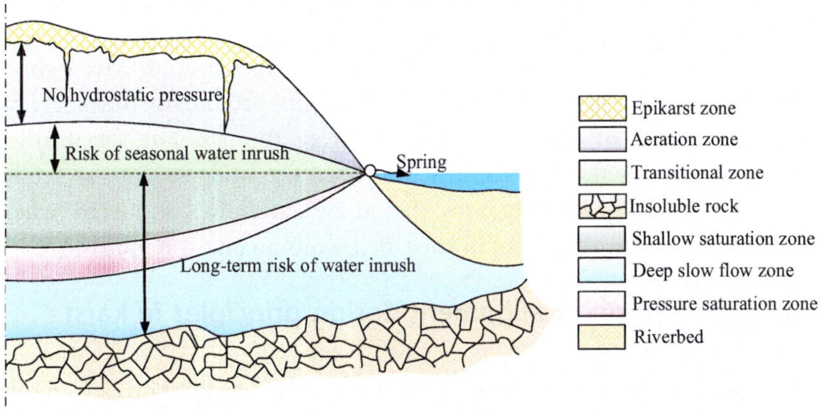

Figure 6.16 Vertical distribution of karst aquifers.

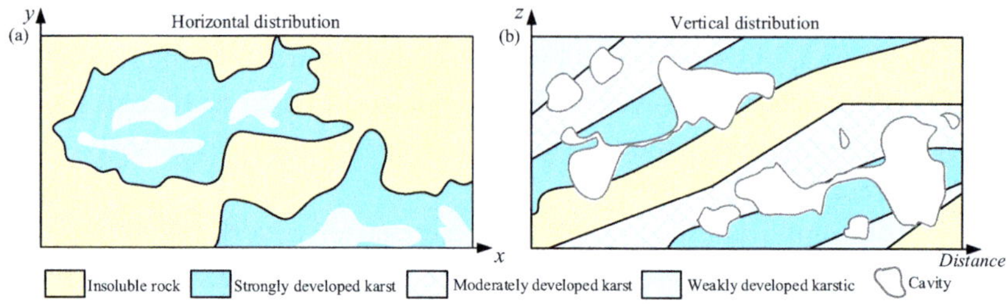

Figure 6.17 Spatial distribution of soluble and insoluble rocks.

significantly affects railway safety. Thus, theoretically, the spatial distribution of karst regions should be investigated before alignment design.

Meanwhile, anticlinal and synclinal structures exhibit distinct hazard characteristics due to groundwater erosion. As shown in Fig. 6.18, a synclinal karst has good catchment conditions. Moreover, the upper part of core forms massive joints and aggravates the erosion by underground water, causing great threats to railway tunnels. The anticlinal core is geologically fragmented or even collapsed to form gullies and valleys, which provides favorable conditions for further water erosion, resulting in poor geologic conditions for construction through the anticlinal core.

In summary, the direct karst hazards, emanating from the coupling of water and rock, mainly pose hazards to underground railway constructions (i.e., tunnels), usually manifested as water inrush, mud outburst, and tunnel roof collapse.

3. Hazards on mountain surface

The frequent occurrence of typical geo-hazards in complex mountainous regions, such as landslides, debris flows and rockfalls limits the feasible regions for railway alignment on the ground. Meanwhile, the water-collecting karst depressions on the ground not only increase the threat of water inrush for railway tunnels, but also further reduce the feasible construction area for bridges, cuts and fills. Accordingly, karst hazards should be carefully considered when designing an alignment in karst-dense mountainous regions.

6.4.3 Description of alignment design principles in karst regions

Railway alignment design principles incorporating karst hazards are closely coupled to other environmental factors, such as landslides, underground

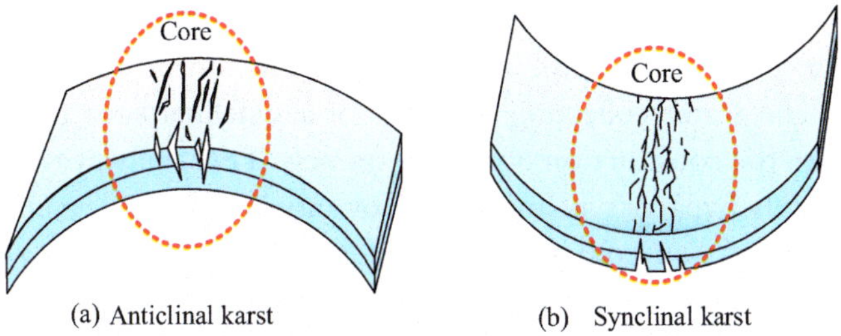

Figure 6.18 Diagrams for (a) anticlinal karst and (b) synclinal karst.

water, and existing tunnels. In this section, an event tree, shown in Fig. 6.19, is constructed to describe the railway alignment design principles in karst regions.

Hence, in order to enhance the railway safety, the feasible region of a railway alignment is described as follows.

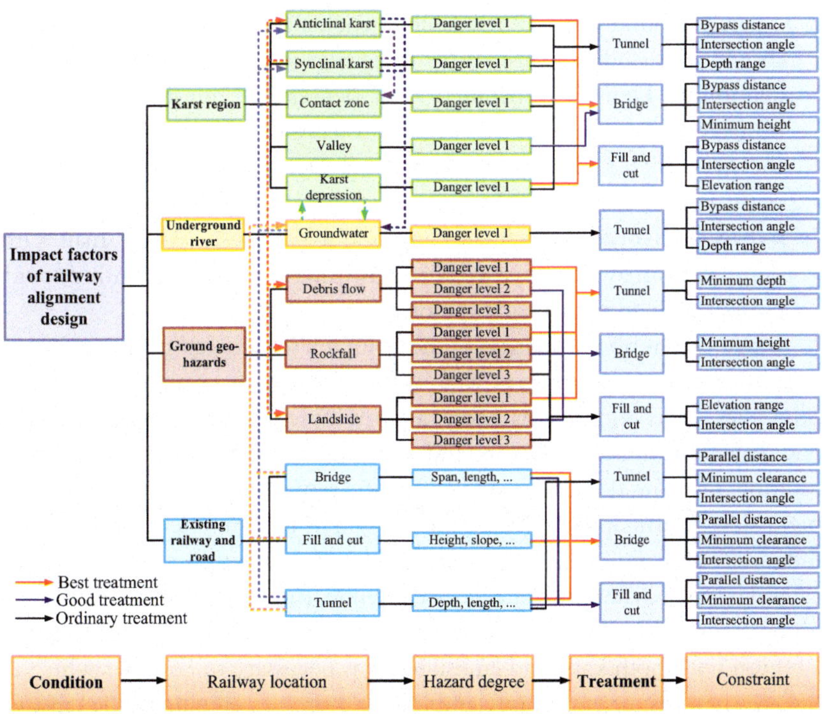

Figure 6.19 An event tree for determining the location of an alignment, where a dashed link represents a coupling relation among specific factors.

1. Considering direct hazards from rock and water
   A karst region is created by the synergistic action of water and soluble rock. The water-collecting structure of a synclinal karst poses a great threat to the construction of railway tunnels. Thus, railway tunnels must completely bypass a synclinal karst. For anticlinal karsts, the engineering condition of non-core areas (see Fig. 6.18) is preferable to a synclinal karst. Therefore, an alignment can traverse non-core anticlinal karst regions. Moreover, the hazard encountered by an alignment varies in different karst regions and changes continuously within a specific region. Additionally, the underground feasible region is strongly coupled with the development of underground water. Thus, when an alignment traverses a karst region, the karst category, as well as distribution of both karst region and underground water must be considered. However, current manual design usually relies on empirical qualitative evaluation, and lacks a more rigorous quantitative evaluation.
2. Considering mountain surface hazards
   Typical ground geo-hazards, such as rockfall, landslide and debris flow should be handled considering the geological vulnerability and susceptibility. Densely-distributed karst depressions are another kind of obstacle for an alignment. Designers usually make judgments based on the size of karst depressions. When a 3-D alignment traverses regions with densely distributed karst depressions, large depressions should be bypassed in order to reduce danger and pollution, while small karst depressions may be traversed through bridges or tunnels with enough clearance in anticlinal karst regions.

Based on the above analysis of the complex terrain and geological characteristics in karst regions, as well as the refinement of alignment design principles, it is found that there are multiple relations among highly-coupled environmental factors and the corresponding treatments. These multiple relations demonstrate the semantic representation of railway alignment design principles in karst regions, which is manifested as the mutual influence of various environmental factors (*conditions*) and the *treatments* caused by their synergistic effects. Meanwhile, these *conditions* have different influences on the choice of *treatments*. For example, if a railway is located in a region with both karst and landslides, it is necessary to consider which of the two hazards, i.e., karst and landslide, is more threatening to the railway. Then, the *treatment* can be determined, which mainly influences the structures (e.g., bridge or tunnel) adopted to traverse this region, and the clearance value (e.g., the minimum span height of a bridge, the minimum depth of a tunnel) for the alignment.

However, the previous RAD-KG builds unidirectional relations from *conditions* to *treatments*. Although a *treatment* may be linked with several *conditions*, these *conditions* are independent of each other, which limits their adaptability to karst hazard assessment. Hence, further enhancements to RAD-KG are necessary for assisting alignment optimization in karst-dense regions.

6.4.4 KaRAD-KG modeling with hypergraph semantics

To solve the above problems, a new semantic network is modeled to represent railway alignment design principles in karst regions.

Specifically, apart from the strong correlation between *condition* and *treatment* entities, different categories of *conditions* also interact with each other, which impacts the determination of preferred *treatments*. For instance, when traversing a debris flow region, the recommended *treatment* is to construct a tunnel. However, if the alignment is also located in a karst region, the karst-related hazards significantly increase for the tunnel. Alternatively, when tunnels are necessary to traverse anticlinal karsts, the feasible region must be carefully delimited based on the spatial distribution of groundwater (see Fig. 6.19).

The above characteristics result in multiple relations among the *condition* and *treatment* entities, where the adoption and preference of the *treatment* entities is related to multiple *conditions* and varies according to the different specific *conditions*. The multiple relations among two categories of entities can be recognized as hyperrelations. In this study, hyperrelations among *condition* and *treatment* entities are shown in Fig. 6.20.

Hence, the original relations among *conditions* and *treatments* in RAD-KG should be enhanced to bidirectional hyperrelations among different *conditions*, different *treatments*, as well as *condition* - *treatment* combinations. Based on these, the properties are improved to reflect the directionality from *treatments* to *conditions*, which is defined as "*Object*". For example, if

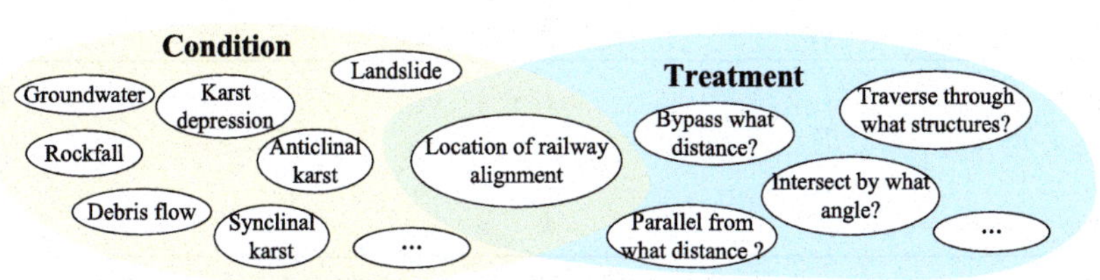

Figure 6.20 Hyperrelation example.

the treatment is manifested as "bypass at a distance of 100m", the "*Object*" is what the alignment should bypass. Moreover, the treatment entities can be merged according to the original property design of RAD-KG. The hyper-relation results in varying relation preferences among entities. Thus, the expression of "*Priority*" can be defined uniformly with the weight of the hyperrelation. For the node (entity) set V ($v \in V$) and the relation set E ($e \in E$), a link I is generated as shown in Table 6.2, all of which construct an adjacency matrix.

For a *treatment* node, the weight of each link is computed as in Eq. (6.30), where N is the sum of different priorities that a *condition* links.

$$\omega_{i,j} = \frac{\text{Priority}_{N-j+1}}{\sum_{n=1}^{N} \text{Priority}_n} \in (0, 1] \tag{6.30}$$

Then, the semantic network structure of KaRAD-KG based on the top-to-bottom ontology modeling method is shown in Fig. 6.21.

Furthermore, as discussed above, different karst regions, i.e., the *condition* entities at the same layer with different properties, impose different threats to railway alignments. To satisfy the distinctions, "*Hazard*" is added to *condition* properties and stored as a user-defined dimensionless specific property to numerically represent the danger magnitude. However, it is difficult to find the spatial distribution of karst at the alignment design stage. Thus, it is usually necessary to infer the feasible region based on human experience when the horizontal boundary of a synclinal karst region is defined. To facilitate alignment optimization, a new property of "*Veto-range*" is added to the specific *condition*, which is explained in Section 5.1. The updated property design is shown in Fig. 6.22.

The retrieval properties must be contained by entities, which construct the corresponding relations among *condition* and *treatment* entities. The inherent properties represent the objective characteristics of entities, such as the level of reserve and height of dam for a *condition* that usually can be

Table 6.2 Example of the hypergraph described by an adjacency matrix.

	e_1	e_2	...	e_j
v_1	$I_{1,1}$	$I_{1,2}$	...	$I_{1,j}$
v_2	$I_{2,1}$	$I_{2,2}$	...	$I_{2,j}$
...	...	...	...	...
v_i	$I_{i,1}$	$I_{i,2}$	...	$I_{i,j}$

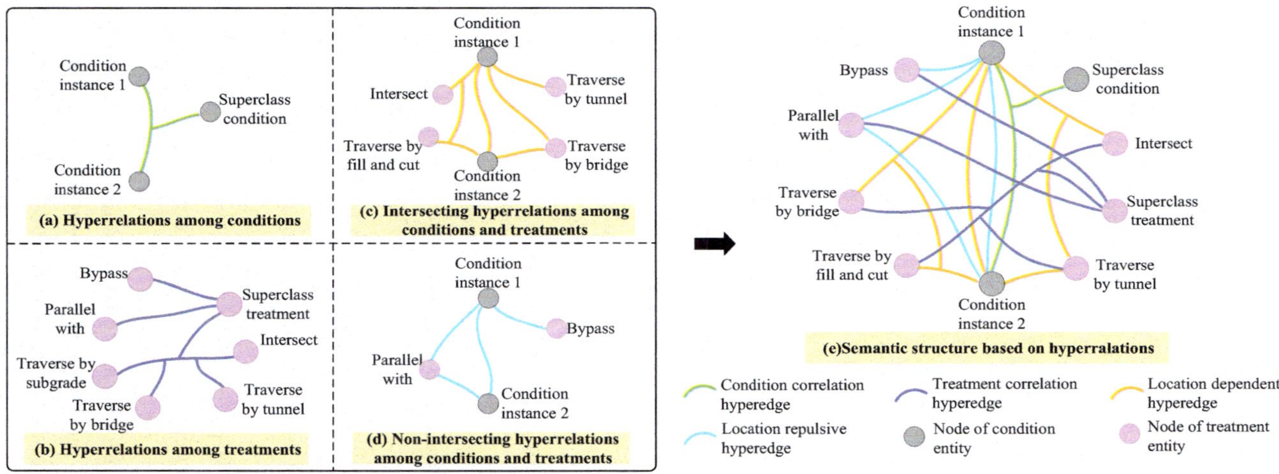

Figure 6.21 Semantic structure of KaRAD-KG.

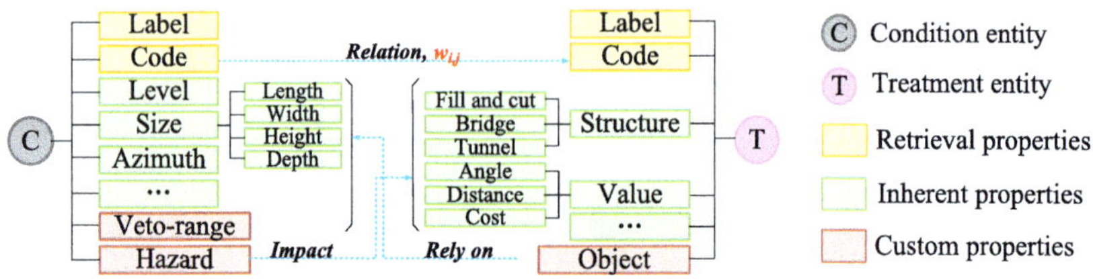

Figure 6.22 Property design for KaRAD-KG.

obtained through an engineering survey, and the required bypass distance of an alignment for a *treatment*. The custom properties are given to the entity and defined by users based on the assessment on inherent properties.

Thus far, the KaRAD-KG for railway alignment design in karst regions is devised. KaRAD-KG effectively enhances the intricate coupling representation of alignment design knowledge, which establishes an autonomous knowledge storage vessel to assist railway alignment optimization in karst regions.

6.5 Conclusion

This chapter addresses the critical issue that geological disaster risk assessment for railway alignment optimization in complex mountainous regions remains largely qualitative (relying on empirical guidelines) during the design stage, while existing research primarily focuses on post-evaluation for operational lines. To resolve this, a comprehensive quantitative geological disaster risk evaluation framework for railway alignments is proposed. First, a CGIM is established by discretizing various types of geographic information (e.g., digital elevation model, geological conditions) into a regular lattice. Based on constraints such as the maximum allowable circuity coefficient (γmax) and maximum gradient (Gmax), the railway's reachable region (R-region) is determined through analyzing horizontal and vertical reachable ranges. Within the R-region, geological hazard zones are classified into three subregions—outbreak regions (O-regions, hosting existing rockfalls, landslides, or debris flows), buffer regions (B-regions, threatened by potential expansion of O-regions), and fuzzy regions (F-regions, facing risks of unknown potential hazards)—using an energy-slope model that simulates energy conversion during soil/rock mass movement to define B-regions. Next, a subregion-specific geological danger evaluation method is developed: O-regions use normalized susceptibility values (derived from environmental factors such as slope angle and NDVI), B-regions adopt normalized kinetic energy (from the energy-slope model, reflecting hazard destructive power), and F-regions apply the Information Value Model (IVM) (assessing susceptibility via factors such as altitude and lithology) to quantify danger. For spatial geohazard assessment, construction vulnerability is evaluated by classifying alignment structures (bridges, tunnels, fills, cuts) based on fill height/cut depth (Δh), assigning vulnerability weights and correction factors (c) to compute final vulnerability; the CGIM grid is extended into 3D voxels to map geological hazard attributes spatially, and the total hazard value of an alignment is obtained by accumulating hazard values of station points converted from the alignment. Additionally, to address fragmented knowledge and limited expression of complex environmental couplings in karst regions, a knowledge-driven system is proposed: the Railway Alignment Design Knowledge Graph (RAD-KG), structured as < condition-relation-treatment > triplets, is enhanced into a Karst Railway Alignment Design Knowledge Graph (KaRAD-KG) using hypergraph semantics. KaRAD-KG captures multi-factor coupling relations, adds properties such as "Hazard" (quantifying danger magnitude) and "Veto-range" (delimiting forbidden areas), and effectively supports alignment optimization in karst regions. Overall, this framework integrates zone classification, danger evaluation, spatial assessment, and knowledge-driven karst analysis, enabling quantitative geological disaster risk evaluation for railway alignment optimization in mountainous (including karst) regions during the design stage.

References

[1] C. Melchiorre, M. Matteucci, A. Azzoni, et al., Artificial neural networks and cluster analysis in landslide susceptibility zonation, Geomorphology 94 (3) (2008) 379–400.

[2] R.L. Michalowski, A. Drescher, Three-dimensional stability of slopes and excavations, Geotechnique 59 (10) (2009) 839–850.

[3] O. Katz, P. Reichenbach, F. Guzzetti, Rock fall hazard along the railway corridor to jerusalem, israel, in the soreq and refaim valleys, Nat. Hazards 56 (3) (2011) 649–665.

[4] S. Yan, S. He, Y. Deng, et al., A reliability-based approach for the impact vulnerability assessment of bridge piers subjected to debris flows, Eng. Geol. 269 (2020) 105567.

[5] B. Palma, M. Parise, P. Reichenbach, et al., Rockfall hazard assessment along a road in the sorrento peninsula, campania, southern Italy, Nat. Hazards 61 (1) (2012) 187–201.

[6] H.W. Weinmeister, Integrated debris flow disaster mitigation-A comprehensive method for debris flow disaster mitigation, J. Mt. Sci. 4 (4) (2007) 293–308.

[7] T. Song, H. Pu, P. Schonfeld, et al., Parallel three-dimensional distance transform for railway alignment optimization using openMP, J. Transport. Eng. 146 (2020) 2473–2907.

[8] W. Chen, L. Fan, C. Li, et al., Spatial prediction of landslides using hybrid integration of artificial intelligence algorithms with frequency ratio and index of entropy in nanzheng county, China, Appl. Sci. 10 (2019).
[9] X. Hu, H. Zhang, H. Mei, et al., Landslide susceptibility mapping using the stacking ensemble machine learning method in lushui, southwest China, Appl. Sci. 10 (2020) 4016.
[10] L.P. Sharma, N. Patel, M.K. Ghose, et al., Development and application of Shannon's entropy integrated information value model for landslide susceptibility assessment and zonation in Sikkim himalayas in India, Nat. Hazards 75 (2015) 1555–1576.
[11] B. Mandal, S. Mandal, Landslide susceptibility mapping using modified information value model in the lish river basin of Darjiling Himalaya, Spatial Inform. Res. 25 (2017) 205–218.
[12] J.J. Wang, Landslide Risk Assessment in Wanzhou County, Three Gorges Reservoir, Doctoral Dissertation, China University of Geosciences, Wuhan, 2015.
[13] D.R. Montgomery, C. Guo, Y. Zhang, Quantitative assessment of landslide susceptibility along the Xianshuihe fault zone, Tibetan plateau, China, Geomorphology 248 (2015) 93–110.
[14] C.J. Van Westen, E. Castellanos, S.L. Kuriakose, Spatial data for landslide susceptibility, hazard, and vulnerability assessment: An overview, Eng. Geol. 102 (2008) 112–131.
[15] M. Färber, F. Bartscherer, C. Menne, Linked data quality of DBpedia, freebase, openCyc, wikidata, and YAGO, Semant. Web 9 (2018) 77–129.
[16] X. Wang, Z. Luo, R. He, et al., Novel medical question and answer system: Graph convolutional neural network based with knowledge graph optimization, Expert Syst. Appl. 227 (2023) 120211.
[17] J. Deng, C. Chen, X. Huang, et al., Research on the construction of event logic knowledge graph of supply chain management, Adv. Eng. Inform. 56 (2023) 101921.
[18] X. Hao, Z. Ji, X. Li, et al., Construction and application of a knowledge graph, Remote Sens. 13 (2021) 2511.
[19] Y. Cui, B. Yang, Application of hypergraph in data mining, Comp. Sci. Rev. (2010) 220–222.
[20] B. Perozzi, R. Al-Rfou, S. Skiena, Deepwalk: Online learning of social representations, in: Proceedings of the 20th ACM SIGKDD International Conference on Knowledge Discovery and Data Mining, 2014.
[21] B. Fatemi, P. Taslakian, D. Vazquez, et al., Knowledge hypergraphs: Extending knowledge graphs beyond binary relations, Statistics (2019).

CHAPTER SEVEN

Environmental model for railway alignment optimization

Abbreviations and corresponding full texts

CEI Carbon Emission Intensity
Cg Geometric Constraints
CGIM Comprehensive Geographic Information Model
Cl Location Constraints
CM C-metric
Cs Structure Constraints
Eb Carbon Emissions from Bridge Construction
EC Carbon Emissions at Construction Stage
EL Life-Cycle Carbon Sink Loss
EM Carbon Emissions at Maintenance Stage
EO Carbon Emissions at Operation Stage
Er Carbon Emissions from Rail Track Production
Es Carbon Emissions from Earthwork Construction
ET Total Railway Life-cycle Carbon Emissions
Et Carbon Emissions from Tunnel Construction
NDVI Normalized Difference Vegetation Index
PSO Particle Swarm Optimization

7.1 Overview

With increasing transportation demand in mountainous regions, railway projects are increasingly intruding into environmentally sensitive areas. Selecting an alignment that is both economical and environmentally responsible can help mitigate ecological damage while controlling construction costs. Existing studies on railway and highway alignment optimization have clear limitations in addressing environmental sustainability. Most models focus solely on cost minimization, while environmental impacts are either neglected or treated crudely—typically measured by the total area of affected sensitive regions, despite the difficulty of accurately delineating these regions and the varying degrees of ecological sensitivity. Furthermore, carbon emissions—an essential component of environmental impacts and a critical concern under the 26th United Nations Climate Change Conference of the Parties (COP26) and national carbon neutrality

Intelligent Railway Alignment Design
ISBN: 978-0-443-45923-8
https://doi.org/10.1016/B978-0-443-45923-8.00012-1

goals—have rarely been incorporated into early-stage alignment planning. Current emissions assessment methods are generally designed for completed projects with detailed construction and operational data, making them unsuitable for preliminary design, and even when cost–emission trade-offs are considered, they usually focus only on localized earthwork processes while overlooking emissions from bridges, tunnels, traction energy use, and long-term maintenance.

To address these gaps, this chapter develops two complementary railway alignment optimization models. The first focuses on traditional environmental objectives, evaluating vegetation loss and soil erosion. The second extends this framework by incorporating life-cycle carbon emissions, thereby improving alignment design to support green and low-carbon development goal. The remainder of this chapter is organized as follows: Section 7.2 presents a study of Multi-objective Railway Alignment Optimization Considering Costs and Environmental Impacts, while Section 7.3 introduces Minimizing Costs and Carbon Emissions in Railway Alignment Optimization: A Bi-objective Model.

7.2 Multi-objective railway alignment optimization considering costs and environmental impacts

In mountainous regions, the natural environment is fragile, with lush vegetation and nature reserves. Selecting an eco-friendly alignment may effectively reduce the environmental impacts of railway construction. However, the costs and environmental impacts of a railway may be conflicting. It is difficult to aggregate these two evaluation criteria into one single-objective function and jointly optimize them, especially if they are not measurable in commensurate units. This section presents our approach to multi-objective railway alignment optimization. The optimization model includes three objectives, i.e., the economic objective of minimizing total costs and two environmental objectives of minimizing vegetation destruction and soil erosion.

7.2.1 Vegetation destruction evaluation

According to the Code for environmental protection design in railway engineering [1], railway alignment should pass through barren land as much as possible to prevent wholesale destruction of vegetation. As a railway alignment traverses different regions, the vegetation growth status and vegetation coverage of the traversed regions vary. To evaluate the damage to the vegetation, existing methods [2,3] usually demarcate some environmentally-sensitive regions before designing the alignment. Then, during the alignment optimization process, the total area of these regions traversed by the alignment is computed. Finally, the computed area is used as the quantitative index to evaluate the damage to the vegetation due to the railway construction. However, in mountainous regions with lush vegetation, it is difficult to demarcate the environmentally-sensitive regions reasonably. The sensitivity of the environment to railway construction in different regions also varies dramatically. Therefore, the environmental impacts cannot be measured accurately.

In this section, we introduce the Normalized Difference Vegetation Index (NDVI) to evaluate the damage to vegetation due to the railway construction. NDVI is the most popular vegetation index which can reflect the vegetation growth status and vegetation coverage of a region [4]. This index is computed with Eq. (7.1) and defines values from −1.0 to 1.0, where negative values are mainly due to clouds, water, and snow. Values close to zero are primarily due to rocks and bare soil, while values close to one reflect dense vegetation:

$$\mathrm{NDVI} = \frac{\rho_{\mathrm{NIR}} - \rho_{\mathrm{RED}}}{\rho_{\mathrm{NIR}} + \rho_{\mathrm{RED}}} \tag{7.1}$$

where ρ_{NIR} is the reflectance value of the near infrared band, and ρ_{RED} is the reflectance value of the red-light band.

The latest NDVI data can be acquired from global datasets, such as SPOT (Satellite Pour l'Observation de la Terre) and MODIS (Moderate-Resolution Imaging Spectroradiometer), which reflect the recent vegetation condition. The acquired NDVI data are stored in the Comprehensive Geographic Information Model (CGIM) [5], a grid-based model that contains all the necessary information for railway alignment optimization, including topography, ground features, unit costs, endpoint coordinates, forbidden zones, major technical standards, and code parameters. The positive NDVI values of all cells traversed by the alignment are summed to obtain the overall NDVI value, which serves as the quantitative evaluation index (as shown in Fig. 7.1). This approach enables accurate assessment of vegetation damage based on up-to-date vegetation conditions.

7.2.2 Soil erosion evaluation

Soil erosion is a major environmental threat to the "sustainability and productive capacity of agriculture [6]". According to the Code for

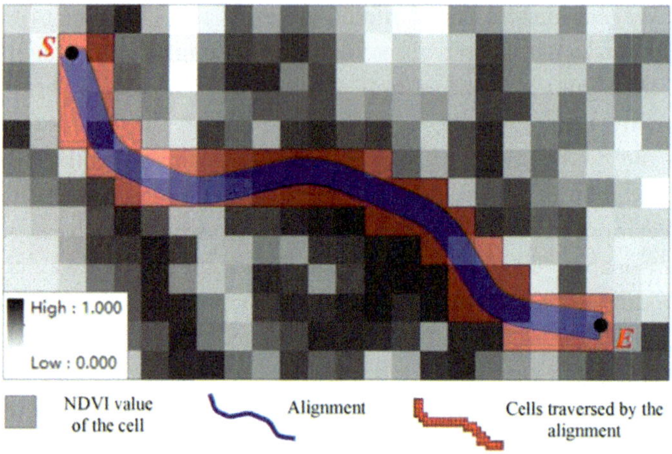

Figure 7.1 Quantitative evaluation index of vegetation destruction.

environmental protection design of railway engineering [1], the soil erosion caused by railway construction should be minimized to conserve the soil and water. Since spoil is the major cause of soil erosion [7], we use the spoil volume as the quantitative index for evaluating the level of soil erosion caused by railway construction. The spoil volume is used here as a proxy variable for soil erosion. The total spoil volume (V_S) is computed with Eq. (7.2) according to the volume of cut sections (V_{SC}), the cut volume of tunnels (V_{TC}), and the volume of fill sections (V_{SF}).

$$V_S = \max\{V_{SC} + V_{TC} - V_{SF}, 0\} \tag{7.2}$$

The volumes of cut sections V_{SC} and fill sections V_{SF} are computed with Eqs. (7.3) and (7.4), respectively, as shown in Fig. 7.2 [8], where A^{3D} represents the 3-dimension alignment, A^{2D} represents the projection of the alignment on the 2-dimension plane (i.e., the horizontal alignment):

$$V_{SC} = \int_0^L \int_{-W_l}^{W_r} \max\{H(s,x) - h(s,x), 0\} dxds \tag{7.3}$$

$$V_{SF} = \int_0^L \int_{-W_l}^{W_r} \max\{h(s,x) - H(s,x), 0\} dxds \tag{7.4}$$

where L is the total length of the alignment, W_l and W_r are the widths of the left and right sides of the subgrade, s represents the distance along the

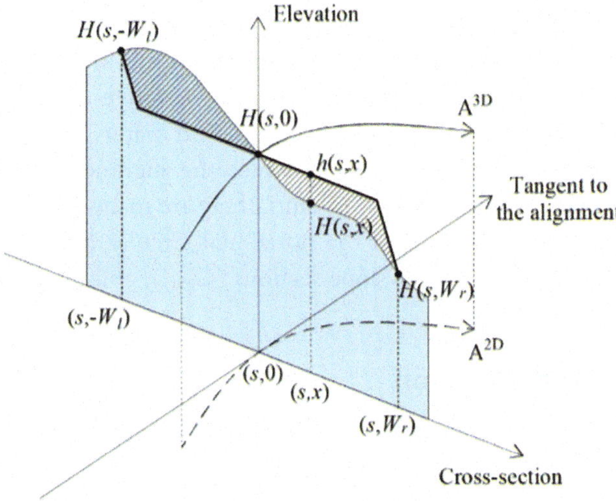

Figure 7.2 Cross section at the mileage of *s*.

alignment, and $s \in [0, L]$. The function $H(.)$ specifies the terrain elevation, and the function $h(.)$ provides the design elevation.

7.2.3 Economic evaluation

Cost is usually the most important factor affecting railway alignment design. In the existing literature on alignment optimization, various economic objective functions including construction cost (C_C) and operating cost (C_O) have been formulated. In this study, the objective functions from our previous study [5] are used. These are expressed in Eqs. (7.5) and (7.6):

$$C_C = \Delta \cdot (C_B + C_T + C_E + C_L + C_R) \tag{7.5}$$

$$C_O = C_A + C_M + C_G \tag{7.6}$$

where "Δ is the capital recovery factor. It is usually set at 0.065 in China [5]." C_B, C_T, and C_E are the construction costs of bridge, tunnel, and earthwork, C_L and C_R are the length-dependent cost and right-of-way cost, while C_A, C_M, and C_G are the operating costs which are sensitive to HPI deflection angle, alignment length, and vertical gradient. Detailed formulations of the objective functions are provided in Chapter 5.

7.2.4 Constraints

One important constraint for the alignment optimization problem is to bypass forbidden zones in an environmentally-sensitive area. The regions of nature reserve, wildlife habitats isolation, and water source protection are usually defined as forbidden zones which the alignment must bypass.

Besides the forbidden zone constraints, there are many other constraints, such as the minimum radius of the horizontal circular curves (R_{min}), the maximum gradient of vertical slope sections (G_{max}), and the crossing constraints [9].

7.2.5 Model formulation

The multi-objective railway alignment optimization model is formulated as shown in Eqs. (7.7) and (7.9):

Objective:

O1: Minimize the overall NDVI value of the cells traversed by the alignment.

$$\text{Min} f_{SV_{\text{NDVI}}}(X, Y, R, K, H) = \sum_{k=1}^{n_c} V_{\text{NDVI}}^{(k)} \tag{7.7}$$

O2: Minimize the total soil erosion volume.

$$\text{Min} f_{V_S}(X, Y, R, K, H) = \max\{V_{SC} + V_{TC} - V_{SF}, 0\} \tag{7.8}$$

O3: Minimize the total economic costs ($C_C + C_O$).

$$\begin{aligned} &\text{Min} f_{T-\text{investment}}(X, Y, R, K, H) \\ &= \Delta(C_B + C_T + C_E + C_C + C_R) + C_A + C_M + C_G \end{aligned} \tag{7.9}$$

7.2.6 Case study

In this section, a railway case is presented to demonstrate the effectiveness of the proposed method. The railway in this case connects S and B, which is a section of the S–T Railway. Through this railway case, the effectiveness of the method is demonstrated by setting different objective weights.

7.2.6.1 Railway case profile

The SB Railway is a Grade I railway, which passes through an environmentally-sensitive region with fragile ecosystems and lush vegetation. The coordinates (in the 3-D Cartesian coordinate system) of the start and end points are S (3292099.157, 32510156.707, 3034.906) and E

(3304500.333, 32476982.557, 2730.593), respectively. The airline distance between the start and end points is 35,416.287 m. The size of the study area is 836 sq. kilometers (38 × 22 km). It is represented with 836 rectangular lattice cells (30 × 30 m). The topography of the study area is shown in Fig. 7.3 and the corresponding NDVI diagram is shown in Fig. 7.4. The symbols "S" and "E" in these figures represent the start and end points of the alignment, respectively. From the NDVI diagram, we can see the NDVI value in more than half the area is nearly one (i.e., red area), which indicates very dense vegetation. The topography data used in this case,

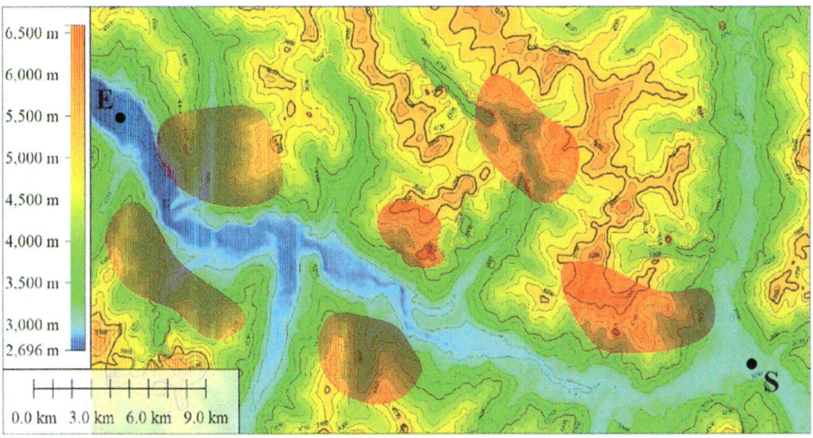

Figure 7.3 Topography data in ASC format.

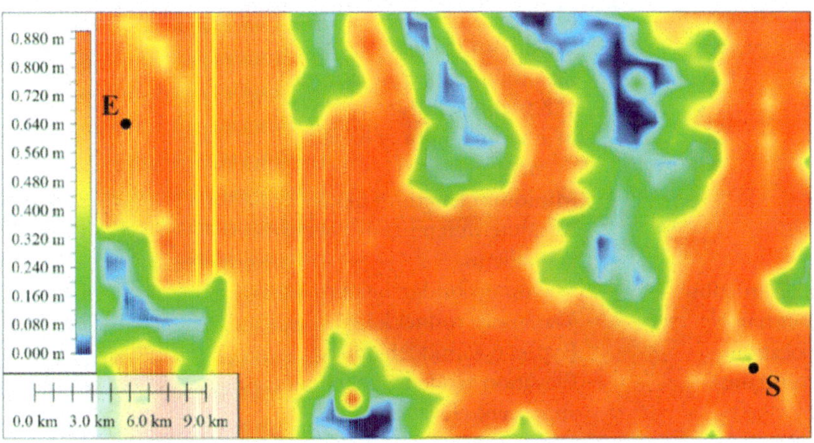

Figure 7.4 NDVI data in ASC format.

whose resolution is 30 m, are downloaded from the Geospatial Data Cloud website (http://www.gscloud.cn). The NDVI data used in this case, whose resolution is 1 km, are downloaded from the Resource and Environment Data Cloud Platform (http://www.resdc.cn).

The main technical standards, constraints, and unit cost information used in this railway case are shown in Tables 7.1 to 7.3.

Since the C-metric (CM) is the most relevant indicator of solution quality in the multi-objective optimization context [10], we introduce CM as the performance indicator to assess which scenario is preferable, as shown in Eq. (7.10).

Table 7.1 Railway major technical parameters.

Major technical standards	Recommendation
Railway grade	Grade I
Design speed	200 km/h
Main line number	Double
Vertical circular curve radius	10,000m
Fill-bridge threshold height	10 m
Cut-tunnel threshold depth	−15 m

Table 7.2 Main constraints.

Classes	Item	Value
Geometric constraints	Minimum radius of curve	2800 m
	Maximum gradient	24‰
	Minimum length of tangent between horizontal curves	120 m
	Minimum length of horizontal circular curve	120 m
	Minimum length of gradient section	250 m
	Absolute difference between adjacent gradients	18‰
Construction constraints	Maximum bridge height	100 m
	Maximum tunnel length	20,000 m
Location constraints	Bypass 6 forbidden zones	
	Elevation tendency: keep increasing alignment elevation	

Table 7.3 Unit costs.

Item	Cost	Item	Cost
Rail track (¥/m)	4000	Filling earthwork (¥/m^3)	18
Right-of-way (¥/m^2)	90	Cutting earthwork (¥/m^3)	20
($L \geq$ 1000m) tunnel (¥/m)	60,000	($H <$ 50) bridge (¥/m)	37,800
($L \geq$ 400m) tunnel (¥/m)	56,000	($H \geq$ 50, $L <$ 500 m) bridge (¥/m)	48,800
($L <$ 400m) tunnel (¥/m)	54,000	($H \geq$ 50, $L \geq$ 500 m) bridge (¥/m)	58,800
One tunnel portal (¥)	250,000	One abutment (¥)	200,000
Unit energy consumption cost (¥/kwh)	2.5	Cost per unit time (¥/h)	7.5
Annual track maintenance and repair cost quota (¥/km)	130,981	Capital recovery factor	5%
Operating cycle (Year)	50		

$$C(s_1, s_2) = \frac{|\{\mathbf{x}_2 \in \mathbf{S}_2 : \exists \mathbf{x}_1 \in \mathbf{S}_1, \mathbf{x}_1 \preccurlyeq \mathbf{x}_2\}|}{|\mathbf{S}_2|} \tag{7.10}$$

where $\mathbf{S}_1$ and $\mathbf{S}_2$ are the solution sets of s_1 and s_2, respectively, $\mathbf{x}_1 \preccurlyeq \mathbf{x}_2$ indicates that solution $\mathbf{x}_1$ dominates or is identical to $\mathbf{x}_2$, C (s_1, s_2) indicates the ratio of the solutions in $\mathbf{S}_2$ that are either dominated by any solution from $\mathbf{S}_1$, or exactly the same as any solution from $\mathbf{S}_1$, to the total number of solutions in $\mathbf{S}_2$If C (s_1, s_2) is larger than C (s_2, s_1) then s_1 is superior to s_2 Table 7.4.

The CM value of each pair among the 16 scenarios is shown in Table 7.5.

According to Table 7.5, s_9, s_{10}, s_{11}, and s_{12} are superior to the other scenarios. The $\boldsymbol{n_p}$ values of these 4 scenarios are all 32 and the value of $\boldsymbol{n_g}$ does not affect the final solution set when the $\boldsymbol{n_g}$ value exceeds 70. Therefore, the values of $\boldsymbol{n_p}$ and $\boldsymbol{n_g}$ are determined as 32 and 70, respectively. All the PSO parameters used for this case are shown in Table 7.6.

Table 7.4 n_p and n_g values of the 16 scenarios.

n_p \ n_g	60	70	80	90
24	s_1 (24,60)	s_2 (24,70)	s_3 (24,80)	s_4 (24,90)
28	s_5 (28,60)	s_6 (28,70)	s_7 (28,80)	s_8 (28,90)
32	s_9 (32,60)	s_{10} (32,70)	s_{11} (32,80)	s_{12} (32,90)
36	s_{13} (36,60)	s_{14} (36,70)	s_{15} (36,80)	s_{16} (36,90)

Table 7.5 C-metric values.

	s_1	s_2	s_3	s_4	s_5	s_6	s_7	s_8	s_9	s_{10}	s_{11}	s_{12}	s_{13}	s_{14}	s_{15}	s_{16}
s_1	1.00	0.97	0.97	0.97	0.70	0.74	0.75	0.75	0.77	0.78	0.78	0.78	0.84	0.84	0.84	0.84
s_2	1.00	1.00	1.00	1.00	0.73	0.77	0.77	0.77	0.78	0.79	0.79	0.79	0.87	0.87	0.87	0.87
s_3	1.00	1.00	1.00	1.00	0.73	0.77	0.77	0.77	0.78	0.79	0.79	0.79	0.87	0.87	0.87	0.87
s_4	1.00	1.00	1.00	1.00	0.73	0.77	0.77	0.77	0.78	0.79	0.79	0.79	0.87	0.87	0.87	0.87
s_5	0.85	0.86	0.86	0.86	1.00	1.00	1.00	1.00	0.82	0.82	0.82	0.82	0.95	0.95	0.95	0.95
s_6	0.80	0.81	0.80	0.80	1.00	1.00	1.00	1.00	0.80	0.76	0.76	0.76	0.89	0.89	0.89	0.89
s_7	0.80	0.81	0.80	0.80	1.00	1.00	1.00	1.00	0.80	0.76	0.76	0.76	0.89	0.89	0.89	0.89
s_8	0.80	0.81	0.80	0.80	1.00	1.00	1.00	1.00	0.80	0.76	0.76	0.76	0.89	0.89	0.89	0.89
s_9	0.88	0.89	0.89	0.89	0.97	0.97	0.97	0.97	1.00	1.00	1.00	1.00	0.95	0.95	0.95	0.95
s_{10}	0.88	0.89	0.89	0.89	0.97	0.97	0.97	0.97	1.00	1.00	1.00	1.00	0.95	0.95	0.95	0.95
s_{11}	0.88	0.89	0.89	0.89	0.97	0.97	0.97	0.97	1.00	1.00	1.00	1.00	0.95	0.95	0.95	0.95
s_{12}	0.88	0.89	0.89	0.89	0.97	0.97	0.97	0.97	1.00	1.00	1.00	1.00	0.95	0.95	0.95	0.95
s_{13}	0.90	0.90	0.91	0.91	0.95	0.96	0.96	0.96	0.87	0.85	0.85	0.85	1.00	1.00	1.00	1.00
s_{14}	0.90	0.90	0.91	0.91	0.95	0.96	0.96	0.96	0.87	0.85	0.85	0.85	1.00	1.00	1.00	1.00
s_{15}	0.90	0.90	0.91	0.91	0.95	0.96	0.96	0.96	0.87	0.85	0.85	0.85	1.00	1.00	1.00	1.00
s_{16}	0.90	0.90	0.91	0.91	0.95	0.96	0.96	0.96	0.87	0.85	0.85	0.85	1.00	1.00	1.00	1.00

Table 7.6 PSO parameters.

Parameters	Value
Swarm size	100
Learning factor	$c_1 = c_2 = 2.0$
Maximum velocity	500 m/iteration
Maximum size of personal best solution set	32
Maximum size of global best solution set	70

7.2.6.2 *Experimental results and analyses*

The total elapsed times of Scenarios 1, 2, 3, and 4 are 29.552 min, 33.168 min, 36.465 min, and 70.183 min, respectively. To normalize the values of the three objectives, the Min–max normalization method is used, as defined in Eq. (7.11):

$$y_i = \frac{x_i - \min x_i}{\max x_i - \min x_i} \tag{7.11}$$

where x_i is the value of the i-th objective and y_i is the corresponding normalized value, min x_i is the minimum value of the i-th objective among all the non-dominated solutions, and max x_i is the maximum value of the i-th objective among all the non-dominated solutions.

To compare the obtained non-dominated solutions of the four scenarios, we combine the non-dominated solutions obtained in Scenario 4 with the non-dominated solutions obtained in the other three scenarios and normalize the objective values based on the maximum and minimum values of the combined solution set. Based on the normalized objective values, the obtained Pareto-optimal fronts in Scenario 4 and the other three scenarios are compared in Fig. 7.5.

As shown in Fig. 7.5, the non-dominated solutions obtained in Scenario 4are more evenly distributed. The results of the other three scenarios are more concentrated around the area where the value of the highly weighted objective is low. Furthermore, in the direction of the highly weighted objective, the non-dominated solutions of Scenarios 1, 2, and 3 lie on a layer outside that of the non-dominated solutions of Scenario 4, which implies that the obtained Pareto-optimal Front is closer to the true Pareto-optimal Front in the direction of the highly weighted objective.

To further compare the non-dominated solutions of the four scenarios, we select the best solution of each scenario according to the preference information, using Eq. (7.12):

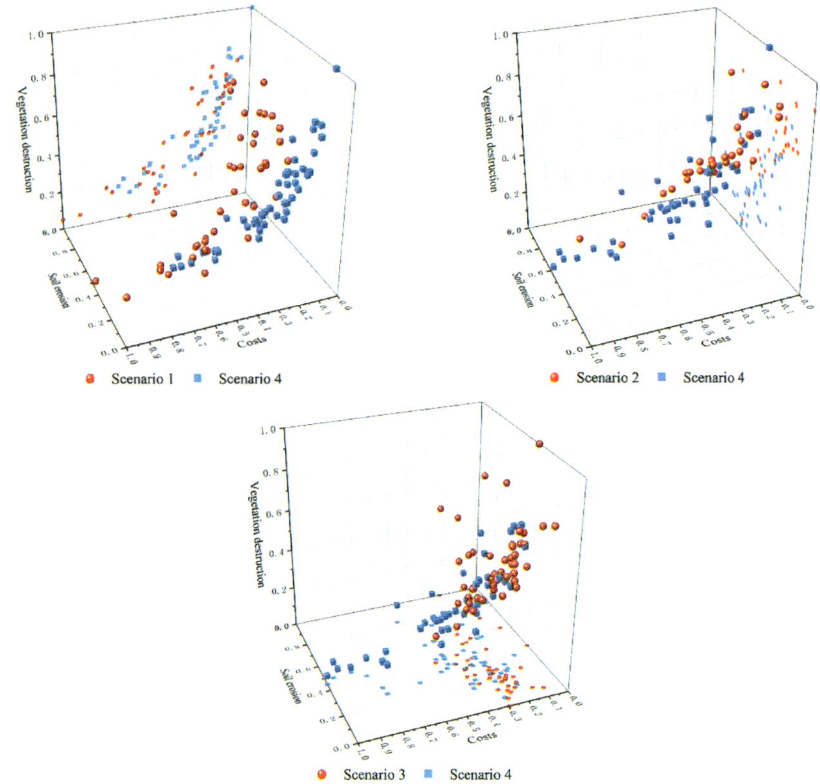

Figure 7.5 Comparison of the post-processed pareto-optimal fronts.

$$CV'^{(j)} = w_1 \cdot \left(C_c'^{(j)} + C_o'^{(j)}\right) + w_2 \cdot SV_{\mathrm{NDVI}}'^{(j)} + w_3 \cdot V_S'^{(j)} \tag{7.12}$$

where $C_c'^{(j)}$ and $C_o'^{(j)}$ are the normalized construction and operating costs of the j-th non-dominated solution, $SV_{\mathrm{NDVI}}'^{(j)}$ is the normalized overall NDVI value of the j-th non-dominated solution, $V_S'^{(j)}$ is the normalized spoil volume of the j-th non-dominated solution, and w_1, w_2, w_3 are the weights of the three objectives, as shown in Table 7.7.

The selected best solutions are termed as Alt1, Alt2, Alt3, and Alt4. The horizontal and vertical alignments of these four solutions are shown in Fig. 7.6. The detailed results are presented in Table 7.8.

Table 7.7 Objective weights.

Scenario no.	Scenario 1	Scenario 2	Scenario 3	Scenario 4
Obj3: Costs	0.1	0.1	0.8	0.4
Obj1: Vegetation destruction	0.8	0.1	0.1	0.3
Obj2: Soil erosion	0.1	0.8	0.1	0.3

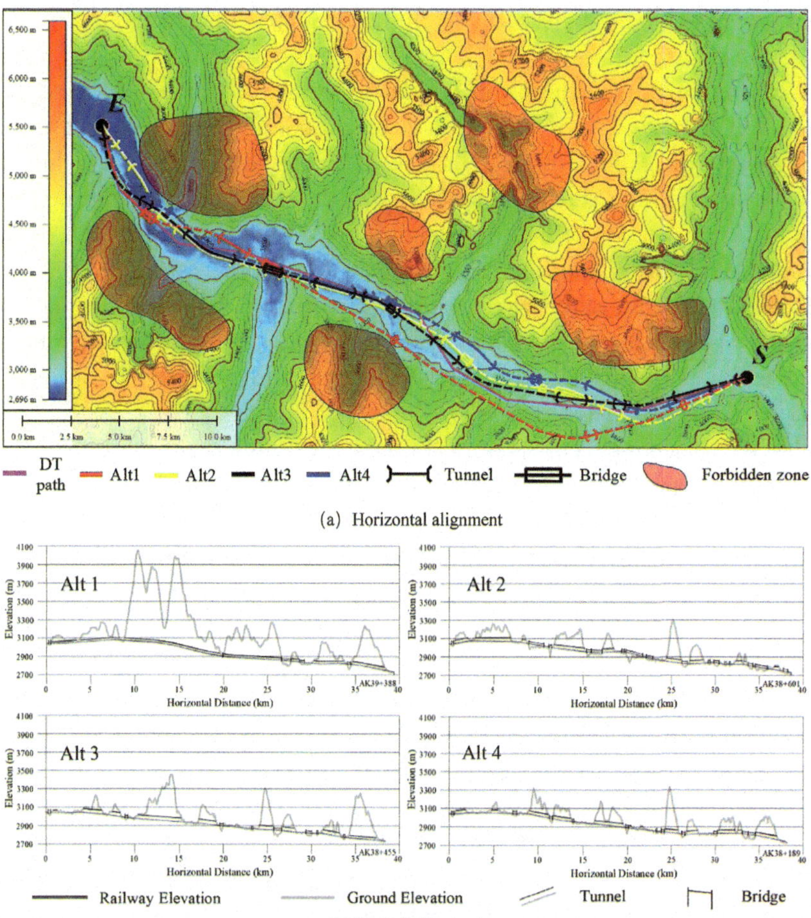

Figure 7.6 Horizontal and vertical alignments of Alt 1, Alt 2, Alt 3, and Alt 4.

Table 7.8 Detailed results for Alt 1, Alt 2, Alt 3, and Alt 4.

	Vegetation destruction (NDVI)	Soil erosion (m^3)	Annual costs (million ¥)	Subgrade length (m)	Bridge length (m)	Tunnel length (m)
Alt1	77.08	5,068,425	29.3846	3361	2442	33,585
Alt2	169.71	3,585,596	24.1246	9125	11,094	18,382
Alt3	210.08	4,392,906	22.9025	12,573	3014	22,868
Alt4	205.52	4,295,591	23.2666	11,149	4186	22,854

By inspecting Fig. 7.6, it can be found that all the pre-specified forbidden zones are bypassed, which demonstrates the effectiveness of the local repair algorithm (LRA).

By reviewing the results of the four alternatives it can be observed that [1]; When the weight of vegetation destruction is highest, the tunnel length of the selected alignment (i.e., Alt 1) is the greatest [2]; When the weight of soil erosion is highest, the bridge length of the selected alignment (i.e., Alt 2) is the greatest [3]; When the weight of cost is highest, the length of cut and fill sections of the selected alignment (i.e., Alt 3) is the greatest [4]; When the weights of the three objectives are similar, the objective function values of the selected best alignment (i.e., Alt 4) are intermediate among these four alternatives. The conflicts among the three objectives considered in this study are analyzed as follows.

The fill sections, cut sections, bridges, and tunnels are the four major structures types of a railway alignment. These four structures types have different impacts on the objectives considered in this study.

(1) The construction costs of fill and cut sections are the lowest. However, serious vegetation damage and soil erosion result if the alignment traverses in the form of fill and cut sections.

(2) The levels of vegetation destruction and soil erosion are much lower if the alignment traverses in the form of bridges. However, the construction costs are higher for bridges than for fill sections when the alignment is close to the terrain surface. Also, the maximum allowable pier height is limited by the current construction technology. This constraint must be satisfied.

(3) The construction costs for tunnels are also higher than for cut sections unless the tunnels are far below the terrain surface. Moreover, building tunnels would produce a vast amount of spoil, which would cause soil erosion. However, the vegetation is not damaged if the alignment traverses through tunnels.

Therefore, if we focus on the economic objective of reducing the costs, the alignment should traverse in the form of relatively shallow fill and cut sections as much as possible. The obtained result of Scenario 3 (i.e., Alt 3) illustrates this point. If soil and water conservation is the main objective, fill and cut sections as well as tunnels are not preferred. The alignment optimization algorithm should seek suitable areas for traversing in the form of bridges. This point is illustrated by the obtained result of Scenario 2 (i.e., Alt 2). If we focus on protecting the vegetation, the alignment should traverse through bridges and tunnels as much as possible. However, since the maximum allowable pier height is limited, it would be difficult to find suitable regions for building bridges, especially in complex mountainous regions as this railway case. Thus, tunnels could be the major structure type of the optimized alignment, which could differ greatly from an alignment whose major structure types are bridges or earthwork segments. The obtained results of Scenario 1 (i.e., Alt 1) illustrate this point.

The above results demonstrate that the proposed method effectively balances economic factors with traditional environmental considerations such as vegetation loss and soil erosion. Based on it, the following section extends the optimization framework to incorporate carbon emissions—an increasingly critical dimension of environmental sustainability—supporting green, low-carbon railway development.

7.3 Minimizing construction costs and carbon emissions in railway alignment optimization

7.3.1 Railway construction cost

The construction cost is important for every railway or road alignment design problem. Detailed formulations for computing it have been provided in Section 7.2.3 and are not repeated here. The next section elaborates the developed carbon emissions model for alignment optimization.

7.3.2 Carbon emissions model

According to the research of the International Union of Railways [11], railway carbon emissions are mostly generated at the construction, operation and maintenance stages during a railway's design life. Therefore, the total carbon emissions formula is given in Eq. (7.13):

$$E_T = E_C + E_O + E_M + E_L \tag{7.13}$$

where E_T is the railway life-cycle carbon emissions; E_C, E_O and E_M are the carbon emissions generated at the railway construction, operation and maintenance stages, respectively; E_L is the life-cycle carbon sink loss caused by vegetation destruction along the railway. Detailed formulations of Eq. (7.13) are specified below.

7.3.2.1 Construction stage

The carbon emissions generated at the construction stage (E_C) are mainly caused by the production of building materials as well as the energy consumption of construction equipment for different railway structures, including bridges, tunnels, earthworks and tracks. They are expressed in Eq. (7.14):

$$E_C = E_b + E_t + E_s + E_r \tag{7.14}$$

where E_b, E_t, E_s and E_r are carbon emissions caused by bridge, tunnel, earthwork and track constructions, respectively.

(1) Carbon emission generated by bridge construction Firstly, E_b mainly consists of the carbon emissions caused by productions of bridge concrete and steel as well as the energy consumption of bridge construction equipment, which are expressed in Eq. (7.15):

$$E_b = g_c \cdot \sum_{i=1}^{N_b} M_c^b(i) + g_s \cdot \sum_{i=1}^{N_b} M_s^b(i) + E_c^b \tag{7.15}$$

where g_c and g_s are concrete (kg/m^3) and steel (kg/kg) carbon emission factors, respectively; N_b is the number of bridges along the alignment; M_c^b and M_s^b are concrete (m^3) and steel (kg) consumptions for bridge construction, respectively; E_c^b represents the emissions caused by bridge construction equipment operation (kg). In this equation, due to limited data at the early railway design stage for alignments, the assumed bridge type is a beam bridge, which is the most widely-used bridge type in practice [12]. The carbon emissions due to complex bridge structures, such as large suspension bridges, need ad-hoc investigations and, in general, are not evaluated at the alignment design stage.

More specifically, the formulas for computing M_c^b, including those for bridge spans (M_c^{bs}), piers (M_c^{bp}) and foundations (M_c^{bf}), are shown in Eqs. (7.16) and (7.19):

$$M_c^b(i) = M_c^{bs}(i) + M_c^{bp}(i) + M_c^{bf}(i) \tag{7.16}$$

$$M_c^{bs}(i) = A_{bs} \cdot L_b(i) \tag{7.17}$$

$$M_c^{bp}(i) = \sum_{n=1}^{N=L_b(i)/l-1} \left[A_{bp} \cdot h_{bp}^n(i)\right] \tag{7.18}$$

$$M_c^{bf}(i) = \sum_{n=1}^{N=L_b(i)/l-1} \left[A_{bf} \cdot h_{bf}^n(i)\right] \tag{7.19}$$

where i is the bridge index; A_{bs}, A_{bp} and A_{bf} are cross-sectional areas of bridge spans, piers, and foundations (m^2), respectively; L_b is bridge length (m); N is the number of bridge piers; l is span length (m); h_{bp}^n is the height of the *nth* bridge pier (m), which is determined by computing the difference between railway design elevation and ground elevation; h_{bf}^n is the height of the *nth* bridge pier foundation (m). The information on bridge spans, piers, and foundations is determined by referring to the Code for Design on Railway Bridges and Culverts [13].

Then, the formulas for computing M_s^b are derived as Eqs. (7.20)–(7.26), considering the usage of longitudinal bars and stirrups on bridge spans, piers, and foundations:

$$M_s^b(i) = \left[M_z^{bs}(i) + M_g^{bs}(i)\right] + \left[M_z^{bp}(i) + M_g^{bp}(i)\right] + \left[M_z^{bf}(i) + M_g^{bf}(i)\right] \tag{7.20}$$

$$M_z^{bs}(i) = A_{bs} \cdot \gamma_z \cdot L_b(i) \cdot \rho_z \tag{7.21}$$

$$M_g^{bs}(i) = A_{bs} \cdot L_b(i) \cdot \gamma_g \cdot \rho_g \tag{7.22}$$

$$M_z^{bp}(i) = \sum_{n=1}^{N=L_b(i)/l-1} \left[A_{bp} \cdot \gamma_z \cdot h_{bp}^n(i) \cdot \rho_z\right] \tag{7.23}$$

$$M_g^{bp}(i) = \sum_{n=1}^{N=L_b(i)/l-1} \left[A_{bp} \cdot h_{bp}^n(i) \cdot \gamma_g \cdot \rho_g\right] \tag{7.24}$$

$$M_z^{bf}(i) = \sum_{n=1}^{N=L_b(i)/l-1} \left[A_{bf} \cdot \gamma_z \cdot h_{bf}^n(i) \cdot \rho_z\right] \tag{7.25}$$

$$M_g^{bf}(i) = \sum_{n=1}^{N=L_b(i)/l-1} \left[A_{bf} \cdot h_{bf}^n(i) \cdot \gamma_g \cdot \rho_g\right] \tag{7.26}$$

where M_z^{bs} and M_g^{bs} are the amounts of longitudinal bars and stirrups used for building bridge spans (kg), respectively; M_z^{bp} and M_g^{bp} are the amounts of longitudinal bars and stirrups used for building bridge piers (kg), respectively; M_z^{bf} and M_g^{bf} are the amounts of longitudinal bars and stirrups, respectively, used for building bridge pier foundations (kg); γ_z and γ_g are the area ratio of longitudinal bars and volume ratio of stirrups (%), respectively; ρ_z and ρ_g are density values of longitudinal bars and stirrups (kg/m^3), respectively.

Lastly, E_c^b is directly related to bridge length and can be computed by Eq. (7.27):

$$E_c^b = e_c^b \cdot \sum_{i=1}^{N_b} L_b(i) \tag{7.27}$$

where e_c^b is an empirical value (kg/m, [14]).

Therefore, the overall formula for computing the carbon emissions in bridge construction is expressed as:

$$\begin{aligned}
E_b = {} & g_c \cdot \sum_{i=1}^{N_b} \left\{ A_{bs} \cdot L_b(i) + \sum_{n=1}^{N=L_b(i)/l-1} \left[A_{bp} \cdot h_{bp}^n(i)\right] + \sum_{n=1}^{N=L_b(i)/l-1} \left[A_{bf} \cdot h_{bf}^n(i)\right] \right\} \\
& + g_s \cdot \sum_{i=1}^{N_b} \left\{ \begin{array}{l} A_{bs} \cdot \gamma_z \cdot L_b(i) \cdot \rho_z + A_{bs} \cdot L_b(i) \cdot \gamma_g \cdot \rho_g + \\ \displaystyle\sum_{n=1}^{N=L_b(i)/l-1} \left[A_{bp} \cdot \gamma_z \cdot h_{bp}^n(i) \cdot \rho_z + A_{bp} \cdot h_{bp}^n(i) \cdot \gamma_g \cdot \rho_g\right] + \\ \displaystyle\sum_{n=1}^{N=L_b(i)/l-1} \left[A_{bf} \cdot \gamma_z \cdot h_{bf}^n(i) \cdot \rho_z + A_{bf} \cdot h_{bf}^n(i) \cdot \gamma_g \cdot \rho_g\right] \end{array} \right\} \\
& + e_c^b \cdot \sum_{i=1}^{N_b} L_b(i)
\end{aligned} \tag{7.28}$$

(2) Carbon emission generated by tunnel construction In this section, the building materials involved in tunnel construction mainly include concrete, anchor bolt and steel mesh. Thus, E_t in Eq. (7.29) can be computed as:

$$E_t = g_c \cdot \sum_{i=1}^{N_t} M_c^t(i) + g_s \cdot \sum_{i=1}^{N_t} M_s^t(i) + E_c^t \quad (7.29)$$

where N_t is the number of tunnels; M_c^t and M_s^t are concrete (m^3) and steel consumptions (kg), respectively; and *Et c* represents the emissions caused by tunnel construction equipment operation (kg).

More specifically, for railway alignment design, the tunnel cross section is generally shaped like a horseshoe, as shown in Fig. 7.7, and the formula for computing M_c^t, which considers the usage of concrete in the tunnel lining, is shown in Eq. (7.30):

$$M_c^t(i) = A_l \cdot L_t(i) \quad (7.30)$$

where i is the tunnel index; L_t is the tunnel length (m); A_l is cross-sectional area of tunnel lining (m^2).

Then, the formulas for computing M_s^t, which considers the usage of anchor bolt and steel mesh (which are generally distributed along the arch wall of the tunnel), is shown in Eq. (7.31):

$$M_s^t(i) = N_{ab}(i) \cdot A_{ab} \cdot l_{ab} \cdot \rho_{ab} + N_{sm}(i) \cdot A_{sm} \cdot l_{sm} \cdot \rho_{sm} \quad (7.31)$$

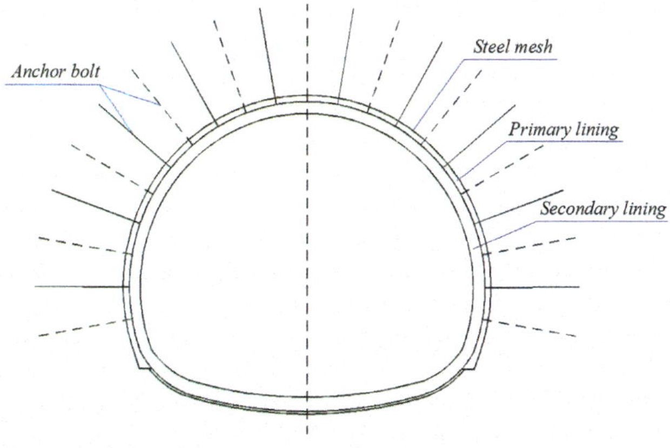

Figure 7.7 Tunnel cross section.

$$N_{ab}(i) = L_w \cdot L_t(i) / s_{ab}^2 \tag{7.32}$$

$$N_{sm}(i) = L_w \cdot L_t(i) / s_{sm}^2 \tag{7.33}$$

where N_{ab} is the number of anchor bolts; N_{sm} represents the converted quantity of steel mesh in a unit area (integer); A_{ab} and A_{sm} are areas of anchor bolt and steel mesh (m^2), respectively; l_{ab} and l_{sm} are anchor bolt and steel mesh lengths (m), respectively; ρ_{ab} and ρ_{sm} are anchor bolt and steel mesh densities (kg/m^3), respectively; L_w is the length of tunnel arch wall (m); s_{ab} and s_{sm} are anchor bolt and steel mesh spacings (m), respectively. It is assumed that cross and longitudinal spacings for anchor bolts (or steel meshes) are the same. Thus, Eqs. (7.32) and (7.33) can be obtained.

Lastly, the formula for computing the emission caused by tunnel construction equipment operation E_c^t (kg) is expressed in Eq. (7.34):

$$E_c^t = e_c^t \cdot \sum_{i=1}^{N_t} L_t(i) \tag{7.34}$$

where e_c^t is an empirical value related to tunnel length (kg/m).

Therefore, the overall formula proposed for computing the carbon emissions caused by tunnel construction is expressed in Eq. (7.35):

$$\begin{aligned} E_t = {} & g_c \cdot \sum_{i=1}^{N_t} [A_l \cdot L_t(i)] \\ & + g_s \cdot \sum_{i=1}^{N_t} \left[L_w \cdot L_t(i) / s_{ab}{}^2 \cdot A_{ab} \cdot l_{ab} \cdot \rho_{ab} + L_w \cdot L_t(i) / s_{sm}{}^2 \cdot A_{sm} \cdot l_{sm} \cdot \rho_{sm} \right] \\ & + e_c^t \cdot \sum_{i=1}^{N_t} L_t(i) \end{aligned} \tag{7.35}$$

(3) Carbon emission generated by earthwork construction In this study, the carbon emissions generated during earthwork construction (E_s) mainly refer to the energy consumed by excavators and rollers that are used to construct cuts and embankments. Therefore, they are directly related to earthwork volumes. Thus, E_s in Eq. (7.14) can be expanded as Eq. (7.36):

$$E_s = g_e \cdot r_e \cdot \sum_{i=1}^{N_s} e_c^s(i) \tag{7.36}$$

where g_e is a factor converting energy consumption into carbon emissions (kg/kwh); r_e is the unit energy consumption rate (%); N_s is the number of earthwork sections; e_c^s is the mechanical energy consumed for earthwork transportation (kwh), which is computed as shown below.

First, the cut and fill volumes ($V_{C/F}$) are computed using Eq. (7.37):

$$V_{C/F} = (K_{i+1} - K_i) \cdot (S_i + S_{i+1}) / 2 \tag{7.37}$$

where K_i and K_{i+1} are the stations of the ith and $(i+1)th$ earthwork sections, respectively; S_i and S_{i+1} are the areas of the ith and $(i+1)th$ earthwork sections (m^2), which are computed using the average-end-area method [15]. Then, the energy consumption for constructing cuts and fills can be expressed in Eq. (7.38):

$$e_c^s(i) = P_C \cdot Q_C \cdot V_C(i)/100 + P_F \cdot Q_F \cdot V_F(i)/100 \tag{7.38}$$

where Q_C and Q_F are the number of required machines for mechanical movement of 100 m^3 cut and fill volumes respectively within unit work time, which is generally set at 8 h according to the Budget Quota for Railway Earthwork [16]; P_C and P_F are the energy consumption of excavators and rollers (kwh), respectively, according to the Rules for Preparation of Machine Movement Cost of Construction Engineering [17].

Therefore, the overall formula for computing the carbon emissions due to earthwork construction is shown in Eq. (7.39):

$$E_s = g_e \cdot r_e \cdot \sum_{i=1}^{N_s} [P_C \cdot Q_C \cdot V_C(i) / 100 + P_F \cdot Q_F \cdot V_F(i) / 100] \tag{7.39}$$

(4) Carbon emission generated by producing rail tracks In this section, three major components of rail tracks are considered in carbon emissions estimation, i.e., the concrete for rail plates and bases, steel for rails, and equipment energy consumptions, which are computed using Eq. (7.40):

$$E_r = g_c \cdot M_c^r + g_s \cdot M_s^r + E_c^r \tag{7.40}$$

where g_c and g_s are concrete (kg/m^3) and steel (kg/kg) carbon emission factors, respectively; M_c^r and M_s^r are concrete (m^3) and steel (kg) amounts, respectively; E_c^r is the emission caused by track construction equipment operation (kg).

More specifically, Eqs. (7.41) and (7.44) are provided for extending Eq. (7.40):

$$M_c^r = A_r \cdot L_r \tag{7.41}$$

$$M_s^r = N \cdot L_r \cdot \rho_r \tag{7.42}$$

$$E_c^r = N \cdot e_c^r \cdot L_r \tag{7.43}$$

where A_r is the total cross section area of rail plates and bases (m^2); L_r is the alignment length (m); ρ_r is the rail unit weight (kg/m); N is the number of railway track lines; e_c^r is an empirical parameter related to track length (kg/m, [14]). Therefore, E_r is ultimately derived from Eq. (7.44):

$$E_r = g_c \cdot A_r \cdot L_r + g_s \cdot N \cdot L_r \cdot \rho_r + N \cdot e_c^r \cdot L_r \tag{7.44}$$

7.3.2.2 Operation stage

In this work, the considered carbon emissions at the railway operation stage are mainly generated by the power supply systems for train traction along the alignment, which are assessed based on the energy consumption of vehicle operation using four steps.

(1) Train traction and resistance analyses

During operation, the train energy consumption is mainly determined by the combined effects of traction and resistance forces [18]. As shown in Fig. 7.8, a typical relation between train traction force and resistance is used for the following explanations.

According to this curve, the relation between traction force (F) and train speed is shown in Eq. (7.45):

$$F = \begin{cases} F_0 - \delta v & 0 < v \le v_0 \\ P/v & v_0 < v \le v_{\max} \end{cases} \tag{7.45}$$

where F_0 is the maximum traction (kN); δ is an empirical parameter which varies with the train type; v is the train speed (km/h); P is the train output power (kw).

Resistance consists of two components, namely the basic and alignment resistances. The basic resistance (w_0) is mainly related to train speed and weight [19]. The additional resistance (w_j) is related to

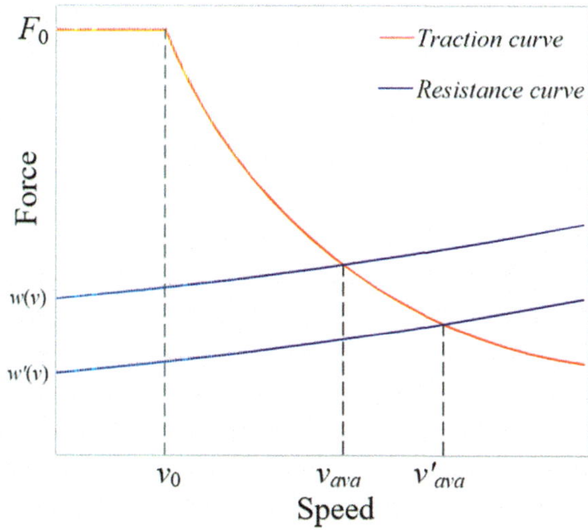

Figure 7.8 Train traction and resistance curves.

alignment conditions and includes three components, namely the gradient, curve and tunnel resistances [20]. They are computed using Eqs. (7.46) and (7.50):

$$\text{Basic resistance }(\text{N}/\text{kN}): w_0 = a + bv + cv^2 \tag{7.46}$$

$$\text{Gradient resistance }(\text{N}/\text{kN}): \text{w}_\text{g} = G \tag{7.47}$$

$$\text{Curve resistance }(\text{N}/\text{kN}): w_r = m/R \tag{7.48}$$

$$\text{Tunnel resistance }(\text{N}/\text{kN}): w_s = n \cdot L_s \tag{7.49}$$

$$w_j = w_g + w_r + w_s \tag{7.50}$$

where a, b and c are empirical parameters determined by train type; v is train speed (km/h); G is alignment gradient (‰); m and n are empirical parameters determined by the railway grade; R is horizontal circular curve radius (m); L_s denotes tunnel length (km). Therefore, the total resistance varies at different alignment sections, as shown in Fig. 7.8.

(2) Alignment division
After obtaining the above train force information, an alignment alternative can be segmented according to the additional resistance (also known as generalized gradient), that is, the alignment sections with the same w_j values can be combined. Besides, a braking position

(S_{bra}) should also be located before the end point to ensure that trains can stop at terminals accurately. It is determined by Eq. (7.51):

$$S_{bra} = L - v_{bra}^2 / (2 \cdot a_{max}) \tag{7.51}$$

where v_{bra} is the braking speed preset by designers (m/s); a_{max} is braking deceleration (m/s^2); L is alignment length (m). Thus, the final alignment division result can be illustrated as in Fig. 7.9.

After dividing the railway alignment into several sections, the maximum allowable speed $V_d(S_i)$ in each section is constrained by both the equilibrium speed V_{max} and the limiting speed V_{lim}. The equilibrium speed V_{max} is determined at the intersection point between the traction and resistance curves in the train traction characteristics. Since the running resistance varies across different sections, the equilibrium speed also differs accordingly. On downhill sections, V_{lim} represents the maximum permissible speed for ensuring operational safety, which can be calculated based on the gradient.

The smaller value between V_{max} and V_{lim} is then taken as the maximum allowable speed $V_d(S_i)$ for each subsection and used in the subsequent train operation simulations.

(3) Train operation simulation
For a detailed introduction to this section, please refer to Chapter 5.

(4) Conversion of energy consumption into carbon emission
Finally, the determined energy consumption is converted into carbon emissions, as shown in Eq. (7.52):

$$E_{opera} = g_e \cdot M_t \cdot d \cdot \sum_{i=1}^{N} E_i(v) \tag{7.52}$$

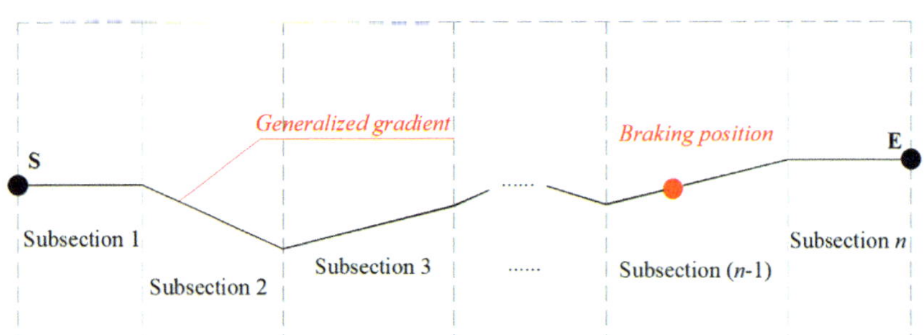

Figure 7.9 Diagram of alignment subsections.

where E_{opera} is annual carbon emissions during the railway operation stage (kg), which will be further modified considering the impacts of carbon emission attenuation; M_t is the average number of daily trains in year t, which is predicted according to the annual passenger flow volume [18]; d is operating days per year; N is the number of alignment segments; $E_i(v)$ is energy consumption of the *ith* alignment segment (kwh); g_e is a factor to convert energy to carbon emission (kg/kwh) [21].

7.3.2.3 Maintenance stage

During operation, railway structures (bridges, tunnels, earthworks and tracks) should be maintained daily and tracks must also be replaced periodically. At this maintenance stage, along with the renovation of equipment and facilities, energy consumption and carbon emissions are still considerable. Thus, the following formulations are proposed to estimate carbon emissions for maintaining these structures.

(1) Structural maintenance

The carbon emissions' generation mechanism at the maintenance stage is similar to that at the construction stage, which also involves a building materials' production factor and an energy consumption factor. In this section, a simplified formula for estimating annual carbon emissions of structure maintenance ($Eb/t/s/r\ M$) is shown in Eq. (7.53):

$$E_M^{b/t/s/r} = e_M^{b/t/s/r} \cdot L_{b/t/s/r} \tag{7.53}$$

where $e_M^{b/t/s/r}$ is an empirical parameter (kg/m) [22] related to structural length $L_{b/t/s/r}$ (m) and defined differently for different structural types ($b=$ bridge, $t=$ tunnel, $s=$ cut/fill, $r=$ rail track). $E_M^{b/t/s/r}$ will be modified considering carbon emissions attenuation as for E_{opera}.

(2) Track replacement

Due to long-term train operations, rail tracks face inevitable damages, especially at curved alignment sections where wheel-track interactions concentrate [23]. To determine the track replacement carbon emissions, the cycle of track repair ($T(R)$), which is closely related to different alignment curve radii (R), should be first computed, as shown in Eq. (7.54):

$$T(R) = M(R)/M \tag{7.54}$$

where $M(R)$ is the critical passing weight of trains (tons) along curve sections with a radius of R; M is the annual passing weight of trains (tons) along an alignment. In the above equation, rails located on curves with different radii have different wear rates and, therefore, different replacement cycles. Moreover, tangents are deemed as special curves with infinite radii.

Then, the formula for computing carbon emissions of track replacement is given by Eq. (7.55):

$$E_M^{r'} = g_s \cdot M_s^r(R) \tag{7.55}$$

where $E_M^{r'}$ is the initial carbon emissions of the rail replacement with a radius of R (kg), which will be further modified considering track replacement periodicity and carbon emission attenuation; $M_s^r(R)$ is the amount of steel used for curve sections with a radius of R (kg), which is related to the curve radius and length; g_s is the carbon emission factor for rail steels (kg/kg).

7.3.2.4 Carbon sink loss estimation

Carbon sink refers to the activity by which carbon dioxide is absorbed through afforestation or vegetation restoration and, thus, can reduce the concentration of greenhouse gases in the atmosphere. In general, three types of land carbon sinks are involved in railway designs, i.e., woodland, grassland and cropland sinks.

Due to railway construction, the land use pattern along the alignment will be changed. Thus, the vegetation may be damaged, which would destroy its ability to absorb carbon dioxide in the study area. In this condition, the carbon sink loss is considered as equivalent to increasing the same amount of carbon emissions throughout the railway life. In this study, the carbon sink loss (E_L) includes that along bridge and earthwork sections, as shown in Eq. (7.56):

$$E_L = \sum_{m=1}^{3} \left[c_m \cdot \left(S_m^s + S_m^b\right)\right] \cdot T \tag{7.56}$$

where c_m is annual carbon sink per unit area (kg/m^2) of the *mth* land cover type, $m = 1$ (woodland), 2 (grassland), 3 (cropland); S_m^s is the railway right-of-way area of earthwork sections (m^2); S_m^b is the requisitioned land area of bridges (m^2); T is the railway economic life (years).

Specifically, S_m^s is computed in Eq. (7.57):

$$S_m^s = \sum_{i=1}^{N_m - 1} \left[(K_{i+1} - K_i) \cdot (W_i + W_{i+1}) / 2\right] \tag{7.57}$$

where N_m is the number of earthwork sections; K_i and K_{i+1} are the stations of the *ith* and $(i+1)th$ earthwork sections, respectively; W_i and W_{i+1} are the widths of the *ith* and $(i+1)th$ earthwork sections (m), respectively.

Then, S_m^b is estimated using Eq. (7.58):

$$S_m^b = L_m^b / l \cdot S_{bp} \tag{7.58}$$

where L_m^b is the length of the *mth* kind of vegetation that is bypassed by bridges (m); l is span length (m); S_{bp} is cross-sectional area of bridge piers (m^2).

Therefore, the formula for estimating the total carbon sink loss is shown in Eq. (7.59):

$$E_L = \sum_{m=1}^{3} \left\{ c_m \cdot \left[\sum_{i=1}^{N_m - 1} (K_{i+1} - K_i) \cdot (W_i + W_{i+1}) / 2 + L_m^b / l \cdot S_{bp} \right] \right\} \cdot T \tag{7.59}$$

7.3.2.5 Carbon emission dynamic analysis

The above method can compute the constant carbon emissions during the railway operation and maintenance stages. However, the long-term carbon emissions during these stages are not invariable and will be affected by many factors, such as national policies, economic development and technological progress. Thus, this section considers the time-varying characteristics of such carbon emissions with a dynamic Carbon Emission Intensity (the carbon emissions per unit GDP, CEI) analysis.

To mitigate the greenhouse effect, the 2015 Paris Agreement set specific emission reduction targets for different countries. Based on the Agreement, Zheng et al. [24] found the change rule of Carbon Emission Intensity (the carbon emissions per unit GDP, CEI) under specific carbon emission reduction target scenarios by combining national carbon emission reduction policies and economic development factors, and proposed six carbon emission scenarios to analyze China's potential emission path from 2011 to 2050.

Referring to that research, this study assumes that annual carbon emissions would continue to increase from the base year (i.e., railway construction year) until they peak in 2030, and then decline in the future, which is called the Peak scenario [24]. In that scenario, the CEI would experience an exponential decay from 2011 to 2050 and can be estimated with an exponential equation as shown in Eq. (7.60) [25]:

$$I_t = I_0 \cdot e^{-b\cdot(t-t_0)} \tag{7.60}$$

where I_t is the CEI of year t; I_0 is the CEI of the base year t_0; b is the attenuation coefficient of CEI.

In response to the Peak scenario, the transportation industry will innovate technology and equipment in real time to improve maximum train frequency and promote energy conservation as well as emission reduction. Therefore, considering the China's carbon peaking strategy and GDP growth trend, this section introduces the carbon emission attenuation rule from this scenario into our model, and the carbon emissions threshold of year t (E_t) can be estimated by Eq. (7.61):

$$\begin{aligned} E_t &= G_t \cdot I_t \\ &= G_t \cdot I_0 \cdot e^{-b\cdot(t-t_0)} \\ &= G_t \cdot E_0 / G_0 \cdot e^{-b\cdot(t-t_0)} \\ &= E_0 \cdot G_t / G_0 \cdot e^{-b\cdot(t-t_0)} \\ &= E_0 \cdot (1+\lambda_t) \cdot e^{-b\cdot(t-t_0)} \end{aligned} \tag{7.61}$$

where E_0 is the initial carbon emissions of the base year t_0; G_t and G_0 are the GDP of years t and t_0, respectively; λ_t is GDP growth rate of year t; b can be computed in the Peak scenario, as detailed in Ref. [24] and is shown in Eq. (7.62).

$$\begin{cases} b \le \ln(1+\lambda_{t_1}) & (t_0 \le t_1 \le 2029) \\ b \ge \ln(1+\lambda_{t_2}) & (2030 \le t_2 \le 2049) \end{cases} \tag{7.62}$$

In theory, to ensure that b has a solution, the following condition should be satisfied, as given in Eq. (7.63):

$$\ln(1+\lambda_{t_2}) \le b \le \ln(1+\lambda_{t_1}) \Rightarrow \lambda_{t_2} \le \lambda_{t_1} \tag{7.63}$$

Thus, in this model, it is assumed that the GDP will successively grow but its annual growth rate will gradually decay to satisfy the above relation. If the above relation is not satisfied, there is no solution to b. In that condition, it is considered that carbon emissions will not decay annually and b is set at 0.

In this study, the carbon emissions at the construction stage are assumed to have a one-time constant value. However, emissions generated at the operation and maintenance stages should be further modified based on Eq. (7.61). Specifically, the modified carbon emissions at the operation stage (E_O in kg) can be formulated as Eq. (7.64):

$$E_O = \sum_{t=t_0+T_C}^{T_f} \left[E_{\text{opera}} \cdot (1+\lambda_t) \cdot \mathrm{e}^{-b\cdot(t-t_0)}\right] \tag{7.64}$$

where E_{opera} is the annual carbon emissions during the railway operation stage (kg) that computed using Eq. (7.52); T_C is the construction period (year); T_f is the time point of railway's long-term stage.

Regarding carbon emissions at the maintenance stage, the track (with a radius of R) replacement time point during the railway' economic life $t(R)$ is first determined by Eq. (7.65):

$$t(R) = t_0 + T_C + m\cdot T(R) \tag{7.65}$$

where t_0 is the base year; m is rail replacement times; $T(R)$ is the cycle of track repair computed by Eq. (7.54).

Then, the carbon emissions at the maintenance stage (E_M in kg) are integrated as follows:

$$\begin{aligned} E_M = & \sum_{t=t_0+T_C}^{T_f} \left[E_M^{b/t/s/r} \cdot (1+\lambda_t) \cdot e^{-b\cdot(t-t_0)}\right] \\ & + \sum_R \sum_{m=1}^{M=T/T(R)} \left\{E_M^{r'} \cdot (1+\lambda_t) \cdot e^{-b\cdot[t(R)-t_0]}\right\} \end{aligned} \tag{7.66}$$

where $E_M^{b/t/s/r}$ denotes the annual carbon emissions for maintaining different infrastructure components (kg, b = bridge, t = tunnel, s = cut/fill, r = rail track), and $E_M^{r'}$ denotes the initial carbon emissions for replacing the rail track (kg); M is the number of rail replacements during a railway's economic life (T).

7.3.3 Overall model formulation

Considering the dimensional inconsistency, it is preferable to treat the construction costs and carbon emissions as multiple objectives rather than add them. Thus, a bi-objective optimization model is formulated as shown in Eq. (7.67):

$$\min \begin{cases} F_{\text{cost}} = C_B + C_T + C_E + C_L + C_R \\ F_{\text{emission}} = E_C + E_O + E_M + E_L \end{cases} \tag{7.67}$$

$$s.t. \quad C_g, C_s, C_l$$

The objective of the overall model is to optimize railway alignments by reducing construction costs and carbon emissions simultaneously. The three types of constraints are explained briefly below.

Geometric constraints (C_g) are mainly imposed to ensure operational safety and maintenance facilitation according to the Code for Design of Railway Line [26], which include the minimum circular curve radius, the minimum curve and tangent lengths, the minimum sloped section length and the maximum gradient.

Structure constraints (C_s) are usually set by considering the construction technology limitations and allowable construction duration, which include the maximum tunnel length and bridge height.

Location constraints (C_l) generally refer to forbidden zones and clearance requirements.

Detailed specifications of these constraints can be found in Ref. [20] and are not repeated here. To solve this model, a bi-objective optimization method is provided in the next section.

7.3.4 Application

7.3.4.1 Case profile

In this section, the NT railway is used as an example to verify the effectiveness of the proposed model and method.

The NT railway is located in a typical mountainous region, as shown in Fig. 7.10. Therefore, to overcome the complex terrain, railway alignments are mainly comprised of expensive bridges and tunnels. However, bridges and tunnels require large amounts of building materials and high energy consumption, which results in high carbon emissions during the construction. Moreover, due to the complex terrain and consequent steep railway

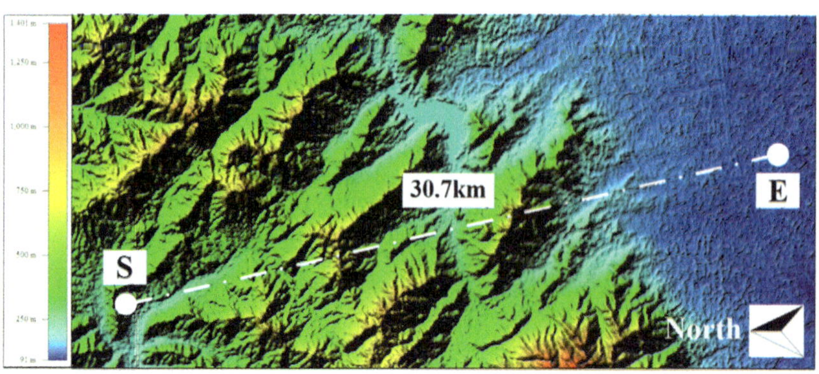

Figure 7.10 Terrain map of the study area.

Figure 7.11 Vegetation map of the study area.

gradients, the energy consumption for train operation as well as the carbon emissions at the operation stage are both high. Besides, as in Fig. 7.11, the railway alignment may traverse large vegetation areas, whose loss leads to serious loss of carbon sink during the railway's life. Therefore, it is highly desirable to optimize construction costs and life-cycle carbon emissions concurrently for this complicated railway case.

All design data for the analysis are provided by the China Railway Eryuan Engineering Group Co. Ltd. The linear distance between two endpoints is 30.7 km. The constraints and design standards are provided in Table 7.9. The unit costs are shown in Table 7.10. The parameters for estimating carbon emissions are presented in Table 7.11. The resolution of

Table 7.9 Constraints and design standards.

Parameter	Value	Parameter	Value
Minimum horizontal tangent length	280 m	Maximum gradient	25‰
Minimum horizontal curve length	280 m	Maximum bridge height	100 m
Minimum circular curve radius	7000 m	Maximum tunnel length	10 km
Minimum sloped section length	900 m	High-speed train type	CRH3
Fill-bridge threshold height	8 m	Design speed	350 km/h
Cut-tunnel threshold depth	12 m	Construction years	4
Operating days per year	365	Economic life years	20

Table 7.10 Unit costs.

Parameter	Value	Parameter	Value
Length-dependent cost (¥/m)	9500	Filling earthwork (¥/m^3)	38
Right-of-way (¥/m^2)	106	Cutting earthwork (¥/m^3)	40
(H $<$ 50 m, L $\geq$ 2500 m) bridge (¥/m)	85,000	(L $\geq$ 6000 m) tunnel (¥/m)	90,000
(H $<$ 50 m, L $\geq$ 500 m) bridge (¥/m)	47,000	(L $\geq$ 3000 m) tunnel (¥/m)	53,000
(H $<$ 50 m, L $<$ 500 m) bridge (¥/m)	35,000	(L $<$ 3000 m) tunnel (¥/m)	37,000
One bridge abutment (¥)	300,000	One tunnel portal (¥)	300,000

Table 7.11 Carbon emission factors and annual carbon sink values.

Category	Carbon emission factor	Source
Rebar	2.34 (kg/kg)	[27]
Steel	2.38 (kg/kg)	
Concrete	385 (kg/m^3)	
Electric power	0.58 (kg/kWh)	http://www.mee.gov.cn
Category	**Annual carbon sink value**	**Source**
Woodland	0.197 (kgCO_2/m^2)	http://www.forestry.gov.cn
Grassland	0.018 (kgCO_2/m^2)	[28]
Cropland	0.103 (kgCO_2/m^2)	[29]

vegetation map is 30 m. This map is obtained from the website of Zenodo (https://zenodo.org). The GDP data from 2021 to 2045 are obtained from the website of the Organization for Economic Cooperation and Development (http://stats.oecd.org). Consequently, the minimum CEI attenuation coefficient b of the Peak scenario is determined as 0.0339.

7.3.4.2 Analysis of results

The model and method were coded using C++ programming language on 64-bit Windows system and run on a computer with Intel (R) Core (TM) i7-7700 CPU @ 3.60 GHz processor, 16 GB of RAM. The computing time of PSO was 6.6 min and the algorithm converged at the 291st generation (i.e., when the Pareto front stays unchanged for 50 successive

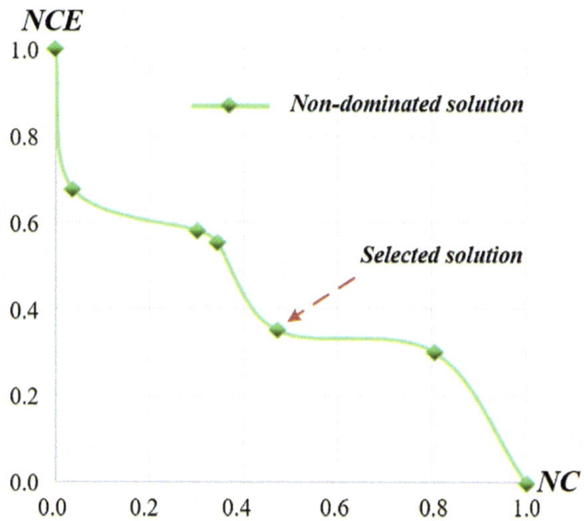

Figure 7.12 Final pareto front.

generations). The final Pareto front is shown in Fig. 7.12. The selected cost-emission balanced solution according to the CDC rationale, which is denoted as A_C hereinafter, is also shown in Fig. 7.13. Then, the best railway alignment manually-designed by experienced human engineers of the China Railway Eryuan Engineering Group Co., Ltd, which is denoted as A_M, is adopted for comparison with A_C. In addition, a least-cost alignment solution considering only the construction cost function is also produced using the PSO, which is denoted as A_S.

The horizontal and vertical alignment comparisons among A_M, A_C and A_S are shown in Figs. 7.13 and 7.14, respectively. The detailed results are presented in Tables 7.12–7.14. From these results, it can be observed that:

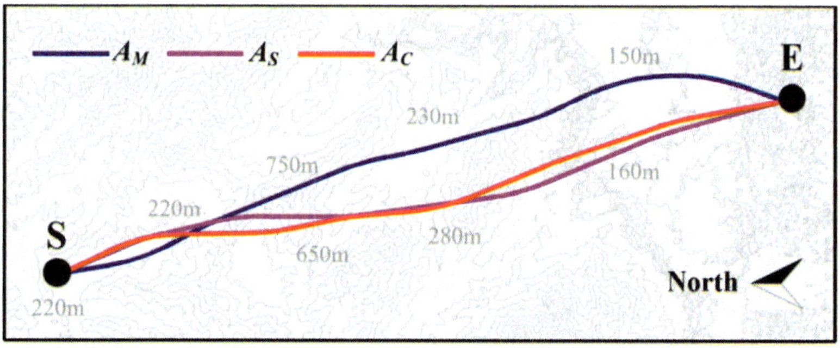

Figure 7.13 Horizontal alignment comparison between A_M, A_C and A_S.

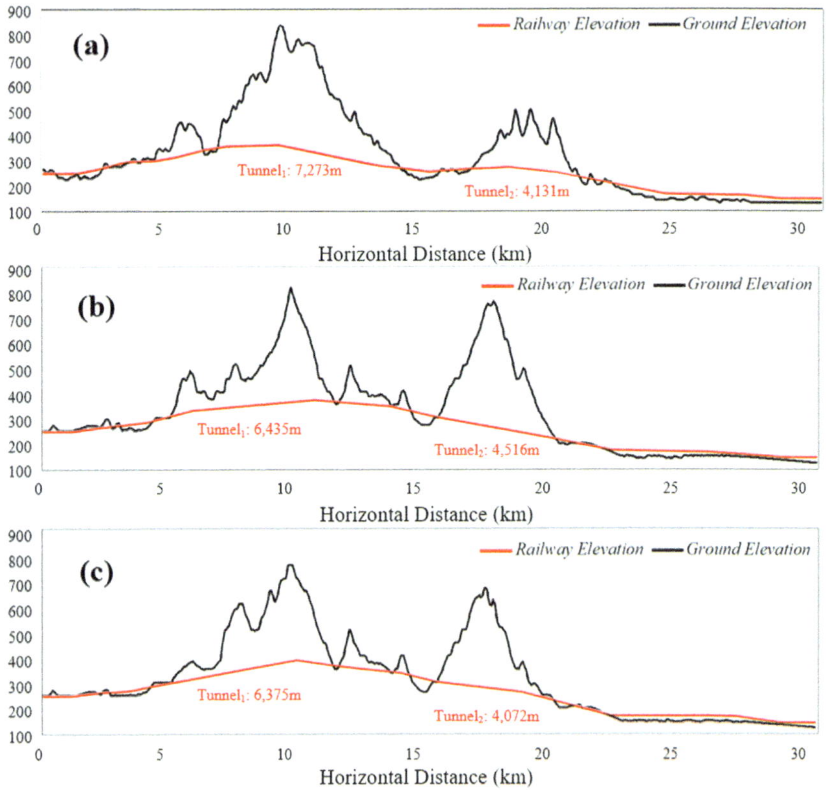

Figure 7.14 Vertical alignment comparison between (a) A_M, (b) A_C and (c) A_S.

(1) From the cost perspective, compared with A_M, A_C reduces the total alignment length by 463 m. In addition, compared to A_M, A_C also reduces the total length of expensive bridges by 3324 m but increases the total tunnel length by 340 m. Moreover, the longest bridge of A_C (i.e., 8295 m) is 112 m shorter than that of A_M (i.e., 8407 m). The longest tunnel of A_C (i.e., 6435 m) is also 838 m shorter than that of A_M (i. e., 7273 m). Thus, by significantly reducing the expensive bridges and shortening both the longest tunnel and the longest bridge, the total construction cost of A_C is 8.0% lower than that of A_M.

(2) From the carbon emission perspective, it can be observed from Tables 7.5 and 7.6 that the carbon emissions are mainly concentrated in the railway construction and operation stages. The total carbon emission of A_C is (1320.0–1264.1)/1320 = 4.2% below that of A_M. More specifically:

Table 7.12 Detailed alignment subsection outputs of A_M, A_C and A_S.

	A_M			A_C			A_S		
ID	L^a	$+G_j^a$	$-G_j^a$	L^a	$+G_j^a$	$-G_j^a$	L^a	$+G_j^a$	$-G_j^a$
0	1380	0.00	0.00	1150	0.00	0.00	1100	0.00	0.00
1	910	20.22	−19.78	1000	15.09	−14.51	1150	12.66	−12.66
2	1050	25.00	−25.00	900	10.26	−9.60	1150	5.78	−5.20
3	1050	3.72	−3.28	1100	12.16	−11.58	6750	18.11	−16.25
4	950	16.11	−15.89	1800	25.00	−25.00	1650	−12.16	13.46
5	1150	25.00	−25.00	4900	8.84	−7.16	2550	−10.71	11.11
6	900	13.54	−13.46	3000	−6.95	8.05	1400	−24.63	24.63
7	2100	3.27	−2.73	1900	−22.24	22.76	3400	−10.27	11.55
8	4100	−19.22	20.78	4700	−17.57	17.97	3600	−25.00	25.00
9	1960	−11.66	12.34	2250	−19.22	20.80	3900	0.17	0.17
10	1550	7.20	−7.20	3800	−1.83	2.17	1100	−3.20	3.20
11	1650	4.03	−3.97	1150	−5.40	5.74	1800	−13.54	13.54
12	1950	−10.52	11.48	2000	−6.87	7.21	1395.69	0.00	0.00
13	4350	−18.57	19.23	1287.53	0.00	0.00			
14	3200	−1.28	1.72						
15	1500	−9.78	10.22						
16	1650	0.00	0.00						

[a] L denotes the sloped section length (m); $+G_j$ denotes the additional gradient from the start to end points (‰); $-G_j$ denotes the additional gradient from the end to start points (‰).

Table 7.13 Detailed engineering outputs of A_M, A_C and A_S.

Objective	Parameter	A_M	A_C	A_S
Cost	Alignment length (m)	31,400	30,937	30,946
	Right-of-way area (m^2)	550,951	377,753	383,549
	Embankment volume (m^3)	258,915	319,431	271,599
	Excavated volume (m^3)	323,032	663,048	486,109
	Bridge number-total length (m)	9-13,746	3-10,422	5-11,508
	Tunnel number-total length (m)	3-13,303	5-13,644	4-12,732
	Total construction cost (million ¥)	2283	2101	2094
	Improvement	**N/A**	**8.0%**	**8.3%**
Carbon emission	Construction stage (kt)	882.2	807.1	819.3
	Operation stage (kt)	303.3	324.1	345.0
	Maintenance stage (kt)	133.5	132.1	127.4
	Carbon sink loss (kt)	1.0	0.8	0.8
	Total carbon emissions (kt)	1320.0	1264.1	1292.5
	Improvement	**N/A**	**4.2%**	**2.1%**

Table 7.14 Detailed carbon emissions results for A_M, A_C and A_S.

	Item	Carbon emission (kt)		
		A_M	A_C	A_S
Construction stage	Bridges	391.7	289.3	331.5
	Tunnels	421.8	432.6	403.7
	Earthwork	29.7	46.9	45.8
	Track	39.0	38.4	38.4
Operation stage	Energy consumption of a train (kwh)	1372	1467	1561
	Carbon emissions per train ($\times 10^5$ kg/year)	2.9	3.1	3.3
Maintenance stage	Bridges	19.5	14.8	16.4
	Tunnels	88.8	91.0	84.9
	Earthwork	2.4	3.8	3.7
	Track	22.8	22.5	22.5
Carbon sink	Woodland (kt)	0.4	0.2	0.2
	Grassland (kt)	0.1	0.1	0.1
	Cropland (kt)	0.5	0.5	0.5

① At the construction stage: The carbon emissions of A_C are (882.2–807.1)/882.2 = 8.5%, i.e., significantly lower than that of A_M. Particularly, the bridge carbon emission per kilometer of A_C is 27.8 kt/km, which is below that of A_M (i.e., 28.5 kt/km). That occurs because the A_C significantly reduces the bridge length and height compared to A_M. Hence, the total bridge length of A_C can be significantly reduced and, therefore, the carbon emissions for bridge construction also decrease.

② At the operation stage: The energy consumption for train operation of A_C is 1466.7 kwh, which exceeds that of A_M (i.e., 1372 kwh). Therefore, at this stage, the carbon emissions of A_C are (324.1–303.3)/303.3 = 6.9% higher than for A_M.

③ At the maintenance stage: The computed carbon emissions of A_C are (133.5–132.1)/133.5 = 1.0% lower than that of A_M, which shows slight improvements. This occurs because, in our model, the maintenance workloads of bridges, tunnels and earthwork sections are closely related to alignment length and the railway life span, which are similar for A_C and A_M. Besides, given that high-speed railways usually adopt a large curve radius (more than 7,000 m in this case), the rail track wear can be effectively reduced and, hence, the corresponding maintenance carbon emissions can also be reduced.

④ Regarding the carbon sink loss, A_C's value (i.e., 0.8 kt) is below that of A_M (1.0 kt). That occurs because A_C bypasses more vegetation areas by appropriately using bridges and tunnels.

Lastly, by considering the least-cost solution generated by a single-objective model (i.e., A_S), it can be seen that A_S has the lowest cost among the three alignments. However, compared with A_C, the cost of A_S only decreases slightly, by (2101-2094)/2101 = 0.3%. Besides, the carbon emissions of A_S increase significantly by (1292.5–1264.1)/1264.1 = 2.2%. This demonstrates the great importance of incorporating carbon emission factors into the alignment optimization. A cost & emissions bi-objective optimization is thus highly desirable for assisting real-world design.

Through the above detailed analyses of output results, the effectiveness of the proposed model and algorithm are verified.

Green alignment design is becoming essential for transportation infrastructure. This chapter introduces a life-cycle carbon emissions framework for railway alignment optimization, integrating construction, operation, and maintenance stages, as well as vegetation carbon sink loss, into a bi-objective cost–emission model solved by an improved PSO algorithm. Application to a real-world case shows that the optimized alignment reduces costs by up to 8% and life-cycle emissions by over 4% compared with manual design, while achieving a better cost–emission balance than the least-cost alternative.

References

[1] Ministry of Railways of People's Republic of China, Code for Environmental Protection Design of Railway Engineering, China Railway Publishing House Co. Ltd, 1998.

[2] A. Maji, M.K. Jha, A multi-objective analysis of impacted area of environmentally preserved land and alignment cost for sustainable highway infrastructure design, Proc. Soc. Behav. Sci. 20 (2011) 966–972.

[3] N. Yang, M.W. Kang, P. Schonfeld, et al., Multi-objective highway alignment optimization incorporating preference information, Transport. Res. C Emerg. Technol. 40 (2014) 36–48.

[4] S.M. Vicente-Serrano, J.J. Camarero, J.M. Olano, et al., Diverse relationships between forest growth and the normalized difference vegetation index at a global scale, Remote Sens. Environ. 187 (2016) 14–29.

[5] W. Li, H. Pu, P. Schonfeld, et al., Methodology for optimizing constrained 3-dimensional railway alignments in mountainous terrain, Transport. Res. C Emerg. Technol. 68 (2016) 549–565.

[6] D. Pimentel, C. Harvey, P. Resosudarmo, et al., Environmental and economic costs of soil erosion and conservation benefits, Sci. 267 (1995), 5201 1117–1123.

[7] H. Yang, L. Lin, Y. He, Soil erosion caused by highway construction in expansive soils districts and its prevention measures, in: Geotechnical Engineering for Disaster Mitigation and Rehabilitation, Springer, 2008, pp. 781–789.

[8] M.E. Vázquez-Méndez, G. Casal, D. Santamarina, et al., A 3d model for optimizing infrastructure costs in road design, Comput. Aided Civ. Infrastruct. Eng. 33 (2018) 423–439.

[9] Ministry of Railways of People's Republic of China, Code for Design of Railway Line, China Planning Press, 2017.
[10] R. Zhang, P.C. Chang, S. Song, et al., Local search enhanced multiobjective PSO algorithm for scheduling textile production processes with environmental considerations, Appl. Soft Comput. 61 (2017) 447–467.
[11] T. Baron, M. Tuchschmid, G. Martinetti, et al., High Speed Rail and Sustainability, Background Report: Methodology and Results of Carbon Footprint Analysis, International Union of Railways (UIC), Paris, 2011.
[12] X. Xie, T.Y. Yang, Performance evaluation of Chinese high-speed railway bridges under seismic loads, Int. J. Struct. Stabil. Dynam. 20 (2020).
[13] China Railway Administration, Code for Design of Concrete Structure of Railway Bridges and Culverts, 2017. TB 10092-2017.
[14] S. Liu, An Environmental Impact Comparative Study of the Beijing-Shanghai High-Speed Railway and Highway Infrastructure Based on LCA, Southeast University, 2015.
[15] P.H. Wright, R.J. Paquette, Highway Engineering, 1987.
[16] China Railway Design Institute, Budget Quota for Railway Earthwork, China Railway Publishing House, 2017.
[17] Research Institute of Standards and Norms Ministry of Housing and Urban-Rural Development, Rules for Preparation of Machine Movement Cost of Construction Engineering, China Planning Publishing House, 2015.
[18] H. Zhang, H. Pu, P. Schonfeld, et al., Railway alignment optimization considering lifecycle costs, IEEE Intell. Transp. Syst. Mag. 14 (5) (2021) 22–40.
[19] Y. Song, W. Song, A novel dual speed-curve optimization based approach for energy-saving operation of high-speed trains, IEEE Trans. Intell. Transport. Syst. 17 (6) (2016) 1564–1575.
[20] T. Song, H. Pu, P. Schonfeld, et al., Simultaneous optimization of 3D alignments and station locations for dedicated high-speed railways, Comput. Aided Civ. Infrastruct. Eng. 37 (4) (2022) 405–426.
[21] H.S. Eggleston, L. Buendia, K. Miwa, et al., 2006 IPCC Guidelines for National Greenhouse Gas Inventories, Institute for Global Environmental Strategies (IGES), Hayama, Japan, 2006.
[22] J. Lin, H. Li, W. Huang, et al., A carbon footprint of high-speed railways in China: A case study of the Beijing-Shanghai line, J. Ind. Ecol. 23 (4) (2019) 869–878.
[23] China State Railway Group Co. Ltd, China Railway Repair Rules (TG/GW 102-2019), 2019.
[24] T. Zheng, J. Zhu, S. Wang, et al., When will China achieve its carbon emission peak? Natl. Sci. Rev. 3 (1) (2016) 8–12.
[25] C. Yue, S. Wang, J. Zhu, et al., 2050 carbon emissions projection for China (carbon emissions and social development, IV), Acta Sci. Nat. Univ. Pekin. 4 (2010) 517–524.
[26] China Ministry of Railways, Code for Design of Railway Line (TB 10098-2017), 2017.
[27] China Academy of Building Research, Standard for Building Carbon Emission Calculation (GB/T 51366-2019), 2019.
[28] L. Zhang, G. Zhou, Y. Ji, et al., Spatiotemporal dynamic simulation of grassland carbon storage in China, Sci. China Earth Sci. 59 (2016) 10 1946–1958.
[29] Y. Yu, Y. Huang, W. Zhang, Modeling soil organic carbon change in croplands of China, 1980–2009, Global Planet. Change. 82 (2012) 115–128.

CHAPTER EIGHT

Railway alignment corridor search methods

Abbreviations and corresponding full texts

3D-MCTS 3D Monte Carlo Tree Search
AHP Analytical Hierarchy Process
CGIM Comprehensive Geographic Information Model
CRITIC Criteria Importance Through Inter-Criteria Correlation
DEM Digital Elevation Model
DT Distance Transform
EECD Exploration-Encouraging Cubic Decay
FCCD Fast Convergence Cubic Decay
FSS Feasible Search Space
GIS Geographic Information System
LD Linear Decay
MCDM Multi-Criteria Decision-Making
MCTS Monte Carlo Tree Search
PSO Particle Swarm Optimization
RTPs Required Traversed Points
RWS Roulette Wheel Selection
SEN Simulation from Extended Nodes
SSN Simulation from the Start Node
UCT Upper Confidence Bounds for Trees

Solving alignment optimization models involves searching for a set of decision variables within a high-dimensional space that satisfy all design constraints while minimizing (or maximizing) the objective function. However, due to the enormous search space, the typically nonlinear and non-differentiable nature of the objective functions, and the presence of numerous complex constraints, especially in mountainous environments, the optimization landscape is often highly rugged and has numerous local optima. As a result, alignment optimization is classified as an NP-complete problem and remains a major research challenge. In practical engineering projects, the design process is typically divided into multiple stages to cope with large spatiotemporal scales and limited computational resources. The first stage, referred to as the "area-to-corridor" stage, involves selecting optimized alignments and corridor candidates across a broad region to reduce the search space and ensure no valuable corridors are overlooked. The second stage, known as the "corridor-to-alignment" stage,

Intelligent Railway Alignment Design
ISBN: 978-0-443-45923-8
https://doi.org/10.1016/B978-0-443-45923-8.00010-8

performs detailed design of the alignment and its associated structures within the selected corridor, focusing on multi-objective collaborative optimization between alignment and infrastructure.

Based on this two-stage framework, this chapter introduces a multi-level intelligent alignment optimization strategy. The current chapter focuses on the "area-to-corridor" stage, introducing efficient methods for alignment–-structure collaborative search of corridor centerlines. The next chapter will address fine-tuned intelligent optimization at the "corridor-to-alignment" stage. Since corridor alignments are typically defined by their centerlines and widths, the key to this stage lies in determining optimized centerlines.

In complex mountainous terrain, where long tunnels and high bridges are essential to connect endpoints, the layout of dominating structures should sometimes be prioritized. These structures often have a decisive influence on construction feasibility, duration, and cost. If neglected, current optimization models may fail to find any feasible solution. Thus, in some cases, the positions of key large-scale structures (e.g., tunnels and bridges) should be fixed first, serving as anchor points that constrain the global alignment. Afterward, alignment sections between these dominating structures are searched using Distance Transform (DT)-based methods. However, these methods are prone to get trapped in local optima due to their greedy search mechanisms, making them less suitable for such complex scenarios.

To overcome these challenges, we adopt the Monte Carlo Tree Search (MCTS) framework, which has proven effective in complex decision-making problems such as AlphaGo. We propose a novel 3D Monte Carlo Tree Search (3D-MCTS) algorithm to address the vast and highly constrained search space of 3D alignment design.

Overall, the details of dominating structure optimization, DT-based alignment sections search, and MCTS methods are presented in the following sections.

8.1 Presetting of dominating structures

8.1.1 Landform recognition

In practice, advanced terrain analysis is an indispensable step in alignment design. Through this procedure, dominating landforms (which refer to the valleys or mountains that alignments cannot avoid traversing) can be recognized in an alignment optimization process. There are various studies devoted to landform classification in the geomorphology field [1].

Different entities (e.g., peak, ridge, slope, valley and flatland) are categorized based on the texture similarity of terrains [2]. For alignment optimization problems, alignment characteristics should be integrated into landform classification through which only valleys and mountains unreachable by alignments need to be recognized. Thus, two Geographic Information System (GIS) procedures, namely alignment spatial reachability analysis and elevation tendency analysis, are implemented to segment the alignment-unreachable ranges. Afterward, a two-pass scan method is employed to identify the dominating landforms. To record the landforms' characteristics, the study area is discretized as a series of grids based on the resolution of digital elevation model (DEM).

8.1.1.1 Segmentation of alignment-unreachable valleys and mountains

In complicated mountainous regions, the structure layouts vary with different vertical alignments. Although their geometric shapes are dissimilar, their common aim is to overcome the elevation difference between two endpoints. Thus, the vertical alignment is expected to be designed quite close to the straight line between the two endpoints. In this regard, the sections with elevations ranging between 100m below or 500m above this vertical straight line are regarded as the preferable range for an alignment, while other sections in the vertical reachable region are defined as potential alignment-unreachable valleys or mountains. Alignments tend to traverse these potential alignment-unreachable valleys or mountains through bridges or tunnels. Therefore, they are combined with initial alignment-unreachable ranges to jointly constitute the alignment-unreachable valleys and mountains, with details provided in Fig. 8.1.

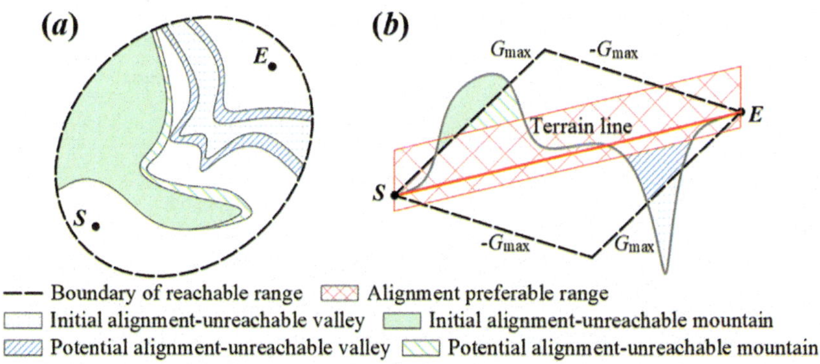

Figure 8.1 (a) Horizontal and (b) vertical views of alignment-unreachable valleys and mountains.

8.1.1.2 Recognition of dominating landforms

Alignment-unreachable valleys and mountains are generally scattered in complicated mountainous regions. Among them, only the regions that entirely separate two endpoints are defined as dominating landforms in this paper. To recognize these dominating landforms, a two-pass scan method [3] used for classifying different connected areas is adopted. Different connected areas can be distinguished after two scanning processes are conducted with an 8-connectedness mask [4]. The steps customized for identifying dominating landforms are listed below:

Step 1: Initializing: The grids in alignment-unreachable regions are defined as foreground grids and others are designated as background grids. A provisional label is specified as 1.

Step 2: First scan: scans from the top left to the right bottom:

1) The provisional label is assigned to the current grid if the grids in its 8-connectedness mask (i.e., the eight surrounding grids that share either an edge or a corner with the current one) [4] are all background grids. The provisional label's value is increased by 1.
2) When there are foreground grids in the mask of the current grid, the minimum provisional label of these foreground grids is assigned to the current grid.
3) All provisional labels assigned to the same connected regions are defined as equivalent labels. Each group of equivalent labels is recorded in an equivalence sequence, as shown in Fig. 8.2a.

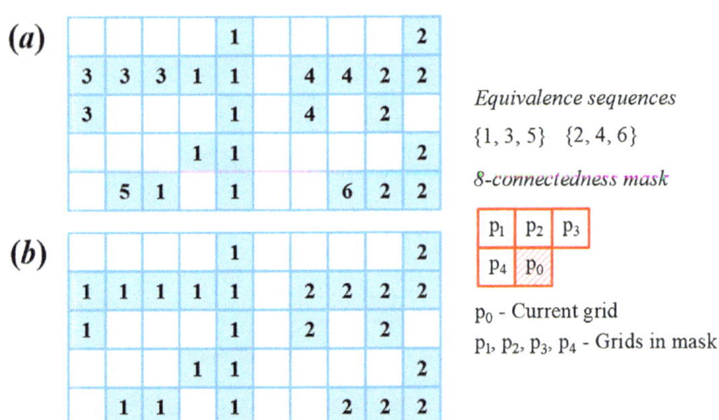

Figure 8.2 The outcomes of: (a) first scanning and (b) second scanning.

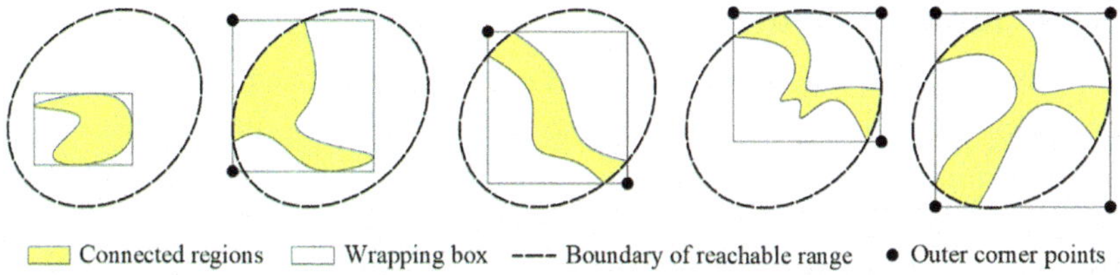

Figure 8.3 Distribution of wrapping boxes and diagonal points for different connected areas.

Step 3: Determining representative labels: The minimum label in an equivalence sequence is designated as the representative label in this connected region.

Step 4: Second scan: Proceeds in the same order as the first scan and relabels each foreground grid by the representative label, as shown in Fig. 8.2b.

After different connected areas are labeled, a rectangular wrapping box for each area can be generated as illustrated in Fig. 8.3. It can be noted that when at least one pair of diagonal points in the wrapping box is outside the railway's horizontally reachable range, the alignment cannot avoid this connected area and hence, the area is designated as a dominating landform Fig. 8.4.

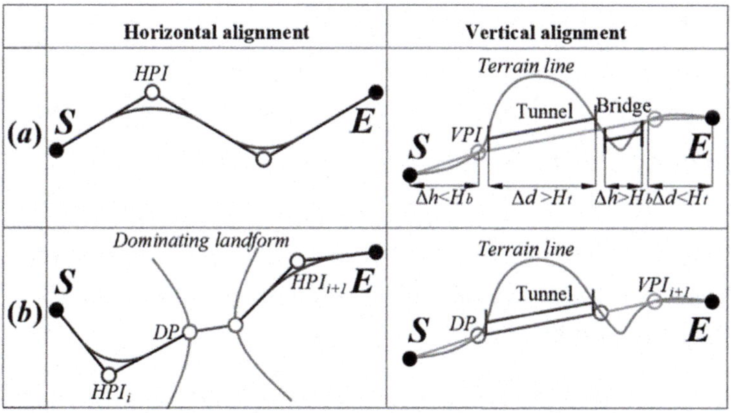

Figure 8.4 (a) Configuring structures after alignment locations in the existing methods, where H_b is the threshold height between fill or bridge, and H_t is the threshold depth between cut or tunnel and (b) presetting structures.

8.1.2 Solution method

To solve the abovementioned bi-level model, a three-stage method is proposed for presetting the dominating structures before optimizing the alignment. Primarily, a scanning process and specified screening operators are designed to generate all the possible point pairs of these structures. Then, after applying a multi-criteria decision-making (MCDM) process, the optimized point pairs of the dominating structures can be selected. Ultimately, the alignment is optimized by a customized bi-objective PSO [5] with the selected point pairs as required traversed points.

8.1.2.1 Structure locations determination

(1) Scanning process

In this process, the whole railway horizontally reachable area is scanned with line segments in various directions to obtain all point pairs crossing the dominating landforms. The parametric formulas of these line segments $\boldsymbol{L}$ (α, i, j) are specified as follows (Eq. 8.1):

$$
\begin{cases}
x_L = \dfrac{x_S + x_E}{2} + \sin\left(f_{SE} + \alpha\right)\cdot w\cdot i + \cos\left(f_{SE} + \dfrac{\pi}{2} + \alpha\right)\cdot w\cdot j \\
y_L = \dfrac{y_S + y_E}{2} + \cos\left(f_{SE} + \alpha\right)\cdot w\cdot i + \sin\left(f_{SE} + \dfrac{\pi}{2} + \alpha\right)\cdot w\cdot j
\end{cases}
$$
$$
i = \left(-\frac{\gamma_{\max}\cdot l_{SE}}{2\cdot w}, \cdots, 1, 2, \cdots, \frac{\gamma_{\max}\cdot l_{SE}}{2\cdot w} \right) \tag{8.1}
$$
$$
j = \left(-\frac{\gamma_{\max}\cdot l_{SE}}{2\cdot w}, \cdots, 1, 2, \cdots, \frac{\gamma_{\max}\cdot l_{SE}}{2\cdot w} \right)
$$
$$
\alpha = \left(-\theta, \cdots, 0, \frac{\pi}{36}, \cdots, \theta \right)
$$

where x_L, y_L = the coordinates of the points in these line segments; x_S, y_S, x_E, y_E = the coordinates of start and end points; f_{SE} = azimuth of the start and end points; w = width of DEM's grid; i, j, α = the $[i + (\gamma_{\max}\cdot l_{SE})/2/w]^{\text{th}}$ point in $[j + (\gamma_{\max}\cdot l_{SE})/2/w]^{\text{th}}$ line segment and the angle between this line segment and the straight line between two endpoints (Fig. 8.5).

(2) Screening operators

There might be several point pairs obtained from above scanning process that do not entirely traverse the dominating landform. These point pairs must be eliminated from the feasible solutions integrated with landform characteristics. Generally, the geometric characteristics of valleys and mountains are different. Valleys are typically caused by surface erosion and thus, they usually have V-shaped or U-shaped profiles in

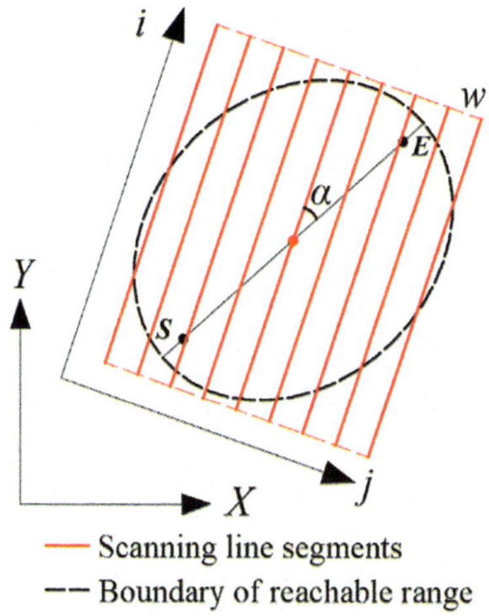

Figure 8.5 Scanning line segments.

longitudinal and quasi-dendritic forms in horizontal, while mountains are formed by rock uplift or orogeny and usually spread widely in a belt-like configuration. Therefore, the point pairs of dominating bridges and tunnels are screened with different operators which consider specific landform features.

For a dominating valley, to preclude point pairs on the dendrite branches, each point pair is disturbed to a certain range. If the line segment formed by this point pair could entirely bypass the valley after disturbance, the corresponding point pair should be eliminated from the possible solutions, as shown in Fig. 8.6a. With the implementation of this bridge-oriented screening operator, all the remaining point pairs that satisfy the aforementioned constraints are stored in an external alternative set.

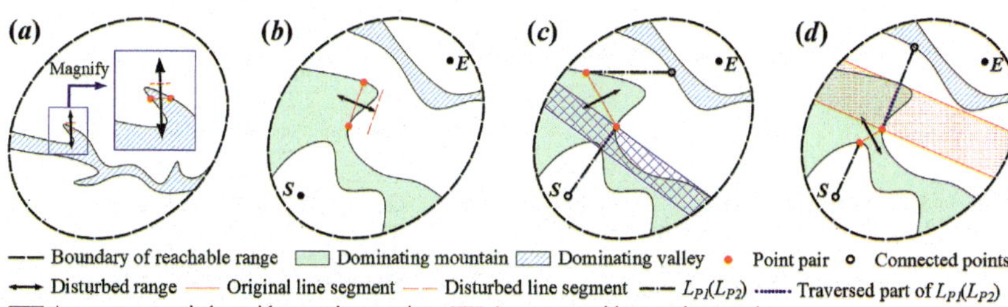

Figure 8.6 (a) Bridge-oriented screening operator; (b), (c) and (d) tunnel-oriented screening operator.

A dominating mountain area can be divided into several irregular belt-like components. Although a point pair may traverse a fraction of the mountain, the circuitous alignments could also be obstructed by other parts of these mountains. Thus, the tunnel-oriented screening operator is described as follows. Primarily, the line segment formed by each point pair is disturbed in a certain range. If the disturbed line segment entirely bypasses the mountainous region, the corresponding point pair is eliminated, as shown in Fig. 8.6b. Otherwise, the point pairs are connected with the start or end point or the DP of another dominating structure. Afterward, the line segments (L_{P1}, L_{P2}) formed by the point pair and their connected points are checked. If L_{P1} or L_{P2} traverse the mountain regions, the traversed part is disturbed within horizontally reachable range (Fig. 8.6c). If the adjusted segment could entirely bypass the mountain after this disturbance (Fig. 8.6d) and the aforementioned constraints are satisfied, the corresponding point pair is recorded to the external alternative set Fig. 8.7.

8.1.2.2 Multi-criteria decision-making analysis

Through the analyses made in Section 3.1.2, there are three categories and seven kinds of factors that significantly influence the layouts of structures. These factors should be aggregated to evaluate the comprehensive fitness of structure layouts. This is intrinsically a hierarchical multi-criteria decision-making (MCDM) problem and thus, it can be solved with an analytical hierarchy process (AHP) combined with a criteria importance through inter-criteria correlation (CRITIC) method for obtaining the criterion weight of each factor.

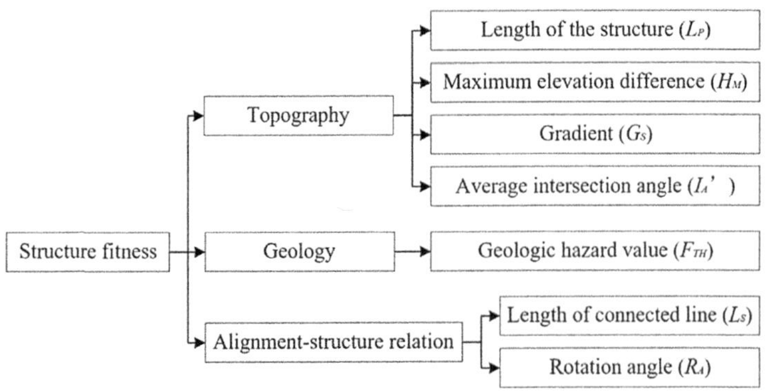

Figure 8.7 The hierarchical structure of the MCDM problem in this study.

Table 8.1 The comparison matrix for all the categories.

Structure fitness	Topography	Geology	Alignment-structure relations	Criterion weight
Topography	1	1	2	0.4
Geology	1	1	2	0.4
Alignment-structure relations	1/2	1/2	1	0.2

After considering the advice of experts from China Railway Eryuan Engineering Group Co. Ltd and China Railway First Survey and Design Institute Group Co. Ltd, the comparison matrix for all categories is obtained and listed in Table 8.1.

Then, the weights of sub-factors in each category are computed based on the specified data's fluctuation and correlation. Supposing the number of dominating structures is N in a study area, then, the structure combination fitness (F_S) can be expressed as follows.

$$\begin{aligned} F_S = & \sum_{q=1}^{N}\sum_{i=1}^{2}\left[W_{Ci}\cdot\sum_{j=1}^{Si}\left(W_{Sij}\cdot x_{qij}\right)\right] + W_{C3}\cdot\sum_{j=1}^{S3}\left(W_{S3j}\cdot x_{Sj}\right) \\ & + W_{C3}\cdot\sum_{j=1}^{S3}\left(W_{S3j}\cdot x_{Ej}\right) + \sum_{q=1}^{N-1}W_{C3}\cdot\sum_{j=1}^{S3}\left(W_{S3j}\cdot x_{qj}\right) \\ & = F_{S1}+F_{S2}+F_{S3}+F_{S4} \end{aligned} \tag{8.2}$$

where W_{Ci} = criterion weight of ith category; Si = number of the sub-factors in ith category; W_{Sij} = sub-criterion weight of jth sub-factor in ith category; x_{qij} = normalized value of jth sub-factor in ith category for qth structure; x_{Sj}, x_{Ej} = normalized value of jth sub-factor in alignment-structure relations category between structure and the start or end points, respectively; x_{qj} = normalized value of jth sub-factor in alignment-structure relations category between two structures. Detailed illustrations are shown in Figs. 8.8a and 8.9.

The point pairs in the external alternative sets can be combined, as shown in Fig. 8.8b, and their F_S are computed with Eq. (8.2). Then, the structure combined form with the maximum F_S value is designated as the optimized combination form of dominating structures. The corresponding point pairs are treated as the required traversed points for the subsequent alignment optimization process.

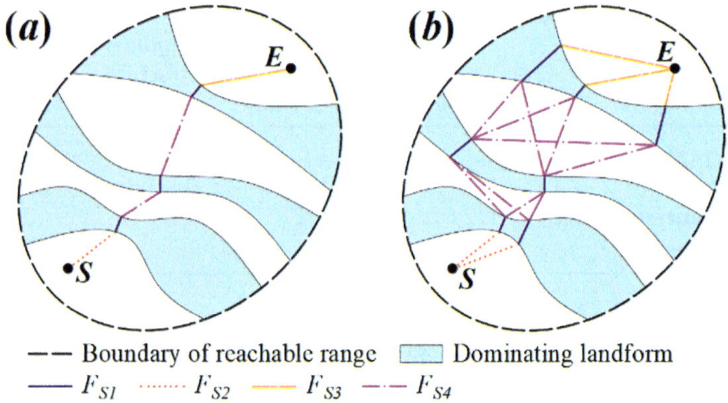

Figure 8.8 Illustrations of: (a) structure combination fitness and (b) structure combined forms.

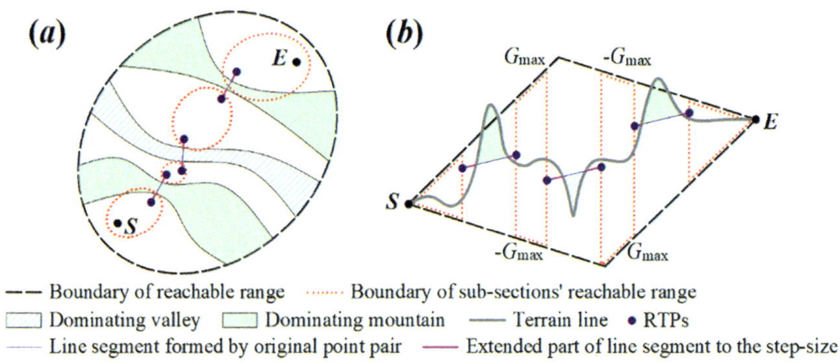

Figure 8.9 (a) Horizontal and (b) vertical demonstrations of RTPs and corresponding sub-sections.

8.1.2.3 Alignment generation

After the locations of dominating structures are determined, the study area is separated into several sub-sections based on the assigned point pairs. Then, the alignment generation is a bi-objective alignment optimization problem incorporating required traversed points (RTPs). Hence it can be solved by DT-based and MTCS methods. In this regard, the endpoints of the newly formed line segment are considered as RTPs. RTPs are assigned as new start or end points of each sub-section, and the stated method is applied for generating optimized alignments. RTPs' design elevation can be adjusted to reduce construction cost and geo-hazard risk concurrently when the search algorithm is applied to each sub-section.

8.2 Distance transform algorithms

8.2.1 Basic flow

Distance transform (DT) algorithms [6] provide a very simple and extremely fast approach for the approximation Euclidean distances from every cell of a rectangular lattice to the nearest cell in a target set. The global distances are approximated by propagating local distances i.e., distances between neighboring cells. The neighboring cells are called a neighboring mask which is described as an adjacency matrix $\boldsymbol{M_d}$. A 5×5 adjacency matrix is shown in Fig. 8.10. The element $d_{r,c}$ in the matrix $\boldsymbol{M_d}$ is the Euclidean distance from the cell (at row r, column c in the neighboring mask) to the center cell. The whole matrix can be divided into two parts: the upper left part and the lower right part, which are, respectively, employed by the forward and backward scan.

Firstly, the distance from each cell to the target is initialized. The value of the target cell is zero, and the others are infinity. Then, a two-pass scan of the lattice data is conducted: a forward scan from the top left to the bottom right, and then a backward scan from the bottom right to the top left. When 1 cell is scanned, the center of the neighboring mask is placed over it, and the local distance $d_{r,c}$ is added to the value of the cell. The new value of this cell is the minimum of all the sums. This process is repeated until no cell value changes. The final value of every cell in the lattice is the minimum distance to the target.

In this process, we assume p is one random cell in the lattice, while $D(p)$, $\Delta R(p)$, and $\Delta C(p)$ are stored in this cell.

$D(p)$ is the minimum distance from p to the target cell.

$\Delta R(p)$ and $\Delta C(p)$ are the incremental row and column movements at p in the corresponding shortest path.

The standard algorithm first initializes the D, ΔR, ΔC value for all cells, as shown in Eq. (8.3):

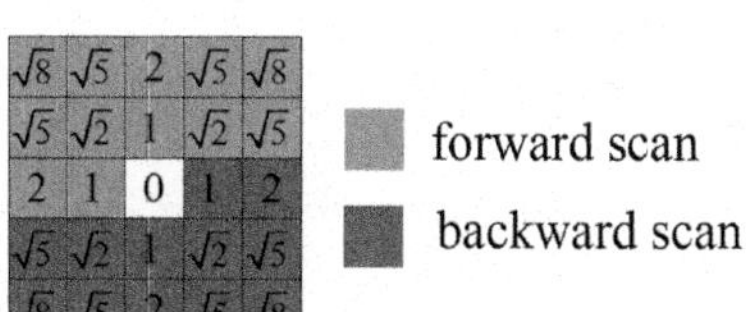

Figure 8.10 Neighboring mask.

$$
\begin{aligned}
&D^0(p) = \begin{cases} 0 & p \in B \\ \infty & p \notin B \end{cases} \\
&\Delta R(p) = 0 \\
&\Delta C(p) = 0
\end{aligned}
\tag{8.3}
$$

where B is the target set and the superscript of D is the iteration.

Then, a two-pass scan of the lattice data is conducted: a forward scan from the top left to the bottom right, and then a backward scan from the bottom right to the top left (Fig. 8.11). When cell p is scanned, its values are refreshed with Eq. (8.4):

$$
\begin{aligned}
&D^t(p) = \min_{r,c}\{D^{t-1}(q) + d_{r,c}\} \\
&\Delta R(p) = r \\
&\Delta C(p) = c
\end{aligned}
\tag{8.4}
$$

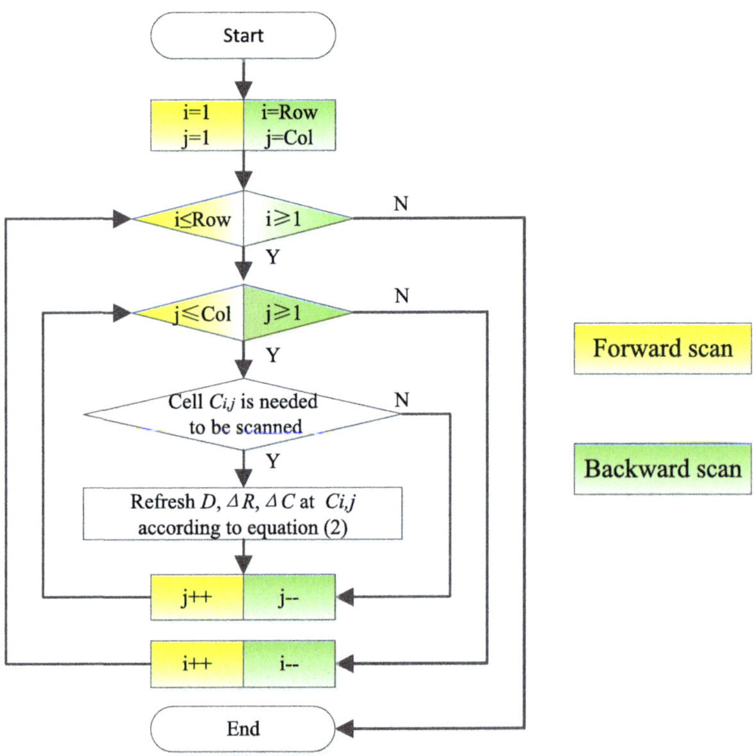

Figure 8.11 Two-pass scan in DT.

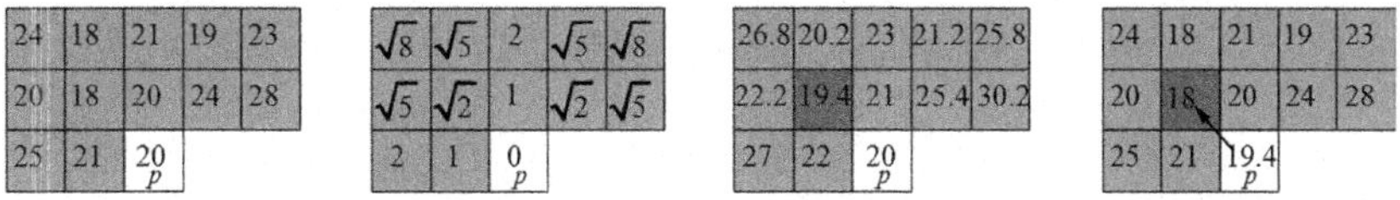

Figure 8.12 Distance refreshment process.

where $D^t(p)$ is the new value at p (i.e., the central cell in the neighboring mask), q is one random cell in the neighboring mask, $D^{t-1}(q)$ is the current distance value at q and $d_{r,c}$ is the local distance from p to q. Fig. 8.12 is a simple case, in which $r = -1$, $c = -1$, the value of $D^{t-1}(q) + d_{r,c}$ get the minimum value so the new $D^t(p) = D^{t-1}(q) + d_{-1,-1}$, $\Delta R\ (p) = -1$, $\Delta C(p) = -1$.

The above two-pass scan is carried out until the value of D at every cell is static. The final $D(p)$ is the minimum distance from p to the nearest target. As shown in Fig. 8.13, the corresponding shortest path can be generated by the $\Delta R\ (p)$ and $\Delta C\ (p)$.

On completion of the two-pass scan, each cell in the resulting lattice will contain the minimum distance to the target point. We represent the distance using the path cost and set the alignment endpoint as the target point. Then, the cell distance value in the resulting lattice is the minimum cost to the endpoint and the corresponding path is the optimized path. In railway alignment optimization, the distance should be replaced by the link cost.

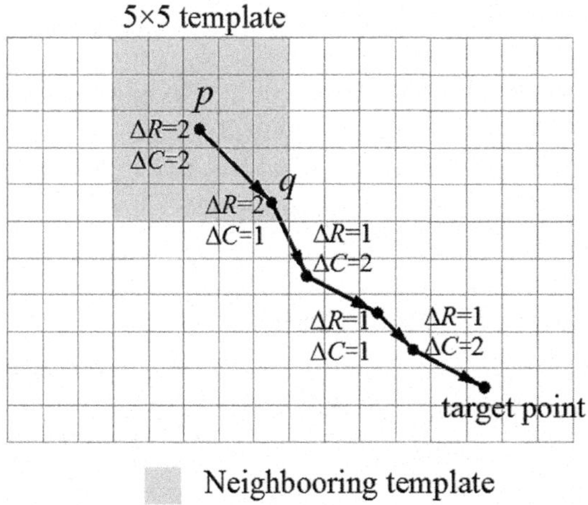

Figure 8.13 Generation of the shortest path from p to the target point.

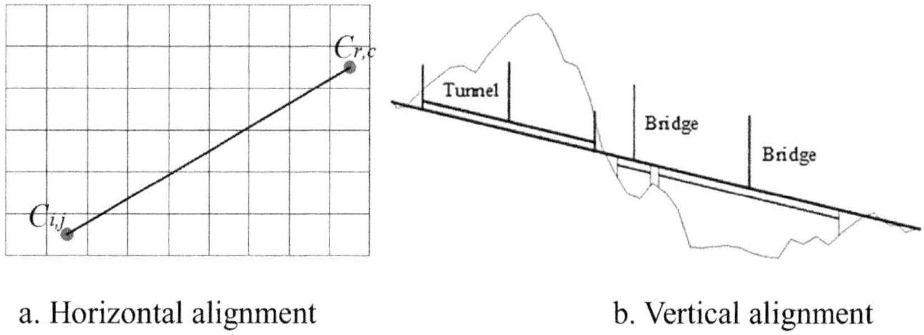

Figure 8.14 Local alignment linked by 2 cells.

8.2.2 Link cost

$C_{i,j}$ is a random cell in the lattice located at row i, column j, whose 3D coordinates are $(X_{i,j}, Y_{i,j}, H_{i,j})$. $C_{r,c}$ is another cell in the lattice, located at row r, column c, whose 3D coordinates are $(X_{r,c}, Y_{r,c}, H_{r,c})$. If these 2 cells are linked, a local alignment can be generated. The horizontal and vertical alignments are shown in Fig. 8.14.

The link cost between $C_{i,j}$ and $C_{r,c}$ is defined as the local alignment cost. Before computing the alignment cost, the locations of railway construction elements including bridges, tunnels, and subgrades should be set. The main procedure is as follows.

(1) Convert the alignment to a set of points (known as stations).
(2) Interpolate the ground elevation of the stations based on DEM.
(3) Compute the fill or cut height (Δh) for all stations.
(4) Auto-set the locations of bridges, tunnels and subgrades according to the stations' Δh.

The cost consists of four parts: 1) construction costs, including bridges, tunnels, earthwork and length-dependent costs; 2) right-of-way costs, which are obtained from the intersection of the alignment with the right of way lattice in the CGIM; 3) operating costs sensitive to the alignment; and 4) environmental impact costs.

8.2.3 Adaptive neighboring mask

There may be no linkable cells in the neighboring mask for some cells during the scan, since multiple constraints apply. These cells are invalid. The path cannot be formed from an invalid cell to the target point. If many such cells exist, there will be no path linking the railway start and end points. To overcome this problem, an adaptive neighboring mask is

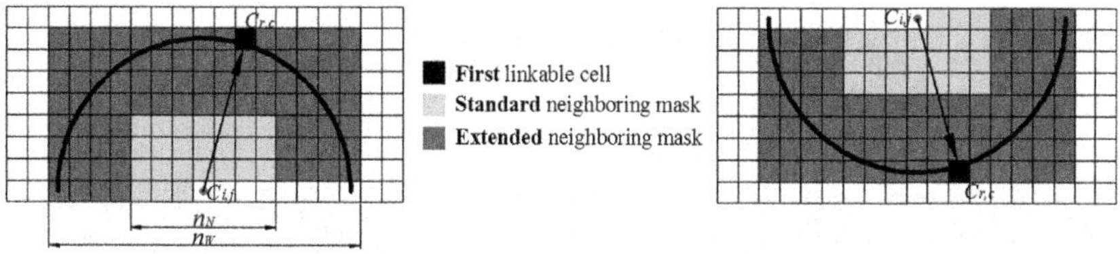

Figure 8.15 Dynamic adaptive neighboring mask.

defined. If there is no linkable cell in the preset standard neighboring mask, the neighboring mask is extended adaptively.

As shown in Fig. 8.15, this neighboring mask includes two parts: inner standard $n_N \times n_N$ neighboring mask and outer extended $n_W \times n_W$ neighboring mask. n_N is a static value which can be determined with De Smith's [6] method. However, n_W is an adaptive value depending on the topography and constraints.

When cell $C_{i,j}$ is scanned, if there are linkable cells in the standard neighboring mask, the extended neighboring mask is not needed. Otherwise, as shown in Fig. 8.15, we traverse the cells from inside out until one linkable cell $C_{r,c}$ is found, which satisfies all the constraints. The extended neighboring mask size n_W is determined by Eq. (8.5):

$$\begin{aligned} n_W &= 2 \times \max(|r-i|, |c-j|) + 1 \\ n_W &< 2 \times L_{\max}/w + 1 \end{aligned} \tag{8.5}$$

where $L_{\max}$ is the admissible maximum distance between two linked cells. We usually set it to be the admissible maximum bridge or tunnel length.

8.2.4 Bidirectional scanning strategy

Setting the railway end point cell as the target, the optimal path from one random cell $C_{i,j}$ to the end point can be generated by a multi-constrained distance transform. If $C_{i,j}$ is the railway start point, this path is the optimal path we are seeking. However, at most one path can be found. Multiple constraints in mountains may even result in no feasible path. Moreover, the designers usually need various possible paths to avoid overlooking valuable candidates. Therefore, we use the bidirectional scanning strategy to generate various paths.

Firstly, the end point cell is set as the target to generate the end distance image DT_E. Similarly, the start distance image DT_S is formed by setting the start point cell as the target. Then, we traverse all the cells in the lattice to generate various 3D paths. The paths from $C_{i,j}$ to the start and end points

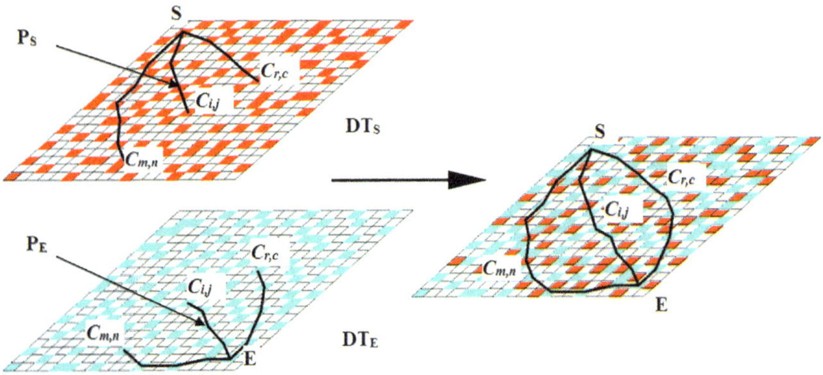

Figure 8.16 Generating paths with bidirectional scanning.

can be found, respectively, by DT_S and DT_E. By linking the above two paths, the integrated path via $C_{i,j}$ is generated (Fig. 8.16).

Using the above strategy, a set of 3D paths can be generated. Then, the paths are sorted by their comprehensive costs.

As stated above, a DT search is capable of identifying optimized corridors within the accessible area. However, it is highly sensitive to grid resolution, with search time increasing rapidly as resolution improves. To overcome this drawback, Monte Carlo Tree Search (MCTS), first proposed by Coulom (2007), provides a promising alternative, as it enables efficient exploration of large search spaces while balancing local and global optimization. The detailed search strategy of the MCTS is presented in the following section.

8.3 3D Monte Carlo tree search method

8.3.1 Basic MCTS

In an MCTS, a search tree is built incrementally by repeatedly performing specific actions from the current node to create new nodes in the tree. The core steps of MCTS are selection, expansion, simulation and backpropagation (Fig. 8.17). The selection step performs a trade-off between global exploration and local exploitation to efficiently search the solution space. Then, in each iteration of MCTS, the expansion step extends one node from the current node on the Monte Carlo tree. During the simulation step, a global reward is estimated to guide the local exploitation process. Finally, the backpropagation step updates the global reward and visited

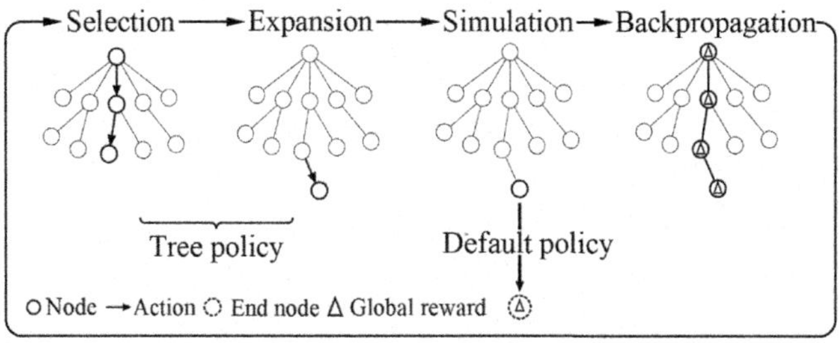

Figure 8.17 Major steps of an MCTS.

times of a parent node, which are then used in the selection step. A step-by-step overview of MCTS process is provided below.

8.3.1.1 Selection

The first step of basic MCTS involves selecting the most promising node on the search tree from which to expand. It is accomplished by using a selection policy, with the most commonly-adopted one being the Upper Confidence Bounds for Trees (UCT) [7].

8.3.1.2 Expansion

After selecting a parent node, MCTS proceeds to expand the node by adding a child node which represents potential search directions and actions from the parent node. Through expansion, the Monte Carlo tree grows and extends continuously to explore the search space.

8.3.1.3 Simulation

MCTS performs a simulation from the expanded node to the end node. The goal of simulation is to estimate the global reward of the current node. Global reward means that, during the MCTS path searching process, the algorithm considers not only the local reward of existing nodes during the selection and expansion steps, but also accounts for the potential rewards from the current node to the end node. By repeatedly executing simulation, MCTS gradually gains insight into different possible trajectories and estimates the expectation of MCTS nodes from the current node to the end node.

8.3.1.4 Backpropagation

The final step of MCTS is backpropagation. In this step, the results obtained from simulation are backpropagated through the tree. The statistical data of all nodes along the tree from the expanded node to the start node are updated based on the simulation results. The backpropagation step aims to improve the estimation of every node's objective function value and visit counts, which are used in the selection phase to guide the global exploration and local exploitation processes.

By iteratively performing these four steps and gradually building the search tree, MCTS can achieve the effective exploration of the study area.

8.3.2 Customized 3D-MCTS

In this study, we propose a 3D-MCTS method to address the railway alignment optimization problem. All essential information, such as the DEM (digital elevation model) data, unit costs, and forbidden area, are stored in the comprehensive geologic information model (CGIM). The data structure of the CGIM is a set of 3D cuboids, known as "voxels." The four steps of 3D-MCTS, namely selection, expansion, simulation, and backpropagation, are all modified in this paper, as shown in Fig. 8.18. More detailed explanations are provided in the following sections.

8.3.2.1 Time-varying decay approach for the balance coefficient

For the selection step, **UCT** balances the trade-off between global exploration and local exploitation by considering the estimated objective function value (exploitation term) and the degree of exploration (exploration term) of each node (Eq. 8.6). By iteratively applying the selection policy, the

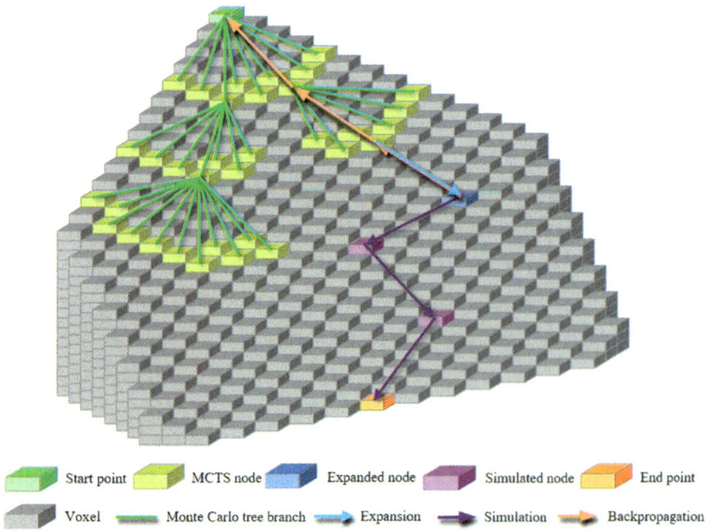

Figure 8.18 Path search diagram of the 3D-MCTS.

algorithm navigates through the search tree toward the most promising regions of the search space:

$$\mathbf{UCT}(v_i, v) = Q_{i,\mathrm{ave}} + C \cdot \sqrt{\frac{\ln N(v)}{N(v_i)}} \tag{8.6}$$

where the first and second terms to the right of the equal sign in Eq. 8.6 are exploitation and exploration terms of node v_i, respectively, $N(v_i)$ and $N(v)$ are the number of times that node v_i and its parent node v have been visited during the search process, respectively, $Q_{i,\mathrm{ave}}$ is the average estimated value of node v_i, and C is a balance coefficient that influences the trade-off between global exploration and local exploitation.

The basic MCTS method relies on a fixed coefficient C in the **UCT** formula (Eq. 8.6). In MCTS, the coefficient C plays a pivotal role by influencing the delicate balance between global exploration and local exploitation. The value of coefficient C directly affects the depth of the search tree. A larger C value encourages broader exploration, which results in a wider but shallower tree. It helps in approaching the globally optimal solution, but at the expense of longer search time. Conversely, a smaller C value emphasizes leveraging local experience, which focuses on promising known Monte Carlo tree branches (Fig. 8.19). It leads to a deeper but narrower tree capable of swiftly generating optimized solutions while also increasing the risk of getting trapped in local optima. Thus, the coefficient C is of utmost importance for effectively seeking optimized solutions within the constraints of computational resources. For 3D alignment optimization, it is crucial to thoroughly explore the search space in the early stage. Then, in the later stage, rapid convergence should be pursued by the program.

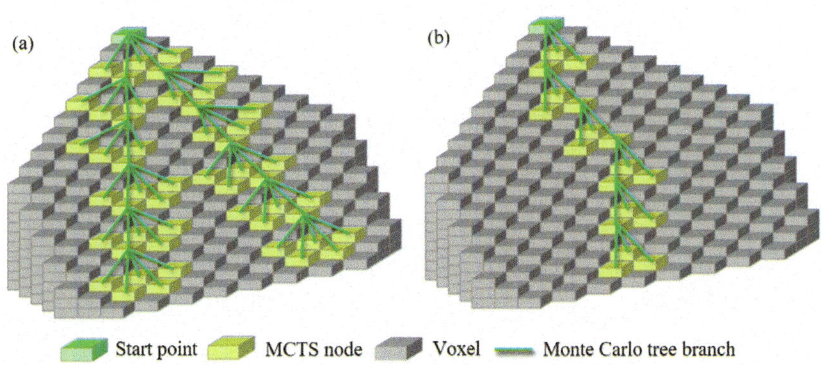

Figure 8.19 The customized selection step with (a) a high or (b) a low value of balance coefficient C.

To this end, we propose a time-varying decay approach for balancing the coefficient C during the search process. According to the variation in decay rate as the search proceeds, the time-varying decay approaches can be categorized into four function types: Constant, Linear Decay (LD), Fast Convergence Cubic Decay (FCCD), and Exploration-Encouraging Cubic Decay (EECD). In addition to the commonly used Constant function, the other three decay functions will be presented sequentially. The linear decay function (LD) for C can be formulated as Eqs. (8.7) and (8.8):

$$C_P = -(C_0 - C_1) \cdot P + C_0 \tag{8.7}$$

$$P = \max\left\{1 - \frac{L_{CE}}{L_{SE}}, 0\right\} \tag{8.8}$$

where P represents the proximity between the current and end nodes, reflecting the search progress, L_{CE} is the linear distance between the current node and the end node, L_{SE} is the linear distance between the start and end nodes, C_P is the time-varying value of C, C_0 and C_1 are the initial and end values of C, respectively.

Besides, a cubic polynomial update function is also devised as an alternative. In this function, the search process is divided into three phases by assuming that the decay amount of the former phase, which does not accumulate across phases, is a_w times that of its latter phase. In this paper, we define n as the decay amount of C_P in the last phase, and m as the ratio of the decay amount between two adjacent stages. Thus, a cubic polynomial decay formulation (Eqs. 8.9–8–13) is designed to pass through four points: $(0, C_0)$, $(1/3, C_0\text{-}m^2n)$, $(2/3, C_1+n)$ and $(1, C_1)$.

$$C_P = a \cdot P^3 + b \cdot P^2 - c \cdot P + d \tag{8.9}$$

$$a = \frac{-9(m-1)^2}{2(m^2+m+1)}(C_0 - C_1) \tag{8.10}$$

$$b = \frac{9(2m-1)(m-1)}{2(m^2+m+1)}(C_0 - C_1) \tag{8.11}$$

$$c = -\frac{11m^2 - 7m + 2}{2(m^2+m+1)}(C_0 - C_1) \tag{8.12}$$

$$d = C_0 \tag{8.13}$$

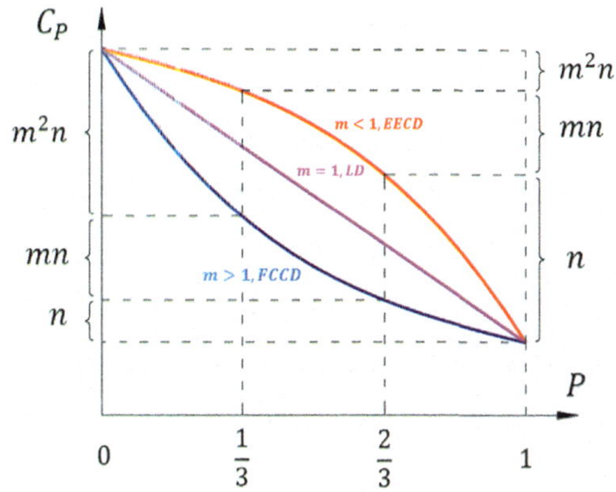

Figure 8.20 Three different decaying functions of balance coefficient C_P.

where the coefficients *a*, *b*, *c*, and *d* in the cubic polynomial decay formulation can be determined as shown in Eqs. (8.10)–(8.13). As seen in Fig. 8.20, when $m = 1$ (the middle curve), C_P degenerates to the linear decay (LD) form; when $m > 1$ (FCCD, the lower curve), the decay rate of C_P gradually decreases, thereby reducing the level of early exploration and promoting final convergence; when $m < 1$ (EECD, the upper curve), the decay rate of C_P gradually increases, allowing more tree node extensions during early exploration to guide the subsequent tree search process. The comparison of the above three decay modes will be discussed in detail in Section 4.3.

In summary, the time-varying selection function proposed for railway alignment optimization can be expressed as Eq. (8.14):

$$\mathbf{UCT}(v_i, v) = -Q_{\min} + C_P \cdot \sqrt{\frac{\ln N(v)}{N(v_i)}} \tag{8.14}$$

where $Q_{\min}$ is the minimum alignment cost from the current node v_i to the end node, C_P is the time-varying balance coefficient.

8.3.2.2 Expansion based on dynamically-determined feasible search space

For MCTS-based 3D alignment optimization, expanding nodes by traversing every voxel in the huge search space would consume considerable computation time and resources. To address this problem, a dynamic expansion operator is proposed for determining feasible search spaces (FSSs) during the search process (Fig. 8.21).

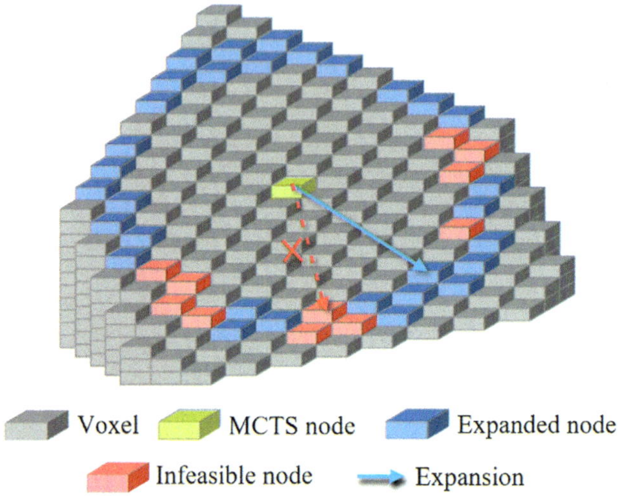

Figure 8.21 The customized expansion step.

For 3D alignment optimization, FSS should be generated for both horizontal and vertical alignments, which will be illustrated below.

(1) Horizontal FSS

In order to control railway construction cost within a reasonable range, the length of an alignment should not be excessive. Typically, the maximum allowable circuity coefficient ($\lambda_{\max}$) is used to control the alignment length. During the path search, the distance from current node to the two endpoints should satisfy Eq. (8.15):

$$\sqrt{(x_S - x_C)^2 + (y_S - y_C)^2} + \sqrt{(x_E - x_C)^2 + (y_E - y_C)^2} \leq \lambda_{\max} \cdot L_{SE} \tag{8.15}$$

where (x_S, y_S), (x_E, y_E), (x_C, y_C) are the coordinates of the start node, end node, and current node, respectively.

As shown in Fig. 8.22, the initial horizontal FSS confines the search space within an ellipse. During the search process, the horizontal FSS is further dynamically narrowed based on the current node. This process can be represented as Eq. (8.16):

$$\sqrt{(x_P - x_C)^2 + (y_P - y_C)^2} + \sqrt{(x_E - x_C)^2 + (y_E - y_C)^2} \leq \lambda_{\max} \cdot L_{SE} - L_{\text{gen}} \tag{8.16}$$

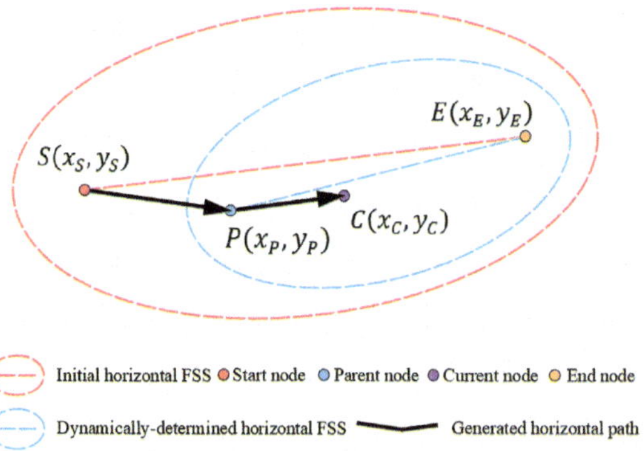

Figure 8.22 Horizontal FSS.

where (x_P, y_P) is the horizontal coordinate of the parent node, L_{gen} is the total length of the generated path.

(2) Vertical FSS

As shown in Fig. 8.23, to satisfy the maximum gradient (G_{max}), the initial vertical FSS is confined within a parallelogram determined by the start and end nodes, as shown in Eqs. (8.17) and (8.18):

$$H_C \leq \min\{H_S + G_{max} \cdot L_{SC}, H_E + G_{max} \cdot \lambda_{max} L_{EC}\} \tag{8.17}$$

$$H_C \geq \max\{H_S - G_{max} \cdot L_{SC}, H_E - G_{max} \cdot \lambda_{max} L_{EC}\} \tag{8.18}$$

where H_S, H_E, H_C are the design elevations of the start node, end node, and current node, respectively, while L_{SC} and L_{EC} denote the linear distance from the start node and end node, respectively, to the current node.

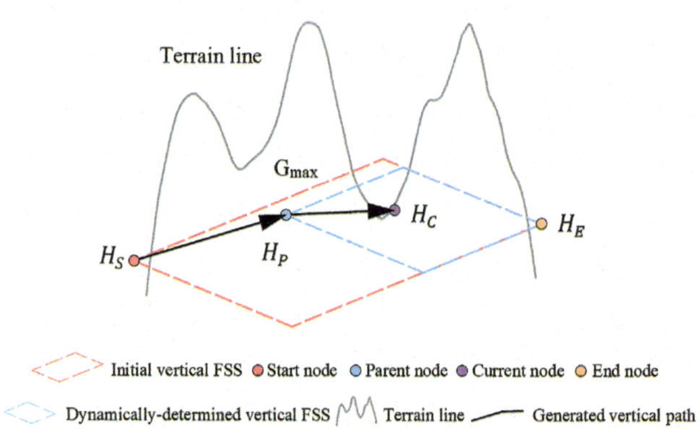

Figure 8.23 Vertical FSS.

After the generation of a new node, the vertical FSS can be gradually narrowed based on the generated path (Eqs. 8.19 and 8.20),

$$H_C \leq \min\{H_P + G_{\max} \cdot L_{PC}, H_E + G_{\max} \cdot \lambda_{\max} L_{EC}\} \tag{8.19}$$

$$H_C \geq \max\{H_P - G_{\max} \cdot L_{PC}, H_E - G_{\max} \cdot \lambda_{\max} L_{EC}\} \tag{8.20}$$

where H_P is the design elevation of the parent node, and L_{PC} is the horizontal linear distance between the parent node and the newly-generated node.

8.3.2.3 Simulation with global reward estimation

The simulation step (Fig. 8.24) is similar to the expansion step that involves dynamic feasibility space determination to exclude infeasible nodes. Additionally, the simulation step requires further customization for 3D alignment optimization. Unlike in the basic MCTS, simulations are performed in two scenarios in 3D-MCTS. First, before executing MCTS, a simulation is initiated from the start node to find feasible paths between the start and end nodes. The construction cost of the simulated path is used as the baseline ($Cost_{\text{Start}}$) for the balance coefficient C, which is referred to as Simulation from the Start Node (SSN). Second, a simulation from an extended parent node is required to evaluate its global reward (the construction cost of the simulation path), which provides a reference for the selection step and is called Simulation from Extended Nodes (SEN).

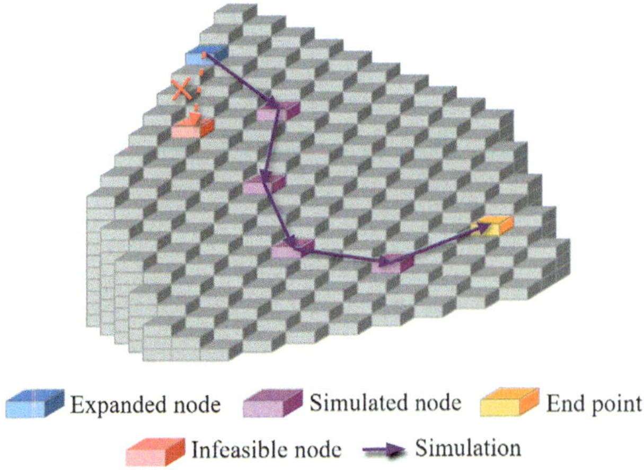

Figure 8.24 The customized simulation step.

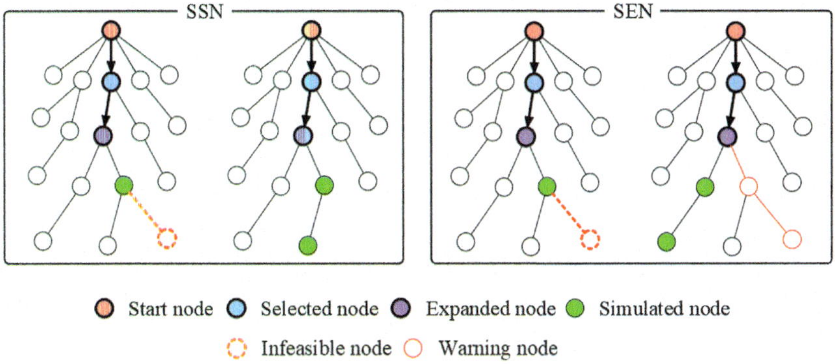

Figure 8.25 Diagrams for SSN and SEN.

For SSN, to ensure the MCTS can find a feasible path from the start node to end node, a path repair strategy is developed (Fig. 8.25). When a simulated child node is infeasible, SSN backtracks to its parent node, and removes the infeasible node from the parent node's FSS. Then, it continues the simulation from the parent node until a feasible path connecting the start and end nodes is found.

SEN, on the other hand, aims to rapidly evaluate the global rewards of the expanded node. Since numerous SEN steps should be executed in 3D-MCTS, to ensure that 3D-MCTS can be completed within a controllable time, we propose a warning strategy combined with a Roulette Wheel Selection (RWS). Similarly to SSN, to efficiently find feasible paths, SEN also adopts a path repair approach. However, when a simulated child node is infeasible, the repair strategy of SEN backtracks from the simulated node to the expanded node instead of its parent node and, then, increases the path repair count of warning nodes (Fig. 8.25) on the infeasible path by 1 (see Section 3.3.4). Afterward, SEN continues the simulation and RWS can determine the child nodes' simulated probability considering their path repair count (Fig. 8.26). The specific steps are as follows:

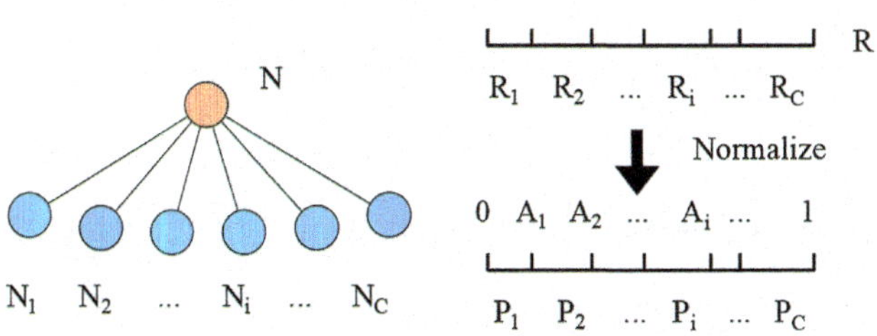

Figure 8.26 Roulette wheel selection used in SEN.

1) For each simulated node, first obtain the path repair count R of its parent node N and R_i of its child node N_i. R_i will be updated at the backpropagation step.
2) Compute the path repair count fraction P_i for each child node, where P_i is defined as the ratio of R_i to R.
3) Compute the cumulative percentage interval A_i of each child node N_i.
4) Generate a uniform random number in the range [0,1], and compare it with the cumulative percentage interval A_i of each child node N_i to find the matched node. Then, the simulated node is selected as the matched child node.

8.3.2.4 Backpropagation

The backpropagation step is responsible for updating the nodes' key information used in the selection step (Fig. 8.27). For 3D alignment optimization, in addition to updating the visited times $N(v_i)$ of node v_i (Eq. 8.21), we must also backpropagate the minimum construction cost and the path repair count. The minimum construction cost is backpropagated as shown in Eq. (8.22).

$$\mathrm{N}\left(v_{i,t+1}\right) = \mathrm{N}\left(v_{i,t}\right) + 1 \tag{8.21}$$

$$Q_{\mathrm{min},t+1} = \min\left\{Q_{\mathrm{min},t}, Q_{\mathrm{simu},t}\right\} \tag{8.22}$$

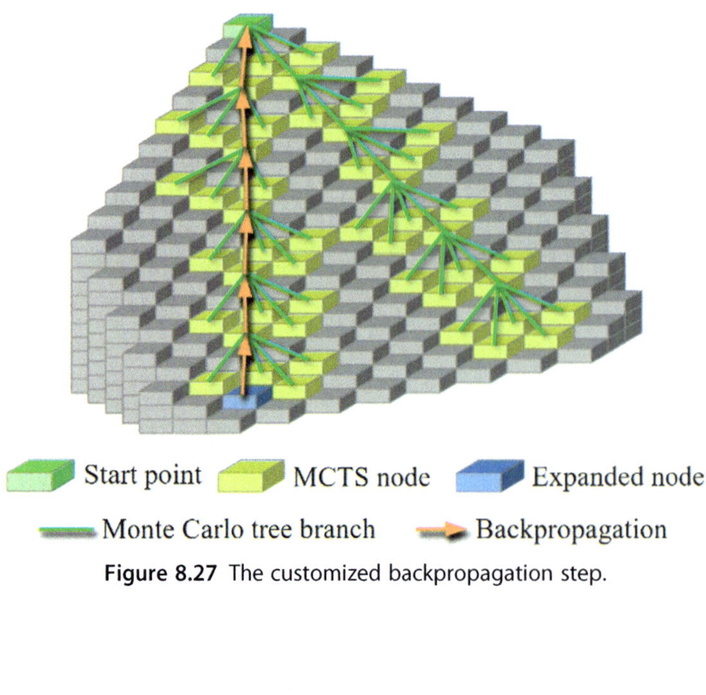

Figure 8.27 The customized backpropagation step.

where $Q_{\min,t}$ is the minimum construction cost of the simulated path from node v_i, $Q_{\text{simu},t}$ is the construction cost of the simulated path from node v_i at the tth iteration.

The backpropagation of the path repair count is used to provide an early warning before the current node reaches an infeasible node that has already been attempted, which is shown in Eq. (8.23):

$$\mathrm{N}_{\text{rep}}\left(v_{j,t+1}\right) = \mathrm{N}_{\text{rep}}\left(v_{j,t}\right) + 1 \tag{8.23}$$

where v_j represents all the simulated nodes (Fig. 8.18) at the tth iteration.

By repeatedly executing the above steps of 3D-MCTS, an optimized 3D corridor from the start point to the end point can be generated.

8.4 Conclusions

This chapter focuses on the "area-to-corridor" stage of the two-stage railway alignment optimization framework, addressing the NP-complete problem of mountainous railway alignment design caused by enormous search space, nonlinear/non-differentiable objective functions, and complex constraints. It proposes three core methods: first, a dominating structure presetting method that identifies dominating landforms (via GIS analysis and two-pass scans) and selects optimal critical structure (tunnel/bridge) point pairs through scanning, screening operators, and MCDM (AHP + CRITIC) to anchor the search; second, adapted Distance Transform (DT) algorithms that compute comprehensive link costs (construction, right-of-way, etc.), use adaptive neighboring masks to solve linkable cell shortages, and adopt bidirectional scanning to generate multiple feasible corridors; third, a customized 3D-MCTS method that optimizes selection (time-varying balance coefficient), expansion (dynamic FSS), simulation (global reward estimation with SSN/SEN), and backpropagation (updating visit counts, costs, and repair counts) to handle large 3D search spaces. These methods together enable efficient identification of constraint-compliant corridor centerlines in complex mountainous terrain.

After obtaining the optimized alignment centerlines (paths) through dominating structure presetting, Distance Transform (DT), and 3D Monte Carlo Tree Search (3D-MCTS), it is necessary to further generate corridor belts.

Taking the optimized path as the reference, the corridor is designed by integrating three key aspects: engineering constraints (e.g., a basic width of

8–12 m for single-track subgrade sections to accommodate track structures and maintenance access, expanded to 15–20 m for tunnel portal sections to account for excavation and slope protection), topographic and environmental boundaries (excluding geohazard-prone zones and ecologically sensitive areas using data from the Comprehensive Geologic Information Model (CGIM)), and flexibility for detailed optimization (narrowed to 8–10 m in gentle terrain sections and expanded to 12–15 m in complex terrain sections). Through path smoothing and segmentation, initial buffer generation, constraint-based clipping, and validity verification, a continuous and feasible belt-shaped region is formed. This corridor belt will serve as the core input for the "corridor-to-alignment" optimization in the next chapter.

References

[1] J. Jasiewicz, T.F. Stepinski, Geomorphons - A pattern recognition approach to classification and mapping of landforms, Geomorphology 182 (2013) 147–156.

[2] T. Matsuura, M. Aniya, Automated segmentation of hillslope profiles across ridges and valleys using a digital elevation model, Geomorphology 177 (2012) 167–177.

[3] L. He, Y. Chao, K. Suzuki, A run-based two-scan labeling algorithm, IEEE Trans. Image Process. 17 (5) (2008) 749–756.

[4] K. Wu, E. Otoo, K. Suzuki, Optimizing two-pass connected-component labeling algorithms, Pattern Anal. Appl. 12 (2009) 2 117–135.

[5] T. Song, H. Pu, P. Schonfeld, et al., Mountain railway alignment optimization considering geological impacts: A cost-hazard bi-objective model, Comput. Aided Civ. Infrastruct. Eng. 35 (2020) 12 1365–1386.

[6] M.J. de Smith, Determination of gradient and curvature constrained optimal paths, Comput. Aided Civ. Infrastruct. Eng. 21 (2006) 1 24–38.

[7] G. Wang, H. Pu, T. Song, et al., A 3D monte carlo tree search method for railway alignment optimization, Appl. Soft Comput. 151 (2024) 111158.

CHAPTER NINE

Searching methods for detailed alignment alternatives

Abbreviations and corresponding full texts

CGIM Comprehensive Geographic Information Model
DT Distance Transform
GA Genetic Algorithm
GPM Gradient Projection Method
HAO Highway Alignment Optimization
HPI Horizontal Point of Intersection
PSO Particle Swarm Optimization
VPI Vertical Point of Intersection

9.1 Overview

The previous chapter elaborated in detail on the intelligent search method for railway corridors during the stage of "region to corridors". In the subsequent "corridors to alignments" stage, the alignment design work is confined within a narrow and long corridor, and the search space is significantly reduced compared to that in the "region to corridors" stage. Therefore, designers can conduct detailed engineering surveys along the corridor to obtain high-precision data and information on features such as terrain and geology, which are used to finalize a refined railway alignment.

Currently, scholars have conducted extensive research on the problem of "region to corridors" alignment optimization. The research team for digital and intelligent railway alignment design at Central South University has successively focused on studies of various alignment optimization algorithms, including the gradient projection method, genetic algorithms, and particle swarm optimization. As a commonly-used mathematical optimization approach, the Gradient Projection Method (GPM) [1] is applied to solve constrained optimization problems. This method computes the gradient of the alignment optimization objective function, takes the gradient descent direction as the alignment search direction, and iteratively approaches the optimized solution under constraints through continuous iterations. Moreover, by introducing appropriate constraint conditions into the alignment optimization process, an alignment which meets the requirements of safety,

Intelligent Railway Alignment Design
ISBN: 978-0-443-45923-8
https://doi.org/10.1016/B978-0-443-45923-8.00005-4

transportation efficiency, and environmental protection can be generated. Meanwhile, the alignment parameters can be optimized.

A genetic algorithm (GA) [2] is an optimization algorithm that roughly simulates the natural evolution process. During the optimization, alignment parameters and alternatives can be represented as chromosomes. Through operations such as selection, crossover, and mutation, the alignment alternatives are continuously optimized to make them adapt to environmental requirements and design objectives. Particle Swarm Optimization (PSO) [3] is an optimization algorithm that simulates the foraging behavior of bird flocks. In this method, alignment alternatives can be represented as particles, and the optimized alignment alternative is found by simulating the changes of particles in the position and velocity. PSO is characterized by rapid global convergence, which is suitable for complex alignment design problems. The application principles of these algorithms will be introduced in detail as follows.

9.2 Improved genetic algorithm

GA was first introduced into the field of alignment optimization by the research group led by Professor Schonfeld from the University of Maryland, USA. It developed a Highway Alignment Optimization (HAO) and proposed a GA-based 3D alignment optimization method [2]. In this model, design variables are abstracted as genes, and alignment alternatives are abstracted as chromosomes composed of genes. Equally-spaced cutting planes are preset between the start and end points of the alignment, and a series of genes are randomly generated on these cutting planes to form a series of initial alignment alternatives. Then, the initial population of alignment alternatives is constructed. The population of alignment alternatives continuously evolves through typical genetic operators such as crossover and mutation, as well as operators customized for alignment optimization, until the algorithm converges to an optimized alignment.

In this method, the distribution and number of control points for the horizontal and vertical alignments are usually consistent, which may result in overlapping horizontal and vertical curves. However, the number of control points for horizontal and vertical alignments in practical design is usually inconsistent. Moreover, to facilitate maintenance, railway alignment design needs to avoid unreasonable overlaps of curves: Thus, overlapping of circular curves and vertical curves should be minimized, while overlapping of transition curves and vertical curves is strictly prohibited.

To address the above issues, an improved GA was developed for 3D railway alignment optimization by Professor Pu's team at Central South University, China [4]. This GA adopts different distributions for the control points of horizontal and vertical alignments, respectively. During the chromosome generation process, horizontal genes are generated first, followed by vertical genes considering horizontal-vertical constraints, and finally the overall gene sequence is formed through adjustment. Corresponding crossover and mutation genetic operators are also designed. The population of alignment alternatives continuously evolves through the designed selection, crossover, and mutation genetic operators until the convergence criteria are met.

9.2.1 Gene coding

It is extremely difficult to directly encode the design variables $\boldsymbol{X}$, $\boldsymbol{Y}$, $\boldsymbol{R}$, $\boldsymbol{L}$ and $\boldsymbol{Z}$ in the optimization model as the genes of the alignment chromosome. Therefore, the GA-based alignment model [5] first presets orthogonal cutting planes along a straight line between a road's start and end points, on which horizontal points of intersection (HPIs) and vertical points of intersection (VPIs) are generated by the GA. The locations of HPIs and VPIs are denoted as genes. Sequences of these genes are combined into an alignment chromosome.

Therefore, a series of cutting planes is generated at certain intervals in the direction perpendicular to the straight line between the start and end points of an alignment. It is assumed that the alignment intersection points lie on these cutting planes, the offset distance of the intersection points relative to the straight line between the start and end points serving as the gene to encode the coordinates of the intersection points. As shown in Fig. 9.1, S and E are the start and end points of the alignment, respectively; $\boldsymbol{e}$ is the unit vector from the start point to the end point; $\boldsymbol{n}$ is the vector obtained by rotating $\boldsymbol{e}$ 90 degrees counterclockwise; the dashed lines are the cutting planes perpendicular to $\boldsymbol{e}$; the distance between adjacent cutting planes is d_H; d_i is the offset of the i-th intersection point relative to the line connecting the start and end points (positive on the right side and negative on the left side). The relation between the coordinates of the intersection points $[X_i, Y_i]^T$ and the offsets d_i is as listed in Eq. (9.1).

$$\begin{bmatrix} X_i \\ Y_i \end{bmatrix} = \begin{bmatrix} X_S \\ Y_S \end{bmatrix} + i \cdot d_H \cdot \begin{bmatrix} X_e \\ Y_e \end{bmatrix} + d_i \cdot \begin{bmatrix} X_n \\ Y_n \end{bmatrix} \tag{9.1}$$

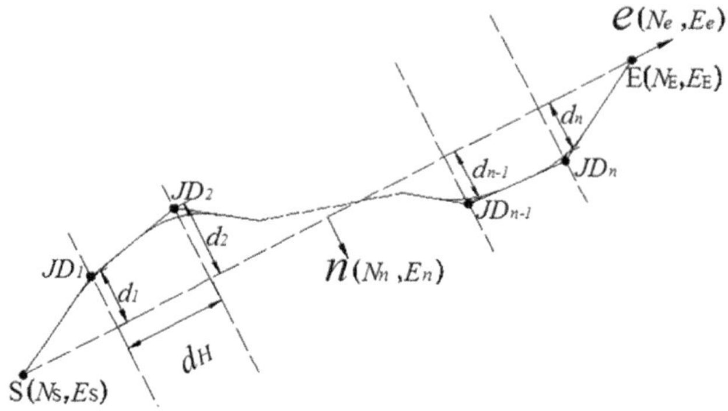

Figure 9.1 Coding of the horizontal intersection points.

where $[X_e, Y_e]^T$ is the unit vector from the starting point to the end point, and $[X_n, Y_n]^T$ is the unit vector perpendicular to the direction of the starting and ending points.

Similarly to horizontal alignment, for the vertical alignment, a series of cutting planes with a normal vector e can first be generated (as shown in Fig. 9.2). The distance d_V between adjacent cutting planes can be set as the required minimum slope length. Set n_V as the total number of cutting planes, the VPIs are respectively denoted as P_1, P_2, ..., P_{nV}. Taking these points as vertical intersection points, a series of slopes can be formed. After that, the design variables $\boldsymbol{L}$ and $\boldsymbol{Z}$ of the vertical alignment sections can be obtained (i.e., the mileage and elevation of the vertical intersection points). The mileage l of the vertical intersection point can be obtained by finding the intersection of the line formed by the horizontal alignment and the projection of the cutting plane on the horizontal, while the elevation z serves as the gene in the GA.

The 3D alignment model of the railway is characterized by complex constraints and a large spatial search range. According to these characteristics, a floating-point encoding method is proposed. The offset distance d_i $(i = 1, 2, \cdots, n_H)$ of the intersection point between the spatial alignment and the cutting plane (projected on the horizontal plane) relative to the airline and the radius R_i $(i = 1, 2, \cdots, n_H)$ are used as horizontal gene encodings, and the elevation z_i $(i = 1, 2, \cdots, n_V)$ of the intersection point between the spatial alignment and the cutting plane (projected on the vertical section) is the vertical gene encoding. The 3D chromosome of the alignment is expressed as Eq. (9.2):

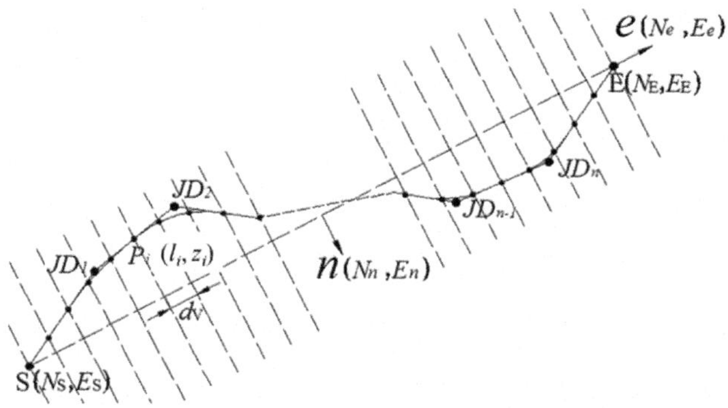

Figure 9.2 Cutting planes of a vertical alignment.

$$\Lambda = \left[\lambda_{i,1}, \lambda_{i,2}, \ldots, \lambda_{i,n}\right] = \left[d_{i,1}, d_{i,2}, \ldots, d_{i,n_H}, R_{i,1}, R_{i,2}, \ldots, R_{i,n_H}, z_{i,1}, z_{i,2}, \ldots . z_{i,n_V}\right]$$
$$d_{i,k} = \lambda_{i,k}, R_{i,k} = \lambda_{i,n_H+k}, z_{i,k} = \lambda_{i,n_H+n_V+k} \tag{9.2}$$

where Λ denotes the chromosome; λ_i represents the gene; $d_{i,1}$, $d_{i,2}$, …, $d_{i,nH}$ are gene segments of horizontal intersection points; $R_{i,1}$, $R_{i,2}$, …, $R_{i,nH}$ denotes the gene segments of horizontal curve radii; $z_{i,1}$, $z_{i,2}$, …, $z_{i,nV}$ represent the gene segments of vertical profiles.

The three gene segments in the alignment chromosome are not completely independent. The selection of HPIs coordinates should match the configuration of horizontal curve radii to ensure that major technical standards such as minimum curve radius, minimum curve length, and minimum tangent length between curves satisfy design requirements. The generation of vertical alignment genes must be based on the horizontal alignment. Therefore, a step-wise encoding method for generating the 3D railway alignment is provided. The general process is as follows:

(1) On the preset cutting planes, HPIs genes are randomly generated on the preset cutting planes, and compute the coordinates of HPIs with Eq. (9.1).

(2) Set the horizontal curve radius for each HPI according to the coordinates of HPIs, the minimum curve radius requirements in *Code for Design of Railway Lines (China Ministry of Railways, 2017, hereinafter referred to as the "Code")*, as well as the geometric position and coupling constraint limits, forming the genes of horizontal curve radii.

(3) Set vertical cutting planes according to the minimum slope length and the computed vertical intersection points.

(4) Determine the elevation of vertical intersection points considering the terrain and constraints including maximum allowable gradient, minimum slope length, minimum clearances between the alignment and crossing rivers, roads, and railways, allowed maximum bridge height and maximum tunnel length.

(5) Merge adjacent sections whose slope differences are within a user-defined threshold, such as 1.1‰, adjust the elevation of vertical intersection points according to requirements in *Code* and elevation control points, and generate the final vertical alignment alternative.

9.2.2 Genetic operator design

Genetic operators in GA include selection operators, crossover operators, and mutation operators. The design of genetic operators is crucial to the implementation effect of GA.

In the GA-based alignment optimization, the selection operator simulates the "survival of the fittest" mechanism in nature, where individuals with higher fitness values (better alignment performance) are more likely to be selected for reproduction, while those with lower fitness values are more prone to elimination. For crossover and mutation operators, various types can be designed. Four types of crossover operators can be adopted, including simple crossover, two-point crossover, arithmetic crossover, and non-uniform heuristic crossover. Three types of mutation operators can be used, such as single-point mutation, multi-point mutation, and straightening mutation, to introduce diversity into the population and explore potential better alignment solutions.

The selection operator is designed using the roulette wheel selection method. The basic design idea is to make the selection probability of each individual proportional to its fitness value. Thus, individuals with higher fitness values have greater chances of being selected. The specific design process is as follows:

① Compute the fitness value $E(i)$ for each individual, where $i = 1, 2 \ldots, Num$ and Num represents the population size.

② Sort the individuals in the population in ascending order of their fitness values.

③ Compute the selection probability Ω_i for individual $P(s)_i$ using Eq. (9.3).

$$P(s)_i = \frac{E(i)}{\sum_{i=0}^{Num} E(i)} \tag{9.3}$$

④ Compute the cumulative probability Ω_i for individual $q(s)_i$ using Eq. (9.4).

$$q(s)_i = \sum_{j=1}^{i} p(s)_j \tag{9.4}$$

⑤ Spin the roulette wheel *Num* times to obtain a new population.
Simple crossover refers to the direct exchange of 3D genes between parent chromosomes at a certain intersection point on the horizontal genes to form two new alignment alternatives. Two-point crossover involves selecting two non-consecutive intersection points on the horizontal genes, and then exchanging the 3D genes between these two points in the parent chromosomes. The main implementation process of two-point crossover is as follows:

① Generate the starting intersection point s_H and ending intersection point e_H for the intended crossover by a uniform random function, as listed in Eq. (9.5).

$$s_H = r_d[1, n_H], \quad e_H = r_d[1, n_H], \quad \text{and} \quad e_H > s_H + 1 \tag{9.5}$$

where $r_d[]$ is the uniform random function, n_H represents the number of horizontal cutting planes.

② Generate the starting and ending vertical intersection points s_V and e_V in the vertical chain slope within the range of these intersection points, as shown in Eq. (9.6).

$$s_V = [(s_H - 1)\cdot d_H/d_V], \quad e_V = [(e_H + 1)\cdot d_H/d_V] \tag{9.6}$$

where d_H represents the distance between horizontal cutting planes, and d_V denotes the distance between vertical cutting planes.

③ Exchange the horizontal genes within the range from s_H to e_H and the vertical genes within the range from s_V to e_V to form two new child chromosomes Λ_i' and Λ_j', as listed in Eq. (9.7).

$$\begin{aligned} \Lambda_i' &= \left[\ldots d_{j,s_H}\ldots d_{j,e_H}\ldots R_{j,s_H}\ldots R_{j,e_H}\ldots z_{j,s_V}\ldots z_{j,e_V}\right] \\ \Lambda_j' &= \left[\ldots d_{i,s_H}\ldots d_{i,e_H}\ldots R_{i,s_H}\ldots R_{i,e_H}\ldots z_{i,s_V}\ldots z_{i,e_V}\right] \end{aligned} \tag{9.7}$$

Arithmetic crossover is a process in which all genes are linearly combined using a certain proportional coefficient to complete the crossover. As shown in Fig. 9.3, let the two parent chromosomes be Λ_i and Λ_j, and the child chromosomes after crossover be Λ'. First, a random number between 0 and 1 is generated through a uniform random function $\omega = r_d[0, 1]$, and then the crossed child chromosomes are obtained through the linear relation $\Lambda' = \omega\cdot\Lambda_i + (1 - \omega)\cdot\Lambda_j$.

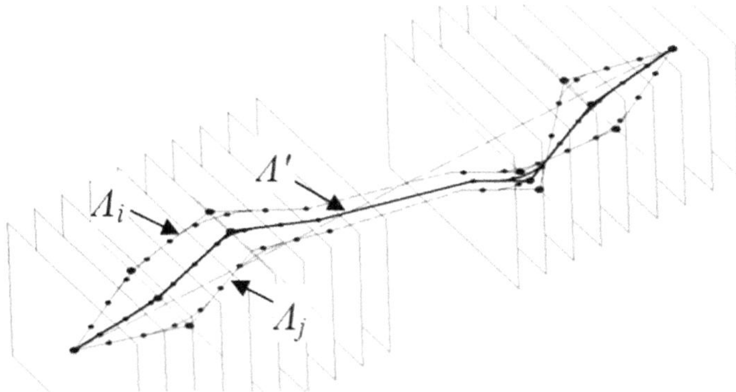

Figure 9.3 Arithmetic crossover three-dimensional space.

Heuristic crossover is similar to arithmetic crossover since it also completes the crossover through linear combination of genes. The difference is that it has a certain degree of heuristic nature. Let the parent chromosomes be Λ_i and Λ_j, with Λ_i being superior. Then, it can be considered that the direction of gene evolution is from Λ_j to Λ_i. Conventional heuristic crossover generates offspring chromosomes by shifting Λ_i in the direction of gene evolution by a certain proportion ω (a uniform random number between 0 and 1). However, in practice, even if Λ_i is globally superior, it may not be superior in all segments. If heuristic crossover is guided solely by the globally-superior direction, it may lack adaptability. To address this defect, a non-uniform heuristic crossover operator can be adopted, where the evolution direction of horizontal and vertical genes is determined by the construction cost values within the range of each intersection point. For any k-th gene in the chromosome, the offspring chromosome is computed as Eq. (9.8). The non-uniform heuristic crossover operator mentioned above is visually demonstrated in Fig. 9.4, which shows its application in three-dimensional space.

$$\lambda_k' = \begin{cases} \omega\left(\lambda_{i,k} - \lambda_{j,k}\right) + \lambda_{i,k}, & \Lambda_i \text{ is superior} \\ \omega\left(\lambda_{j,k} - \lambda_{i,k}\right) + \lambda_{j,k}, & \Lambda_j \text{ is superior} \end{cases} \tag{9.8}$$

Mutation operators start with single-point mutation: randomly select a horizontal intersection point on the alignment and set it as $m_H, m_H = r_d[1, n_H]$. Then, perform mutation operations on each gene in the corresponding horizontal gene segment $d_{m_H}, R_{m_H-1}, R_{m_H}, R_{m_H+1}$ and vertical

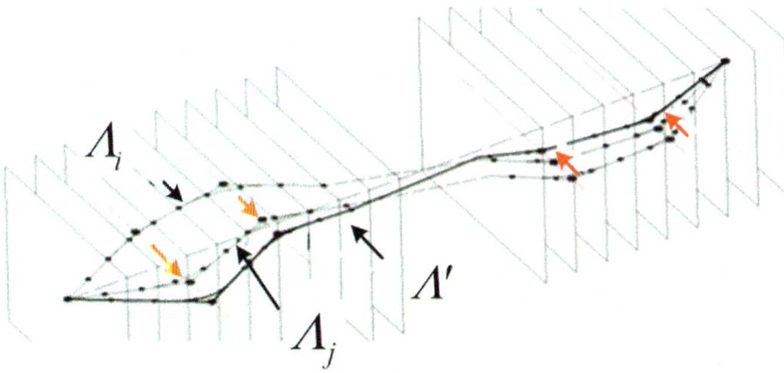

Figure 9.4 Non-uniform heuristic crossover in three-dimensional space.

gene segment $[z_{sv}...z_{ev}]$ in sequence, with the mutation process needing to satisfy coupling constraints. The start and end vertical intersection points s_V, e_V of the vertical slopes corresponding to this intersection range can be computed using Eq. (9.6). Assume that the allowable interval of any gene λ_i under constraint conditions is $[\lambda_{\min}, \lambda_{\max}]$. When topographic conditions permit, short and straight railway alignments can increase the operating speed of trains, thereby enhancing the carrying capacity without increasing the train departure interval. Thus, the modified allowable interval λ_i' is determined using a normally-distributed random number $N(\overline{\lambda}, \sigma)$, with λ_i as the expectation $\overline{\lambda}$. Meanwhile, to ensure that most genes are within the allowable value range, the standard deviation is set to 1/3 of the maximum absolute value of the lower and upper limits (i.e., $\lambda_{\min}$ and $\lambda_{\max}$) of the allowable interval. The computation is shown in Eq. (9.9).

$$\lambda_k' = N\left(\lambda_i, (\max(\lambda_{\max} - \lambda_{\min}))^2/9\right), \quad \lambda_{\min} < \lambda_i' < \lambda_{\max} \tag{9.9}$$

The mutation process of single intersection points in three-dimensional space, as described above, is visually illustrated in Fig. 9.5.

Multi-point mutation involves performing mutation operations on gene segments corresponding to multiple consecutive intersection points. The specific implementation method is to first determine the gene segments to be mutated using Eqs. (9.5) and (9.6), and then perform mutation operations using Eq. (9.9).

The purpose of straightening mutation is to mutate several consecutive intersection points onto a straight line, which can delete some intersection points to form a short and straight alignment scheme. First, determine the start and end intersection point numbers s_H, e_H and the start and end chain

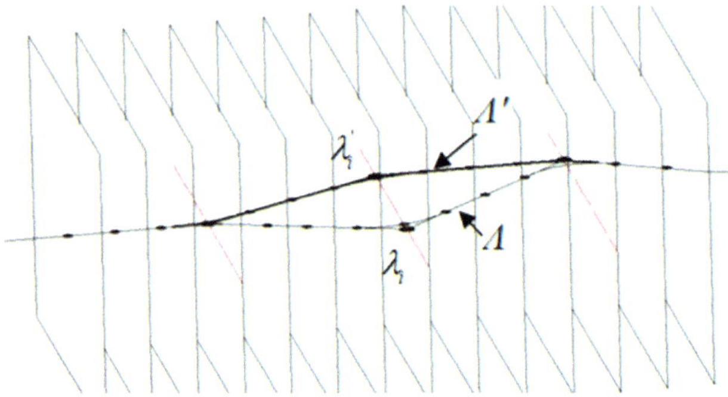

Figure 9.5 Mutation of single intersection points in three-dimensional space.

slope change point numbers s_V, e_V to be mutated using Eqs. (9.5) and (9.6). Then compute the mutated value of any gene d_{k_H} in $[d_{s_H}...d_{e_H}]$ according to Eq. (9.10). The horizontal curve radius gene segment $[R_{sH} \ldots R_{eH}]$ is all set to 0.

$$d'_{k_H} = d_{s_H} + (k_H - s_H + 1) \times (d_{e_H+1} - d_{s_H-1}) / (e_H - s_H + 2) \tag{9.10}$$

After generating the horizontal gene segment, the vertical gene segment $[z_{sV} \ldots z_{eV}]$ can be determined using the method described in Section 9.2.1 (Fig. 9.6).

9.2.3 Algorithm process

(1) Input initial parameters, including the number of genes on a chromosome and the number of initial populations.
(2) Encode for the genes of 3D alignment and generate initial alignment alternatives.
(3) Evaluate the individual fitness of alternatives in the population and compute the selection probability.

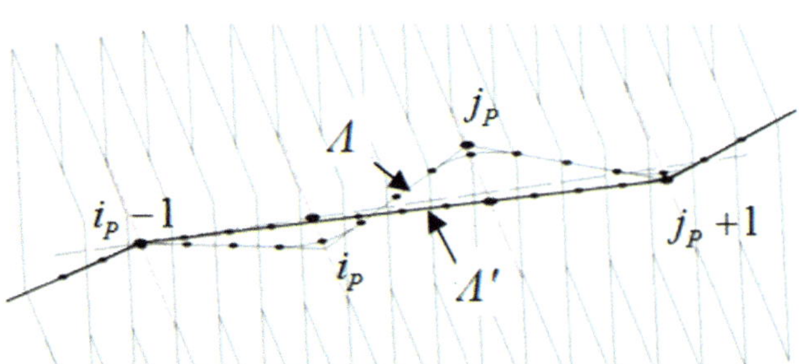

Figure 9.6 Straightening mutation in three-dimensional space.

(4) Use the selection operator to select alignment alternatives for the next generation and evolution.
(5) Generate child alternatives by crossover and mutation operations.
(6) After merging all child alternatives with the original alternatives of the current population, sort the alternatives according to their fitness values. To keep the number of alignment alternatives in the population *Num* unchanged, select the top *Num* alternatives with high fitness values to form a new group of alignment alternatives. Additionally, the best alternative from each generation is always saved into the new generation, so the fitness value never worsens through successive iterations.
(7) Determine whether the convergence criterion is met. If not, return to step (4); if met, stop the optimization.

9.2.4 Convergence criterion

There are generally two termination criteria for GAs:
(1) Maximum number of iterations: If a maximum number of iterations is set, the algorithm should stop when it reaches this number.
(2) Population stability: The change in population fitness should be monitored. When the change in population fitness is less than a threshold value for a specific number of successive generations, the algorithm is considered to have converged.

To prevent the algorithm from converging prematurely or falling into a local optimum, a combination of the two criteria is adopted, i.e., setting the maximum number of iterations as N. In each iteration, the change in population fitness is detected. If the change in population fitness is below a preset threshold and the current number of iterations is less than or equal to N, the population is considered to have stabilized, and the algorithm stops. If the current number of iterations exceeds N, the algorithm is considered to have not reached a stable state, and the iteration process continues until the change in population fitness is better (i.e., less) than the set threshold.

9.2.5 Case study

This approach has been applied to many real-world railway alignment designs. One of the most difficult railway sections in China's mountainous regions is used as a case study in this book.

9.2.5.1 Case profile

This railway is a Grade I railway whose design speed is 120 km/h. It is 1629 km long and winds through mountains. The cumulative height it climbs exceeds 14,000 m. The size of the study area is 910 km^2 (35 × 26 km). It is represented with a 1188 × 854 grid with a resolution of 30 m. The elevation difference between the start and end points is 1307.650 m, whereas the straight-line distance is only 32,598.289 m. The natural terrain gradient $G_T = 40.1‰$, which greatly exceeds the maximum allowable gradient $G_{\max} = 20‰$. This is classified as a severe gradient alignment region. The alignment must then be circuitous. Besides, bridges and tunnels become the major constructions in this region filled with high mountains and deep valleys (as shown in Fig. 9.7). The bridge heights and tunnel lengths should stay within the allowed limits.

The main constraints to be satisfied are listed in Table 9.1.

The SS4 locomotive is selected for designing this railway. The locomotive and design train's tonnage are respectively 184 and 3500 tons.

According to the *Code for Design of Railway Lines (China Ministry of Railways, 2017)*, the minimum vertical circular curve radius is 10,000 m for Grade I railways. The subgrade-bridge boundary height and subgrade-tunnel boundary depth are respectively 8 m and −12 m. The global unit costs are shown in Table 9.2. Different unit costs can be set in special regions.

9.2.5.2 Optimization processes and results

The program is performed on a workstation with Intel Xeon E5620 2.4 G CPU, 4 GB of RAM. Forty-five promising paths are generated after 4748 s of bidirectional DT scanning. Then, these paths are quickly sorted according to their comprehensive costs. The first (cost = ¥354.783 million) and the fourth (cost = ¥391.633 million, close to manual alignment) paths are shown in Fig. 9.8. The program selects the best path for generating a corridor and then searches for an optimized alignment with an improved GA [4].

In practical alignment design, there are some complex factors (e.g., social, political), which cannot be quantified, but may influence the alignment alternative decision. Hence, besides the best one, other paths are very valuable to designers in finding desirable alternatives. Designers can select any path to generate a corridor and optimize an alignment in it.

We choose the best path to generate one railway corridor whose width is 2000 m and set 16 cutting planes on the key points (in Fig. 9.9).

Then the optimized alignment is searched with our improved GA. The population size is 200. The maximum number of generations is 200. After 200 generations, or if the annual comprehensive cost of the best alternative does not decrease in 20 successive generations, the evolution is stopped.

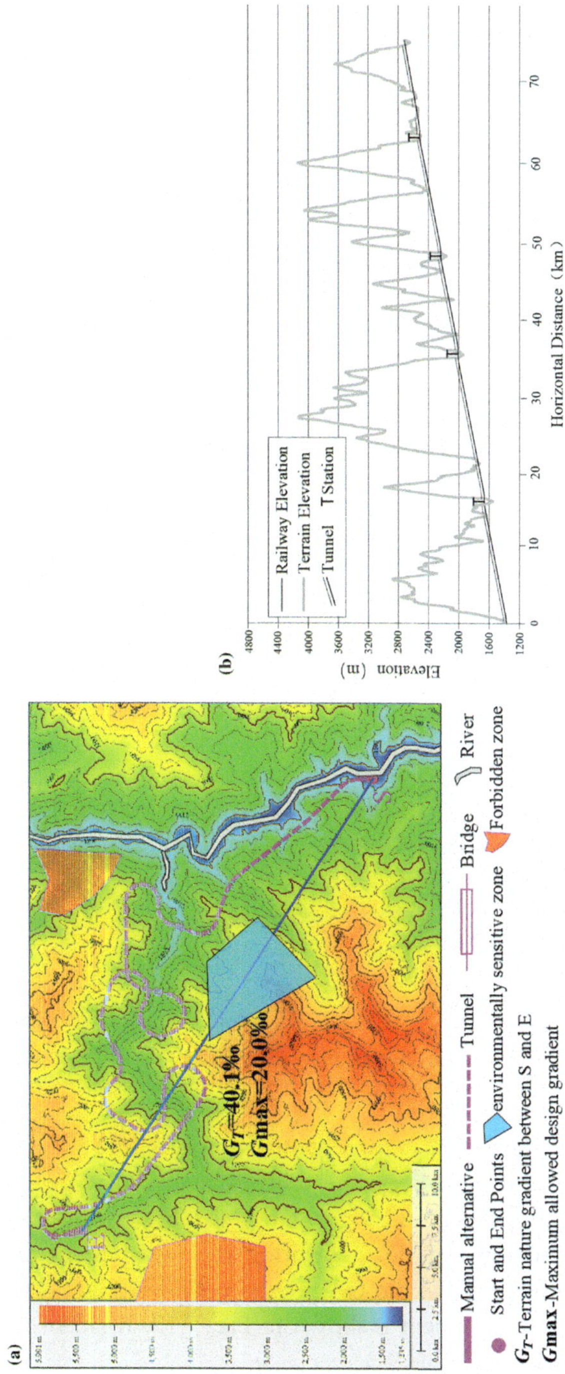

Figure 9.7 Manually designed alignment. (a) Horizontal alignment. (b) Vertical alignment.

Table 9.1 Main constraints.

Classes	Item	Value
Geometric constraints	Minimum radius of curve	800 m
	Maximum gradient	20‰
	Minimum length of tangent between horizontal curves	60 m
Construction constraints	Minimum length of horizontal circular curve	60 m
	Minimum length of slope section	350 m
	Algebraic difference between adjacent gradients	12‰
	Maximum bridge height	80 m
	Maximum tunnel length	16,000 m
	Stations must not overlap with tunnels	
Location constraints	Bypass two forbidden zones	
	If crossing river, clearance $\geq$6 m	
	Elevation tendency: Keep increasing alignment elevation	

Table 9.2 Unit costs.

Item	Cost
Rail track (¥/m)	4000
Right of way (¥/m^2)	90
Filling earthwork (¥/m^3)	18
Cutting earthwork (¥/m^3)	20
($H < 50$) bridge (¥/m)	37,800
($H \geq 50$, $L \geq 500$ m) bridge (¥/m)	58,800
($H \geq 50$, $L < 500$ m) bridge (¥/m)	48,000
One abutment (¥)	200,000
($L \geq 1000$ m) tunnel (¥/m)	60,000
($L \geq 400$ m) tunnel (¥/m)	56,000
($L < 400$ m) tunnel (¥/m)	54,000
One tunnel portal (¥)	250,000
Penalty cost for environmentally sensitive zone (¥/m^2)	50,000

In this case, the annual comprehensive cost reaches ¥325.446 million at the 122nd generation and keeps this value in 20 successive generations. The best objective function in every generation can be seen in Fig. 9.10.

The computation time for generating 200 initial alternatives is 117 s. The entire elapsed time in the GA process is 4275 s (Fig. 9.11).

The final best alignment is shown in Fig. 9.12. It satisfies all the constraints, as verified in Table 9.3.

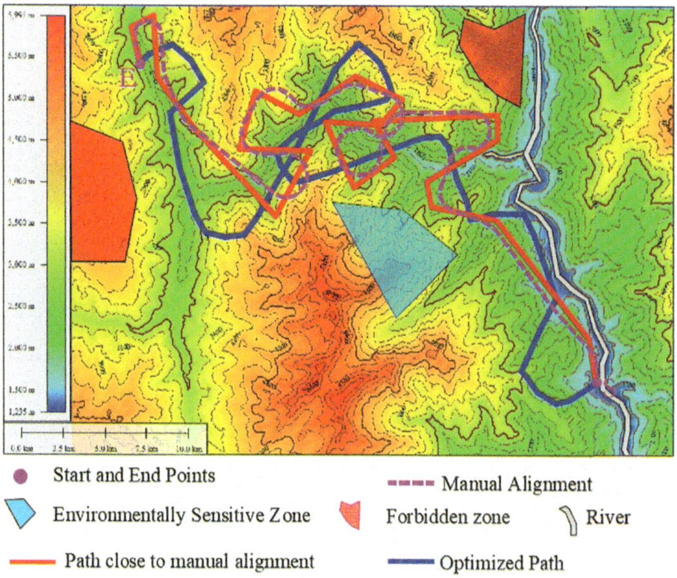

Figure 9.8 Optimized and close-to-manual alignment paths generated by the DT introduced in Chapter 8.

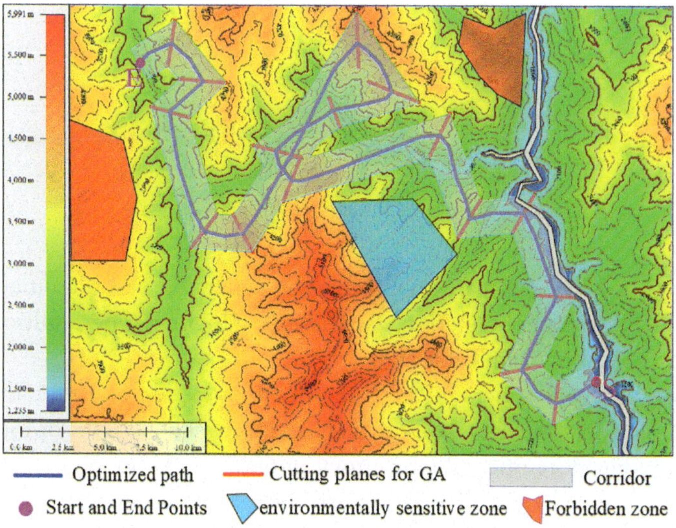

Figure 9.9 Corridor and preset cutting plane for GA.

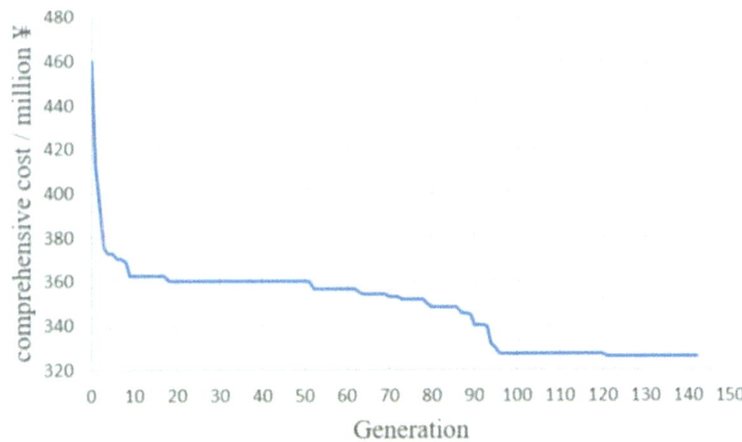

Figure 9.10 Comprehensive cost of the best alternative in every generation.

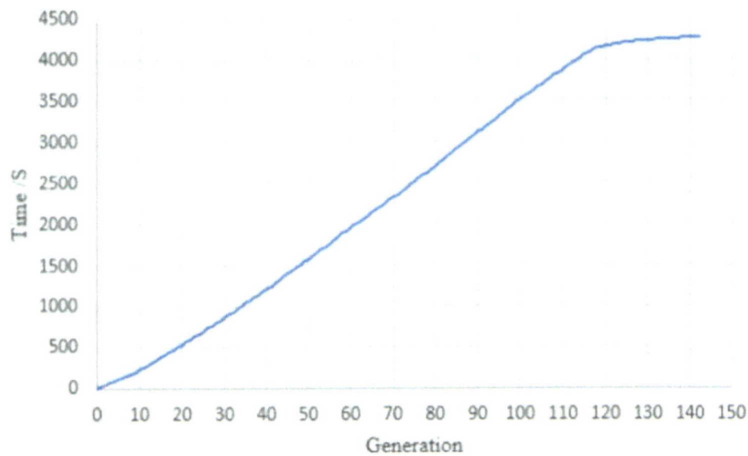

Figure 9.11 Elapsed time in GA.

Afterward, we compare the computer-generated best alignment with the manual solution obtained by experienced designers. The resulting characteristics are shown in Table 9.4.

The computer-generated best alignment is 2487 m shorter than the manual one. Its length-related costs are reduced. The lengths of long tunnels are reduced significantly, by a total of 6567 m. Although the embankment volume and right-of-way area are increased and bridges are 1384 m longer, the annual comprehensive cost is 6.8% lower because the long and expensive tunnels are reduced significantly. These results show that

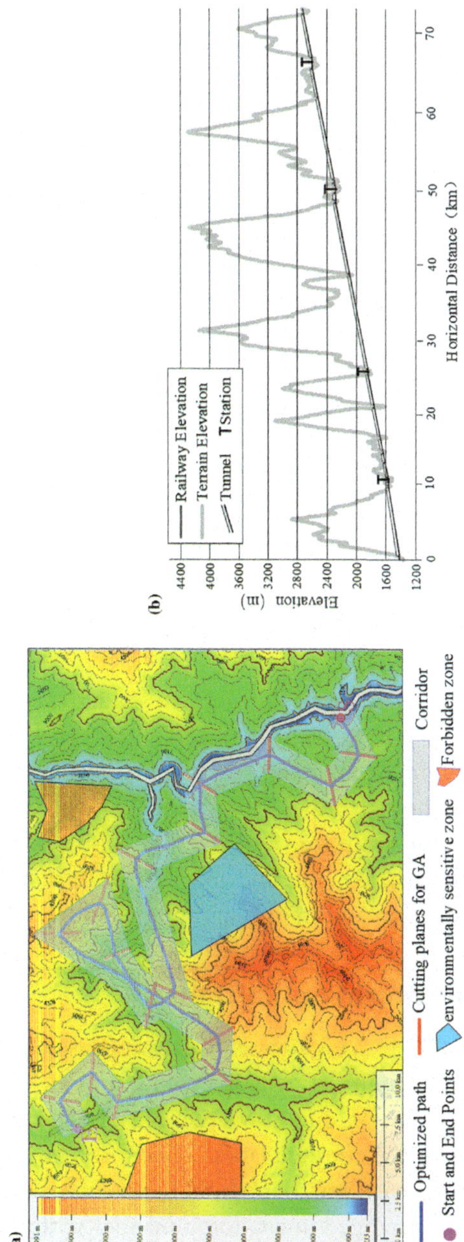

Figure 9.12 Computer-generated best alignment. (a) Horizontal alignment. (b) Vertical alignment.

Table 9.3 Check of constraints.

Constraints	Actual value	Allowed value	Check
Minimum circular radius	800 m	≥800 m	√
Minimum tangent length between horizontal curves	207.3 m	≥60 m	√
Minimum length of horizontal circular curve	208.4 m	≥60 m	√
Maximum gradient	20‰	≤20‰	√
Minimum length of slope section	700 m	≥350 m	√
Algebraic difference between adjacent gradients	12‰	≤12‰	√
Maximum bridge height	42.5 m	≤80 m	√
Maximum tunnel length	12,255 m	≤16,000 m	√
Stations overlapping with tunnels	No	No	√
Forbidden zones	Bypassed	Bypassed	√
If cross the river, clearance ≥6 m	No crossing	≥6 m	√
Elevation tendency	Increase	Increase	√

this approach can improve upon the alternatives designed by experienced designers.

Besides the computer-generated best alignment, other alternatives can also be generated in the remaining 44 corridors. This method can automatically find not only one optimized alignment but also various promising alternatives. These alternatives, offering various trade-offs, can be considered by designers in evaluating and improving their alignments.

9.3 Particle swarm optimization

A review of GA process reveals that from the generation of initial intersection points, through the genetic evolution process, to the completion of algorithm evolution and convergence, all alignment intersection points are restricted to the cutting planes. Therefore, the genetic algorithm is essentially a discrete space search method. In practice, alignment intersection points should be arranged in a tailored manner within the continuous spatial domain of the 3D corridor, rather than on cutting planes with fixed spacing. Consequently, the aforementioned GA performs searches within a

Table 9.4 Computer-generated best alignment compared with manual work.

Item	Manual work	Computer-generated best alignment
Length/m	74,962	72,475
Right of way/m^2	98,992	152,555
Embankment volume/m^3	95,771	187,795
Excavated volume/m^3	58,509	46,842
($H \geq 50$ m) bridge/number-combined length (m)	–	–
($H < 50$ m) bridge/number-combined length (m)	9–3499[a]	9–4883[a]
($L \geq 1000$ m) tunnel/number-combined length (m)	8–69,957[a]	9–63,390[a]
($L \geq 400$ m) tunnel/number-combined length (m)	–	2–939[a]
($L < 400$ m) tunnel/number-combined length (m)	1–244[a]	1–325[a]
Total construction costs/million ¥	4662.610	4375.658
Annual construction costs ($\Delta = 0.065$)/ million ¥	303.070	284.418
Annual operating cost/million ¥	46.214	41.028
Annual comprehensive cost/million ¥	349.284	325.446
Annual comprehensive cost saving/million ¥	–	23.838
Ratio of annual comprehensive cost saving	–	6.8%

[a]Number before hyphen indicates number of bridges or tunnels; number after hyphen indicates combined length.

limited discrete subspace, which may easily lead to the omission of valuable alignment alternatives. Particle Swarm Optimization (PSO) is a heuristic algorithm based on swarm intelligence, whose main advantages include suitability for solving continuous space optimization problems, fast convergence speed, and good robustness. Based on the obtained alignment corridor, this section introduces an alignment spatial search method based on stepwise particle swarm optimization.

9.3.1 Basic principle

Particle Swarm Optimization (PSO) is an evolutionary computation technique based on swarm intelligence proposed in 1995, inspired by the migration and flocking behaviors of bird flocks during foraging [6]. To solve an optimization problem using the PSO algorithm, it is necessary to first generate a set of initial feasible solutions and abstract them as particles in a multi-dimensional space. The spatial dimension is determined by the number of decision variables of the problem to be solved, and each feasible solution is represented by the position of each particle in the multi-dimensional space. All particles simulate the "flight" (i.e., iterative evolution) of a bird flock in the study area in order to find the optimized position. During the "flight" process, each particle is endowed with fitness and velocity characteristics. The fitness characteristic is computed according to the objective function value of the problem to be solved, while the velocity characteristic is comprehensively determined by the "flight" direction and velocity magnitude of each particle. In the iterative process, each particle updates its velocity characteristic based on its current velocity, as well as the individual best particle position (**pbest**) and the global best particle position (**gbest**) in the current swarm. The latest position of the particle is computed based on its current position and the updated velocity, with the formulas as Eqs. (9.11) and (9.12).

$$\begin{aligned}\mathbf{Vel}_i(t+1) = {} & w\cdot\mathbf{Vel}_i(t) + C_1\cdot r_1(t)\cdot[\mathbf{gbest}(t) - \mathbf{Pos}_i(t)] \\ & + C_2\cdot r_2(t)\cdot[\mathbf{pbest}_i(t) - \mathbf{Pos}_i(t)]\end{aligned} \tag{9.11}$$

$$\mathbf{Pos}_i(t+1) = \mathbf{Pos}_i(t) + \mathbf{Vel}_i(t+1) \tag{9.12}$$

where t denotes the current number of iterations; w is the inertia weight; $\mathbf{Vel}_i(t)$ and $\mathbf{Vel}_i(t+1)$ represent the current velocity characteristic and the updated velocity characteristic of the i-th particle in the population, respectively; $\mathbf{Pos}_i(t)$ and $\mathbf{Pos}_i(t+1)$ denote the current position and the updated position of the i-th particle in the population, respectively; C_1 and C_2 are the cognitive constant and social constant, respectively. C_1 controls the speed at which an individual particle moves toward its historical best position (individual experience), while C_2 controls the speed at which an individual particle moves toward the global best position (social experience); typically, $C_1 = C_2 = 2$; r_1 and r_2 are random numbers distributed in the range [0, 1]; $\mathbf{gbest}(t)$ is the optimized position of all particles in the entire population up to the current iteration number t, i.e., the global best position; $\mathbf{pbest}_i(t)$ is the optimized position of the i-th particle up to the current iteration number t, i.e., the personal best position. Fig. 9.13 illustrates the evolutionary process of particle swarm optimization in a two-dimensional space:

In Eqs. (9.11) and (9.12), the inertia weight parameter w is used to control the degree to which the position of a particle after iteration is influenced

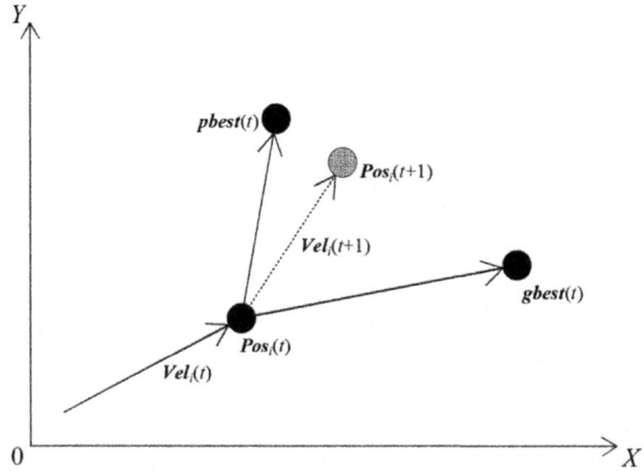

Figure 9.13 The evolutionary process of PSO in 2D space.

by its position before iteration. During the iteration process, if the value of w is too large, particles will exhibit significant velocity fluctuations in the search process, which may easily lead to the omission of some potential solutions and result in a relatively slow convergence speed of the algorithm. Conversely, if the value of w is too small, the velocity fluctuations of particles during the search will be reduced, which may cause the algorithm to stagnate prematurely near a locally optimized solution. Studies have shown that the inertia weight typically takes a value within the interval [0, 1], and that gradually reducing the inertia weight during the iteration process helps balance the relation between global search and local exploration. This enables the algorithm to search more effectively in the solution space and eventually converge to the optimized solution. Specifically, the cubic polynomial formula has a large descending gradient in the early stage of iteration, which facilitates sufficient exploration of the solution space and avoids missing feasible solutions. In the later stage of iteration, as the search space gradually approaches the neighborhood of the optimized solution, the inertia weight slows down its descent gradient accordingly, which helps enhance the local search capability. Therefore, adopting the cubic polynomial inertia weight update strategy can more fully explore the neighborhood space of the optimized solution, which is beneficial for improving the quality of the alternative. The weight update formula is Eq. (9.13).

$$w_t = -2.25 \times \frac{t^3}{7T^3} + 6.75 \times \frac{t^2}{7T^2} - 8 \times \frac{t}{7T} + 1 \tag{9.13}$$

where t and T denote the current number of iterations and the total number of iterations, respectively (Fig. 9.14).

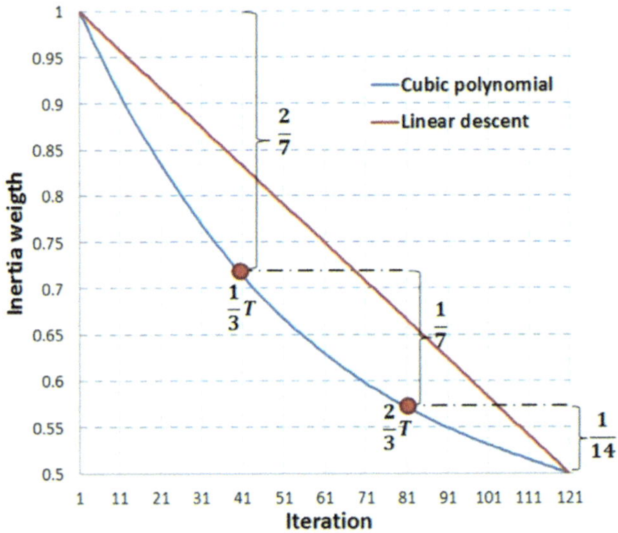

Figure 9.14 Cubic polynomial inertia weight update strategy.

Furthermore, the particle velocity $\mathbf{Vel}_i(t+1)$ is subject to the maximum velocity constraint in Eq. (9.11). The "velocity limitation" criterion is intended to restrict the maximum single-step velocity of particles during each iteration, thereby preventing particles from "flying out" of the solution space and confining them within the feasible domain. The "velocity limitation" criterion can be expressed as Eq. (9.14).

$$\mathbf{Vel}_i(t+1) = \begin{cases} \mathbf{Vel}_i(t+1) & \text{if } \mathbf{Vel}_i(t+1) < \mathbf{V}_{\max} \\ \mathbf{V}_{\max} & \text{if } \mathbf{Vel}_i(t+1) \geq \mathbf{V}_{\max} \end{cases} \tag{9.14}$$

where, $\mathbf{Vel}_i(t+1)$ represents the "computed velocity" of the i-th particle obtained from Eq. (9.11); $\mathbf{V}_{\max}$ denotes the maximum allowable velocity in each iteration. The "velocity limitation" can restrict the single-step displacement of each particle, thereby preventing the algorithm from diverging.

9.3.2 Initial particle swarm generation method

When using PSO to solve the alignment optimization problem, the horizontal and vertical alignments are abstracted as particles in a multidimensional space. Each position through which a particle passes represents an alignment alternative. The initial position of the particle represents the initial alignment alternative. The alignment optimization is achieved by continuously searching for the best particle position in the multi-dimensional space based on the initial particle positions.

The quality of initial alignment alternatives significantly affects the quality and efficiency of PSO. The alignment corridor obtained by the bidirectional generalized DT algorithm described in Chapter 8 can provide a scientific basis for the initial distribution of alignment control points, thereby generating high-quality initial alignment alternatives. Therefore, an initial PSO method based on the alignment corridor can be adopted. Since the distribution and number of horizontal and vertical alignment control points are usually inconsistent, and since overlaps of horizontal and vertical curves should be avoided as much as possible, a stepwise initialization strategy is adopted. That is, horizontal initialization is performed first, and after generating the horizontal alternative, vertical initialization is carried out along the horizontal alignment to generate the initial 3D alignment alternative.

9.3.2.1 Horizontal alignment initialization

Most existing heuristic optimization methods distribute horizontal alignment control points by dividing cutting planes, and realize the initialization of horizontal alignments by randomly setting the positions of horizontal control points on the cutting planes. However, since this method restricts horizontal alignment control points to cutting planes, it reduces initial freedom and often fails to create effective initial horizontal alignment alternatives in mountainous regions with complex terrain and geological conditions. To address this issue, the concept of the butterfly-shaped area template is proposed to initialize horizontal alignments. Compared with initialization on cutting planes, this approach can provide sufficient initial freedom, significantly increase the initial success rate, enhance the diversity of initial alternatives, and offer more evolutionary momentum for subsequent optimization.

(1) Generation of butterfly-shaped area

Based on the obtained railway corridor, a separate local polar coordinate system is established at each key point of the DT path. The origin and axis of each coordinate system are the corresponding key point and angular bisector, respectively. The total length of every axis is fixed at $2L$, where L equals the value of maximum velocity, and the midpoint of this alignment segment is the origin. Afterward, all the coordinate axes are rotated clockwise and counter-clockwise for 45 degrees so that the butterfly-shaped areas as shown in Fig. 9.15 are scanned. This butterfly-shaped area serves as the movement area for the particle

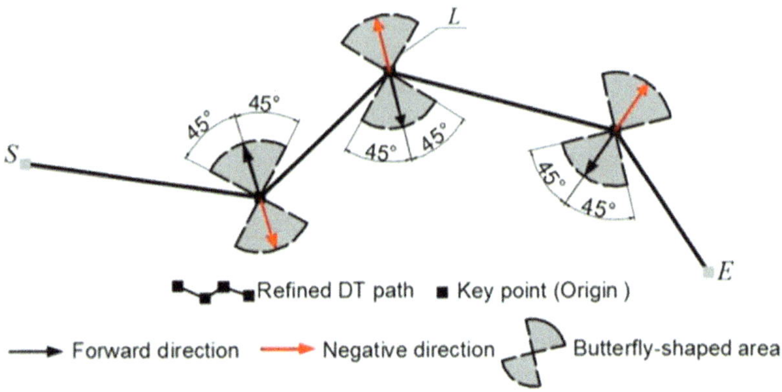

Figure 9.15 Generates butterfly-shaped areas based on line corridor belts.

swarm in the search space. Compared with the previous (GA) method that restricts the points of intersection to cutting planes, it has a larger and more flexible activity space.

(2) Position of the initial horizontal intersection point

After generating the butterfly-shaped areas, the unit vector of the angular bisector (**UnitVec**) at each key point is first computed according to the vector relation between two adjacent connecting paths at each key point on the railway corridor, as listed in Eqs. (9.15–9.18).

$$\mathbf{P}_i\mathbf{P}'_{i-1} = \frac{\mathbf{P}_i\mathbf{P}_{i-1}}{|\mathbf{P}_i\mathbf{P}_{i-1}|} = \begin{pmatrix} V_{x_{i-1}}/\sqrt{V_{x_{i-1}}^2 + V_{y_{i-1}}^2} \\ V_{y_{i-1}}/\sqrt{V_{x_{i-1}}^2 + V_{y_{i-1}}^2} \end{pmatrix} \tag{9.15}$$

$$\mathbf{P}_i\mathbf{P}'_{i+1} = \frac{\mathbf{P}_i\mathbf{P}_{i+1}}{|\mathbf{P}_i\mathbf{P}_{i+1}|} = \begin{pmatrix} V_{x_{i+1}}/\sqrt{V_{x_{i+1}}^2 + V_{y_{i+1}}^2} \\ V_{y_{i+1}}/\sqrt{V_{x_{i+1}}^2 + V_{y_{i+1}}^2} \end{pmatrix} \tag{9.16}$$

$$\mathbf{Vec}_i = \mathbf{P}_i\mathbf{P}'_{i-1} + \mathbf{P}_i\mathbf{P}'_{i+1} = \begin{pmatrix} V_x \\ V_y \end{pmatrix} \tag{9.17}$$

$$\mathbf{UnitVec}_i = \begin{pmatrix} V_x/\sqrt{V_x{}^2 + V_y{}^2} \\ V_y/\sqrt{V_x{}^2 + V_y{}^2} \end{pmatrix} = \begin{pmatrix} Unit\,V_x \\ Unit\,V_y \end{pmatrix} \tag{9.18}$$

where, $\mathbf{P}_i\mathbf{P}'_{i-1}$ and $\mathbf{P}_i\mathbf{P}'_{i+1}$ are the unit vectors of the lines connecting the i-th key point P_i with the adjacent path vertices respectively; $\mathbf{Vec}_i$ is the angular bisector vector of the two unit vectors; and $\mathbf{UnitVec}_i$ is the unit vector of the angular bisector at the i-th key point, as shown in Fig. 9.16.

Then, the offset l from the initial horizontal point of intersection to the horizontal point of intersection of the railway corridor and the deflection angle α are randomly generated. After that, the offset vector $\mathbf{Vec}'$ from the horizontal point of intersection of the railway corridor to the initial horizontal point of intersection is computed with Eqs. (9.19–9.21).

$$\mathbf{Vec}_i' = n \times \begin{pmatrix} UnitV_{x_i} \cdot \cos\alpha - UnitV_{y_i} \cdot \sin\alpha \\ UnitV_{x_i} \cdot \sin\alpha + UnitV_{y_i} \cdot \cos\alpha \end{pmatrix} = \begin{pmatrix} V_{x_i}' \\ V_{y_i}' \end{pmatrix} \tag{9.19}$$

$$n = \begin{cases} 1, & if\ m \geq 0.5 \\ -1, & if\ m < 0.5 \end{cases} \tag{9.20}$$

$$m = rand(0, 1) \tag{9.21}$$

where $\mathbf{Vec}_i'$ is the offset vector from the i-th horizontal point of intersection of the railway corridor to the initial horizontal point of intersection; parameters n and m are used to determine the direction of the random offset vector, as shown in Fig. 9.16.

Finally, the coordinates of the initial horizontal point of intersection are computed according to the offset l and the offset vector $\mathbf{Vec}'$, as Eq. (9.22).

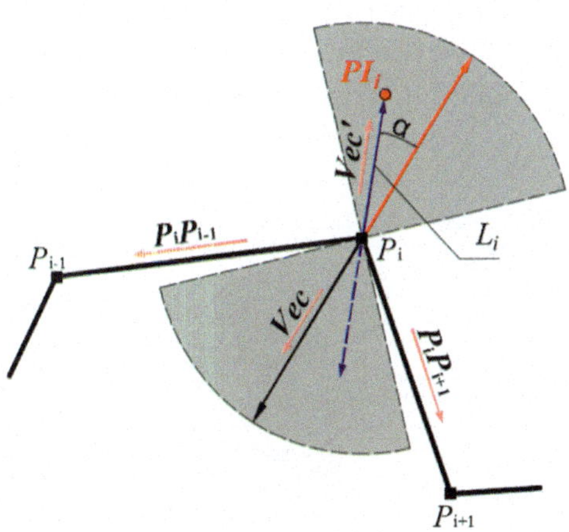

Figure 9.16 Initial horizontal intersection position (n = −1).

$$\begin{cases} PI_{i,x} = P_{i,x} + L_i \cdot V'_{x_i} \\ PI_{i,y} = P_{i,y} + L_i \cdot V'_{y_i} \end{cases} \tag{9.22}$$

where $P_{i,x}$ and $P_{i,y}$ are the coordinates of the i-th horizontal point of intersection on the railway corridor; $PI_{i,x}$ and $PI_{i,y}$ are the coordinates of the i-th initial horizontal point of intersection generated based on the railway corridor.

After obtaining the positions of the initial horizontal points of intersection, the initial horizontal alignment alternatives can be generated by configuring circular curves and transition curves, as shown in Fig. 9.17.

9.3.2.2 Vertical alignment initialization

The vertical ground line of the area traversed by the railway can be derived based on the constructed CGIM. The slope sections of the vertical railway alignment should be flexibly set according to the terrain undulation. Moreover, shorter and more numerous slope sections result in better adaptation to the ground line and reduce the railway construction costs. To obtain a high-quality initial vertical alignment alternative, the initialization is carried out according to the principle of "minimizing the sum of squared fill and cut volumes" between the railway and the ground surface. The details are as follows:

First, the feasible region for vertical points of intersection (VPIs) of the vertical railway alignment is delineated to ensure that the generated alternative satisfies all the vertical constraints, as shown in Fig. 9.18(a).

Then, the ground line is modified according to the feasible region for VPIs, and vertical cutting planes are set at intervals of the minimum slope section length (L_{Smin}) to determine the stations of VPIs, as shown in Fig. 9.18(b). The minimum slope section length is used as the interval

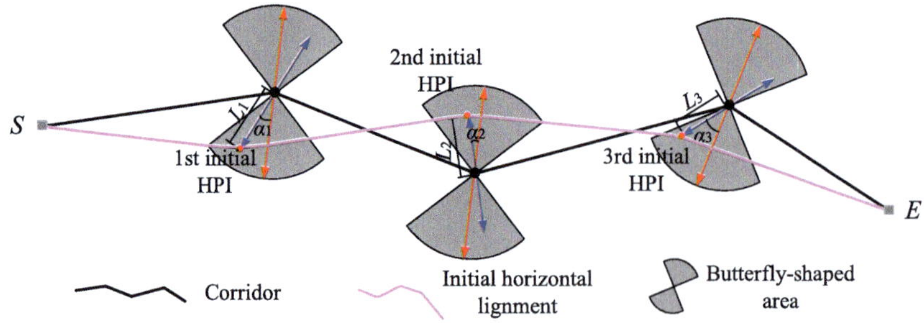

Figure 9.17 Horizontal initialization.

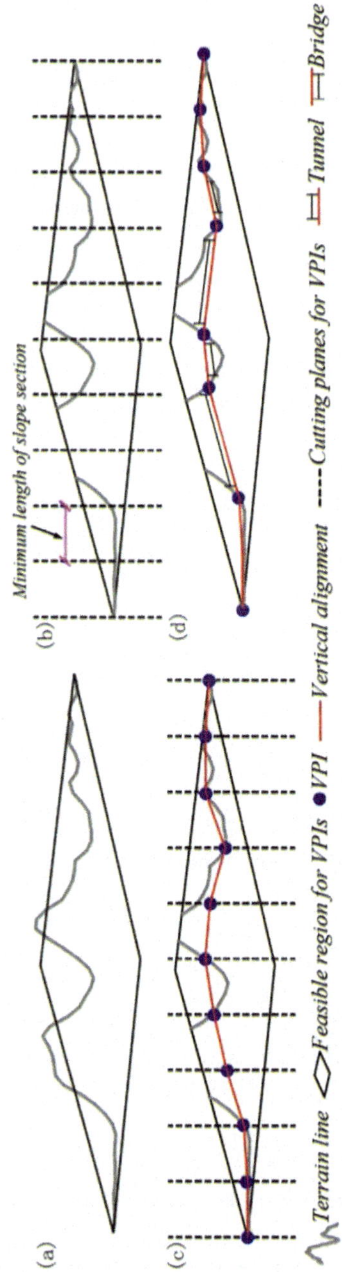

Figure 9.18 Vertical initialization.

because that shorter slope sections can better adapt to terrain changes for reducing the earthwork volume. For two adjacent VPIs, minimizing the sum of squared fill and cut volumes between the railway and the ground surface at each station between them, can be expressed as Eq. (9.23). Let m denote the number of stations between the (i-1)-th and i-th VPIs; G_P represents the theoretical gradient value; K_{Ej} is the station of the jth station; K_i is the station of the i-th VPI; H_i is the design elevation of the i-th VPI; and H_{Ej} is the ground elevation at the j-th station.

$$\min f(G_p) = \min \sum_{j=1}^{m} \left[G_p \times (K_{Ej} - K_i) + H_i - H_{Ej}\right]^2 \tag{9.23}$$

By finding the partial derivative of G_p according to Eq. (9.23), the gradient value G_p that minimizes $f(G_p)$ can be obtained by Eqs. (9.24) and (9.25).

$$\frac{df(G_p)}{dG_p} = 2 \times G_p \times \sum_{j=1}^{m} (K_{Ej} - K_i)^2 + 2 \times \sum_{j=1}^{m} (K_{Ej} - K_i) \times (H_i - H_{Ej}) \tag{9.24}$$

$$G_p = -\frac{\sum_{j=1}^{m} (K_{Ej} - K_i) \times (H_i - H_{Ej})}{\sum_{j=1}^{m} (K_{Ej} - K_i)^2} \tag{9.25}$$

Then, the G_p computed with the above equation must be revised by considering the maximum gradient constraint g_{max}, as shown in Eq. (9.26).

$$G_p = \begin{cases} G_p, & |G_p| < g_{max} \\ \pm g_{max}, & |G_p| \geq g_{max} \end{cases} \tag{9.26}$$

Based on the gradient G_p and the elevation of the starting point of the railway, the elevation of the vertical point of intersection (VPI) of the railway vertical alignment can be computed with Eq. (9.27), as shown in Fig. 9.18(c).

$$H_{i+1} = H_i + G_p \times L_{smin} \tag{9.27}$$

When $i = 0$, H_i is the elevation of the starting point of the railway; L_{smin} is the minimum length of the slope section.

Finally, slope sections with an algebraic difference of gradients below a given threshold are merged, and slope sections with an algebraic difference of gradients exceeding the maximum allowable value are adjusted. The positions of bridges and tunnels are determined through an automatic bridge and tunnel setting method. Then, the initialization of the railway vertical alignment is completed, as shown in Fig. 9.18(d).

9.3.3 Stepwise particle swarm optimization strategy

After the initialization, the best spatial alignment location is searched based on the initialized particle swarm. At this stepwise optimization stage, we must point out that all horizontal particles in our swarm have the same number of positions as the number of HPIs of each alignment is uniform, but different vertical particles in the same swarm can have different numbers of positions. That occurs because the numbers of VPIs of different alignments may differ since the vertical alignments differ along different horizontal alignments. Therefore, we do not establish a unified quantitative standard for VPIs but generate VPIs dynamically. The basic flow of stepwise optimization is as follows:

Step 1: Compute the 2-D velocities and positions of each horizontal particle using PSO operators (Eqs. 9.11 and 9.12)) and generate horizontal particle swarm;

Step 2: Interpolate the individual terrain line of each horizontal particle by using a DEM and, similarly to Section 9.3.2, generate vertical cutting planes with minimum slope length (L_{smin});

Step 3: As Fig. 9.19 shows, obtain the horizontal coordinates of each cutting plane according to its station, denoting the corresponding horizontal point as P1. Then construct projective lines from P1 to the gbest and the pbest, denoting the intersection point as P2 (P2 may not be a VPI of the gbest or the pbest). Finally compute the design elevation of P2 according to the elevation of starting point, P2's station and vertical slopes of the gbest or the pbest in successive steps, and name it "projection elevation".

Step 4: Use the projection elevation as the vertical position (velocity) of the gbest and the pbest to determine the elevation of each VPI on the cutting planes by adopting PSO rules (Eqs. 9.11 and 9.12).

Step 5: Refine the vertical alignment and generate the integrated 3D alignment. Then, compute its comprehensive cost.

After one iteration, all the newly-generated alignments in the swarm are sorted according to their comprehensive costs. Then, the pbest swarm and gbest particle are updated. In this stepwise approach, the horizontal and vertical particles of the alignment are not completely independent: the update

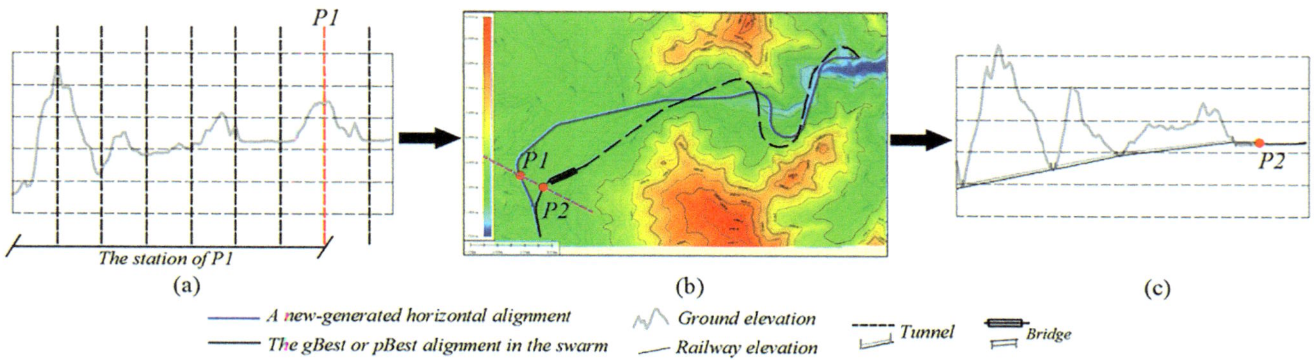

Figure 9.19 Determining the projection elevation: (a) Compute the station of each cutting plane *P1*; (b) obtain the horizontal coordinates of *P1* and determine the horizontal projection point *P2*; (c) compute the design elevation of *P2* and set it as the "projection elevation".

of HPIs must take into account the elevation difference, earthwork, tunnel and bridge structures, complex vertical constraints of vertical alignment; the generation of vertical particles must be based on horizontal iterative results. Clearly, this approach satisfies the requirements of railway alignment design and can well adapt vertical alignments to specific topography.

9.3.4 Dynamic determination of feasible domain in vertical section

In complex mountainous regions, vertical alignment constraints such as maximum gradient and maximum allowable bridge height are important factors affecting the railway alignment. To ensure that the vertical alignment satisfies the constraints during the alignment search process, a method for dynamically generating the feasible region of vertical intersection points can be adopted. As shown in Fig. 9.20, first, the maximum and minimum elevations that can be reached within the range of the start and end points of the alignment are determined. That is, a quadrilateral region is formed starting from the start and end points with the maximum gradient respectively. This region is the elevation range of the alignment, and all vertical intersection points must be within this region. Then, the quadrilateral region is adjusted according to the elevation control points to determine the initial elevation feasible region.

Within the initial elevation region, the position of the first vertical intersection point is determined. After identifying that, the first vertical

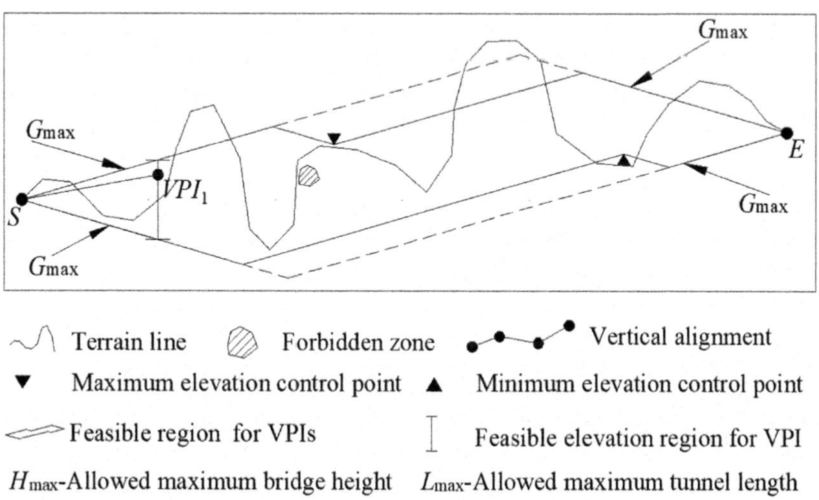

Figure 9.20 Feasible area of initial elevation.

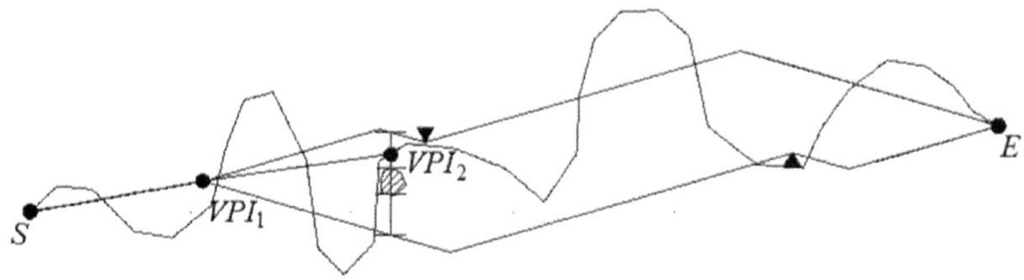

Figure 9.21 Feasible areas for adjusting the elevation.

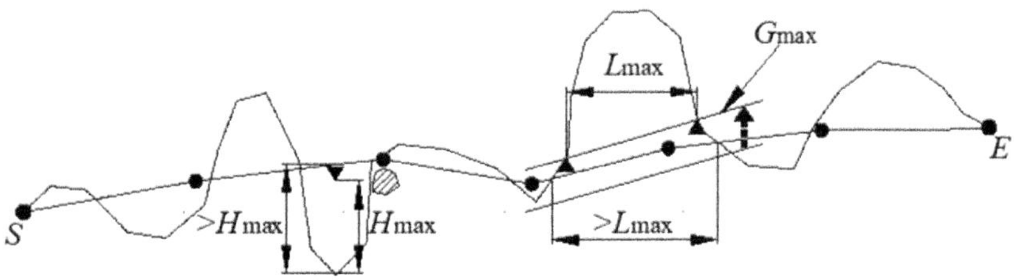

Figure 9.22 Added elevation control points for restricting high bridges and long tunnels.

intersection point is set as the starting point, and the elevation feasible region is readjusted, as shown in Fig. 9.21. This process is repeated until the positions of all vertical intersection points are determined.

The above method can ensure that the generated vertical alignment satisfies the constraints of maximum gradient, elevation control point, and vertical forbidden zones. However, it cannot guarantee that the vertical alignment alternative satisfies the maximum bridge height and tunnel length constraints. To address this, the following method can be adopted:

Step 1: Generate the vertical alignment alternative and check for the presence of overly high bridges and overly long tunnels, as shown in Fig. 9.22.

Step 2: If there is a bridge whose elevation exceeds its maximum allowable height, set a maximum elevation control point at the location with the lowest terrain elevation. The height of this elevation control point is the sum of the terrain elevation and the maximum allowable bridge height, as shown in Fig. 9.22.

Step 3: If there is a tunnel exceeding the maximum allowable length, its gradient is adjusted to the maximum allowable gradient to reduce the tunnel length. When the maximum tunnel length is reached, set two minimum elevation control points at the two portals of the tunnel, as shown in Fig. 9.22.

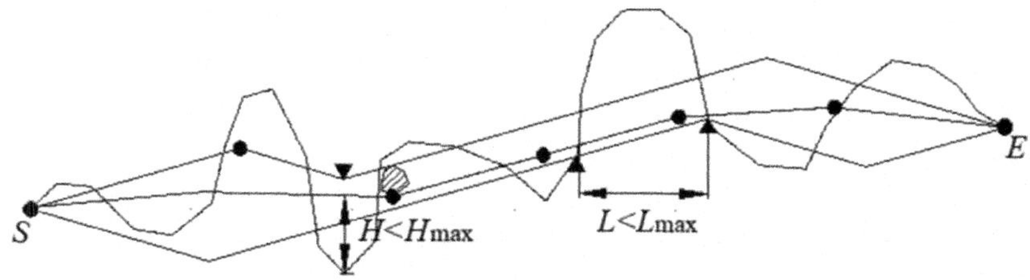

Figure 9.23 Regenerates the feasible area for elevation.

Step 4: The feasible region for the vertical alignment is regenerated based on the new elevation control points, as shown in Fig. 9.23.

Step 5: If the feasible region is non-empty, the vertical alignment alternative within this region is generated; otherwise, the horizontal alignment alternative is infeasible and must be regenerated.

Through the above method, the 3D alignment can be searched in continuous space, and an alignment alternative with the lowest comprehensive cost that satisfies various constraint requirements can be obtained.

9.3.5 Search strategy integrating genetic operators

Some important parameters are involved in the particle swarm optimization algorithm, such as inertia weight and acceleration constant. Inappropriate parameter settings may cause particles to be overly concentrated near the local optimized solution, losing the global search ability. Especially when facing multiple optimization objectives, this further aggravates the complexity of the problem, and thus no optimized solution may be obtained. This section addresses the issue of congestion in the decision variable space of the particle swarm optimization algorithm. It adopts a particle swarm optimization method that integrates genetic operators during the search of spatial variables. By using genetic operators, the diversity of each generation of population alternatives in the decision variable space is enhanced, thereby solving the problem of premature convergence of the particle swarm optimization algorithm.

Section 9.2.2 has elaborated on two types of genetic operators, namely crossover operators and mutation operators, whose effect characteristics are significantly different: Crossover operators have significant optimization capabilities at the beginning of evolution, but their effect weakens toward the end of evolution. In contrast, the effect characteristics of mutation operators are exactly opposite to those of crossover operators. They have slight optimization effect at the beginning of evolution but produce a significant optimization effect at the end of evolution. Therefore, crossover operators and mutation operators are applied to different stages of particle swarm search. Specifically, crossover operators are used after each PSO iteration to explore the search variable space more fully, while mutation operators are used when the evolution is slow.

9.3.5.1 Crossover operators for each iteration

Step1: After each iteration, compare the objective function value of $CurObj_i$ of each $pbest_i$ ($1 < i < N$, where N is the swarm size) with the previous iteration's value $PreObj_i$, as computed as Eq. (9.28).

$$precision_i^{(j)} = \frac{\left| CurObj_i^{(j)} - \text{Pre}\,Obj_i^{(j)} \right|}{\text{Pre}\,Obj_i^{(j)}} \tag{9.28}$$

where, $precision_i^{(j)}$ is the change rate of the j-th objective function value at the position of the i-th pbest particle; $CurObj_i^{(j)}$ is the j-th objective function value at the position of the i-th pbest particle in the current generation; $\text{Pre}\,Obj_i^{(j)}$ is the j-th objective function value at the position of the i-th pbest particle in the previous generation.

If the change rates of all objective function values at the position of a pbest particle are below the set threshold, the railway alignment alternative represented by the position of this pbest particle will be added to the "selection pool".

Step 2: If the number of alternatives in the "selection pool" is less than 2, exit this module and proceed to the next iteration; otherwise, randomly select two schemes from the current "selection pool" and proceed to Step 3.

Step 3: Randomly select a crossover operator to apply to the two selected schemes, and judge whether the newly generated offspring individuals are better than their parent individuals. If the offspring individuals are better, update the corresponding parent individuals in the "selection pool".

Step 4: Repeat Steps 2 and 3, and end this module when it reaches $0.5N$ times (where N is the number of particles in the population). The offspring individuals generated after crossover are saved as the corresponding pbest particle positions for the next generation of particle swarm iterative evolution.

9.3.5.2 Mutation operator fusion module

When the position of the gbest particle remains unchanged for N1 consecutive generations, the mutation operator will be applied to update the position of the gbest particle in subsequent iterations; when the position of the gbest particle remains unchanged for N2 consecutive generations, a random mutation module will be entered to randomly perform mutation operations on the schemes in the current population. If a better scheme is generated through the mutation operator, the algorithm continues. If no better gbest particle position is generated after 0.5N attempts, it is considered that convergence has been achieved and the iterations are terminated.

Through the above method, the diversity of population schemes in each generation in the variable space can be improved during the particle swarm search process, and the ability of particles to jump out of local optimized solutions can be enhanced, thereby solving the premature convergence tendency of the particle swarm optimization algorithm.

9.3.6 Particle position decision-making method

The position of the pbest particle is mainly used to guide the evolutionary direction of the particle itself and must be able to guide the particle to evolve toward improving the alternative quality. Therefore, the decision-making of the pbest particle position should focus on the alternative quality. The multi-objective decision-making method based on the tournament idea can select the multi-objective comprehensive optimized alternative from the perspective of quality. Thus, the tournament idea is adopted for pbest particle position decision-making.

A storage archive is established for each particle to store all non-dominated solutions obtained by each particle up to the current generation, thereby constructing an individual non-dominated solution set for each particle. The maximum number of solutions that can be stored in each particle's individual non-dominated solution set is set as N_p. Then, the objective function values of each non-dominated solution are compared in sequence, and the better-performing solution after comparison is recorded as one victory. The victory value of each objective function in each non-dominated solution is computed through the tournament equation, and the ranking value of each non-dominated solution is computed through the ranking equation. The ranking value is taken as the fitness of the pbest particle, where a solution with a larger ranking value dominates a solution with a smaller ranking value. The Pareto dominance relation is sorted in descending order of the ranking values. When the number of solutions exceeds N_p, the solutions ranked after N_p are deleted.

It should be noted that when computing the ranking value using the ranking equation, if the victory value of a certain indicator is 0, its ranking value is 0. The goal in the distance transform scanning process is to find the multi-objective comprehensive best local path; therefore, paths with excessively poor objective functions have low value and can be directly discarded. However, in the particle swarm search process, the role of the pbest particle position is to guide the evolution of the population. During the alignment search, there may be an alternative that performs very poorly in one aspect but extremely well in other aspects. In such cases, this scheme also has the value of guiding the population evolution, especially when the objective function with poor performance has a smaller weight. To solve this problem, the following ranking equation can be adopted as Eq. (9.29).

$$R(a) = \sum_{j=1}^{m} w_j \cdot T^{(j)}(a, \mathbf{\Phi}_i) \tag{9.29}$$

where $R(a)$ is the ranking value of alternative a; m is the number of evaluation indicators; w_j is the weight value of the j-th indicator specified by designers, $T^{(j)}(a, \mathbf{\Phi}_i)$ is the total number of times that alternative a is better than other alternatives in the non-dominated solution set $\mathbf{\Phi}_i$ in terms of the j-th indicator. This ranking equation can avoid the situation where the ranking value of an alternative is 0 due to the poor performance of a certain objective function, thus preventing the loss of alternatives with high value.

The decision-making process for pbest particle positions is as follows:

Input: non-dominated solution set $\mathbf{\Phi}_i$ of the i-th particle individual
Output: the best alternative of the i-th particle individual

$N_p \leftarrow$ the maximum number of non-dominated solution set $\mathbf{\Phi}_i$
$N_o \leftarrow$ the number of objective fuctions
for $a \leftarrow 0$ **to** N_p, **do**
  for $j \leftarrow 0$ **to** N_o, **do**
    for $b \leftarrow 0$ **to** N_p, **do**
      if $f^{(j)}(a) \geq f^{(j)}(b)$ **then**
        $t^{(j)}(a, b) = 0$
      if $f^{(j)}(a) < f^{(j)}(b)$ **then**
        $t^{(j)}(a, b) = 1$
      end if
    end for

$$T^{(j)}(a, \mathbf{\Phi}_i) = \sum_{\forall b \in \Phi_i \wedge a \neq b} \frac{t^{(j)}(a, b)}{(|\mathbf{\Phi}_i| - 1)}$$

  end for

$$R(a) = \sum_{j=1}^{N_o} w_j \cdot T^{(j)}(a, \mathbf{\Phi}_i)$$

end for
R $\leftarrow$ sort[a, $R(a)$]//sort in descending order according to R(a)
$S \leftarrow$ the top-ranked alternative
return S

The position of the gbest particle is mainly used to guide the evolutionary direction of all particles in the population, and must improve the quality of the alternatives' quality and distribution uniformity on the non-dominated frontier. To meet this requirement, a gbest particle position decision method is introduced below, which is driven by the synergy of tournament and crowding distance.

First, a global non-dominated solution storage archive is constructed to store all non-dominated solutions searched by all particles in the population up to the current generation, thereby constructing a global non-dominated solution set. The maximum number of solutions that can be stored in the global non-dominated solution set is N_g. Then, the distance of each solution in the global non-dominated solution set from other solutions in the multi-dimensional solution space is computed with Eq. (9.30), and a crowding distance function (Eq. 9.31) is formulated to compute the crowding distance of each solution. The solutions in the global non-dominated solution set are sorted according to their crowding distances, with solutions having lower crowding distances ranking higher, resulting in sequence R1. When the number of solutions exceeds N_g, the solutions ranked after N_g are deleted.

Then, the tournament algorithm is used to sort the solutions in the global non-dominated solution set, obtaining sequence R2. Finally, based on the roulette wheel selection idea [7], a probability decision equation (Eq. 9.32) that comprehensively considers R1 and R2 is formulated. The computed probability value is taken as the fitness of the gbest particle, where a solution with a higher probability value dominates a solution with a lower one. The solutions are sorted in descending order of their probability values according to the Pareto dominance relation. Finally, the top-ranked solution is selected as the global optimized solution:

$$D(a,b)=\sqrt{\sum_{j=1}^{m}\left(\frac{f^{(j)}(a)-f^{(j)}(b)}{f^{(j)}_{\max}(\boldsymbol{\Phi})-f^{(j)}_{\min}(\boldsymbol{\Phi})}\right)^{2}} \tag{9.30}$$

where, $D(a,b)$ is the distance between solution a and solution b in the global non-dominated solution set within the multi-dimensional solution space; m is the number of objective functions; $f^{(j)}(a)$ is the j-th objective function value of solution a in the global non-dominated solution set; $f^{(j)}(b)$ is the j-th objective function value of solution b in the global non-dominated solution set; $f_{\max}^{(j)}(\mathbf{\Phi})$ is the maximum value of the j-th objective function in the global non-dominated solution set; $f_{\min}^{(j)}(\mathbf{\Phi})$ is the minimum value of the j-th objective function in the global non-dominated solution set.

$$\rho_a = \frac{1}{\left((1/k)\sum_{r=1}^{k} D_a^{(r)}\right)} \tag{9.31}$$

where ρ_a is the crowding distance value of solution a in the global non-dominated solution set; k is a constant representing the top k solutions closest to solution a in the global non-dominated solution set; $D_a^{(r)}$ is the distance between solution a and the r-th closest solution to it in the global non-dominated solution set. P_a is the probability decision value of solution a in the global non-dominated solution set, whose computation refers to Eq. (9.32).

$$P_a = \frac{2(|\mathbf{\Phi}| + 1) - R1_a - R2_a}{|\mathbf{\Phi}|(|\mathbf{\Phi}| + 1)} \tag{9.32}$$

where $R1_a$ is the position of solution a in sequence R1 after sorting by crowding distance; $R2_a$ is the position of solution a in sequence R2 after sorting by the tournament algorithm; $|\mathbf{\Phi}|$ is the number of global non-dominated solutions.

In sum, the gbest particle decision-making process is as follows:

Input: the global non-dominated solution set $\mathbf{\Phi}$
Output: global optimized solution

$N_g \leftarrow$ the maximum number of solutions in the global non-dominated solution set $\mathbf{\Phi}$
$N_o \leftarrow$ the number of objective functions
$f_{\max}^{(j)}(\mathbf{\Phi}) \leftarrow$ the maximum value of the j-th objective function of all solutions in the global non-dominated solution set $\mathbf{\Phi}$
$f_{\min}^{(j)}(\mathbf{\Phi}) \leftarrow$ the minimum value of the j-th objective function of all solutions in the global non-dominated solution set $\mathbf{\Phi}$
for $a \leftarrow 0$ **to** N_g, **do**
  for $b \leftarrow 0$ **to** N_p, **do**
    for $j \leftarrow 0$ **to** N_o, **do**
      $d^{(j)}(a,b) = \frac{f^{(j)}(a) - f^{(j)}(b)}{f_{\max}^{(j)}(\mathbf{\Phi}) - f_{\min}^{(j)}(\mathbf{\Phi})}$
  end for
  $D(a,b) = \sqrt{\sum_{j=1}^{N_o} d^{(j)}(a,b)^2}$
 end for

—cont'd

$R \leftarrow \text{sort}[a, D(a, b)]$ //arrange in ascending order according to $D(a, b)$

$\rho_a = \frac{1}{\left((1/k)\sum_{r=1}^{k} D_a^{(r)}\right)}$ //Compute the congestion degree

end for

$R1 \leftarrow \text{sort}[a, \rho_a]$ // Arrange in ascending order according to ρ_a

for $a \leftarrow 0$ **to** N_g, **do**

  for $j \leftarrow 0$ **to** N_o, **do**

    for $b \leftarrow 0$ **to** N_g, **do**

      if $f^{(j)}(a) \geq f^{(j)}(b)$ **then**

        $t^{(j)}(a, b) = 0$

      if $f^{(j)}(a) < f^{(j)}(b)$ **then**

        $t^{(j)}(a, b) = 1$

      end if

    end for

$$T^{(j)}(a, \boldsymbol{\Phi}_i) = \sum_{\forall b \in \Phi_i \wedge a \neq b} \frac{t^{(j)}(a, b)}{(|\boldsymbol{\Phi}_i| - 1)}$$

  end for

$$R(a) = \sum_{j=1}^{N_o} w_j \cdot T^{(j)}(a, \boldsymbol{\Phi}_i)$$

end for

$R2 \leftarrow \text{sort}[a, R(a)]$ //Arrange in descending order according to $R(a)$

for $a \leftarrow 0$ **to** N_g, **do**

$P_a = \frac{2(|\boldsymbol{\Phi}| + 1) - R1_a - R2_a}{|\boldsymbol{\Phi}|(|\boldsymbol{\Phi}| + 1)}$

end for

$S \leftarrow$ solution with the max P_a value

return S

The above methods can select the pbest and gbest particle positions for each generation by considering both the quality and the uniformity of the non-dominated frontier distribution according to the designers' preferences, which can balance the conflict resolution and equilibrium among different objectives in optimizing railway alignments.

9.3.7 Case study

The algorithm introduced above has been applied to the alignment design of a railway project in the mountainous regions of western China.

9.3.7.1 Railway case profile

The railway applied in the case study is a Grade I railway that crosses a region with much rise and fall of terrain. The airline distance between start and end points is 34,032.709 m and the elevation difference between two points is 1041.971 m. Therefore, the natural terrain gradient can be computed as Eq. (9.33).

$$G_T = 1041.971/34032.709 = 30.62‰ \tag{9.33}$$

Since $G_T = 30.62‰$ exceeds the maximum allowed design gradient $G_{max} = 20‰$, it is a severe gradient alignment region. Therefore, the alignment must be circuitous to allow the necessary elevation changes using acceptable gradients. Bridges and tunnels become the major constructions, accounting for over 90% of the overall length of the alignment.

For this case, the essential data, including the topography, satellite images, land cover, coordinates of two endpoints, forbidden areas, unit costs, code provision parameters and railway major technical standards of this study are all provided by the China Railway Eryuan Engineering Group Co. Ltd. Aerial survey and satellite remote sensing imagery analysis are used to collect topographic information, satellite images and land cover. These key geospatial data are formatted in standard file types for subsequent analysis, as illustrated in Fig. 9.24. Other data are provided by alignment designers directly. The railway major technical parameters of this project are shown in Table 9.5. The main constraints to be satisfied are listed in Table 9.6. The global unit costs are shown in Table 9.7. Different unit costs can be set in special regions.

9.3.7.2 Optimization results and analyses

The program runs on the HP Z600 workstation (Intel Xeon E5506 2.13 G processor, 4 GB RAM, 500 GB Hard disk). We first build a CGIM covering the whole study area with a resolution of 90m for the DT search. After 37 min of bidirectional DT scanning, 63 promising paths are generated. The best path with the lowest comprehensive cost is selected for generating a corridor and butterfly-shaped areas. Then we build a CGIM which just covers the promising corridor with a resolution of 2.5 m, to optimize the alignment with the stepwise & hybrid PSO.

(1) Comparison with manual solution and existing methods

To fully illustrate the effectiveness of our algorithm, we expand our study by comparing the stepwise & hybrid PSO-generated best alignment with the best solutions obtained by (1) experienced alignment designers in the China Railway Eryuan Engineering Group CO. LTD, (2) a genetic algorithm and (3) a non-stepwise PSO. In this experiment, we use the parameters shown in Table 9.8. The entire elapsed time in the PSO process is 98 min. It satisfies the convergence criteria at the 899th iteration.

Figure 9.24 (a) Topography data in ASC format; (b) satellite images in TIF format; and (c) land cover data in ASC format.

Table 9.5 Railway major technical parameters.

Major technical standards	**Recommendation**
Railway grade	Grade I
Design speed	160 km/h
Main line number of tracks	Single-track
Vertical circular curve radius	10,000 m
Type of traction	Electric traction
Type of locomotive	SS4 locomotive
Locomotive tonnage	184 tons
Design train tonnage	3500 tons
Effective length of the receiving-departure track	650 m
Train operation control mode	Auto-control
Traffic command mode	Centralized traffic control
Subgrade-bridge threshold height	10 m
Subgrade-tunnel threshold depth	−15 m

Table 9.6 Main constraints.

Geometric constraints	Minimum radius of curve	1600 m
	Maximum gradient	20‰
	Minimum length of tangent between horizontal curves	130 m
	Minimum length of horizontal circular curve	130 m
	Minimum length of slope section	350 m
	Absolute difference between adjacent gradients	15‰
Construction constraints	Maximum bridge height	100 m
	Maximum tunnel length	19,000 m
	Terminals are not overlapping with tunnels	
Location constraints	Bypass 1 forbidden zone and 2 environmentally sensitive zones	
	If crossing river, clearance $\geq$6 m	
	Elevation tendency: Keep increasing alignment elevation	

The horizontal alignment results of the four methods are compared in Fig. 9.25. For the vertical alignment, the design results of each method are separately illustrated in Figs. 9.26–9.29.

The final best alignment satisfies all the constraints, as verified in Table 9.9. The resulting characteristics are shown in Table 9.10.

Compared with the manually obtained alignment, the best stepwise & hybrid PSO-generated alignment is shorter by 3701.12 m. Its

Table 9.7 Unit costs.

Item	Cost	Item	Cost
Rail track (¥/m)	4000	Filling earthwork (¥/m^3)	20
Right-of-way (¥/m^2)	75.1	Cutting earthwork (¥/m^3)	24
($L \geq 9500$ m) tunnel (¥/m)	76,000	($H < 50$) Bridge (¥/m)	35,800
($L \geq 7000$ m) tunnel (¥/m)	72,000	($H \geq 50$, $L < 500$ m) Bridge (¥/m)	39,000
($L \geq 4500$ m) tunnel (¥/m)	68,000	($H \geq 50$, $L \geq 500$ m) Bridge (¥/m)	43,200
($L \geq 2000$ m) tunnel (¥/m)	51,000	One abutment (¥)	200,000
($L < 2000$ m) tunnel (¥/m)	40,000	Penalty cost for environmentally sensitive zone (¥/m^2)	80,000
One tunnel portal (¥)	180,000	Capital recovery factor Δ	0.065

Table 9.8 PSO experimental parameters.

Parameters	Value
Swarm size	100
Maximum velocity	500 m/iteration
Accelerator coefficients	2:2
Total number of iterations	1000

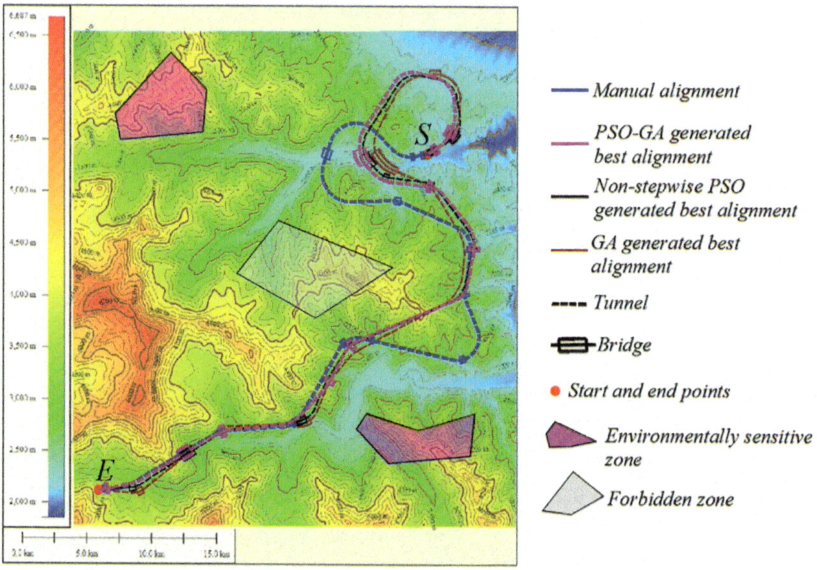

Figure 9.25 Comparison of PSO-generated horizontal alignment with manual work and existing methods.

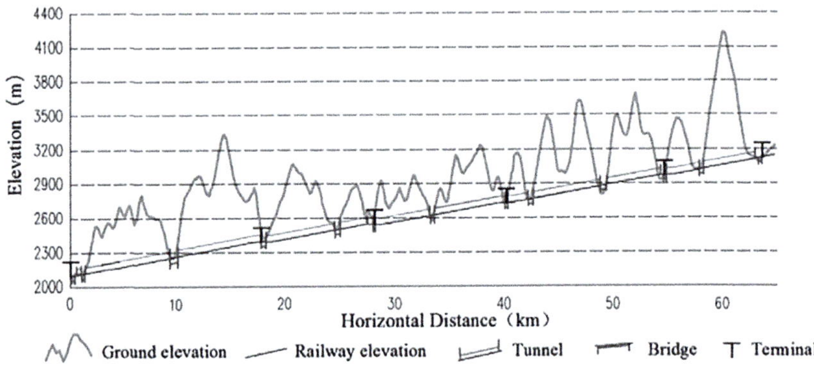

Figure 9.26 Manually designed vertical alignment.

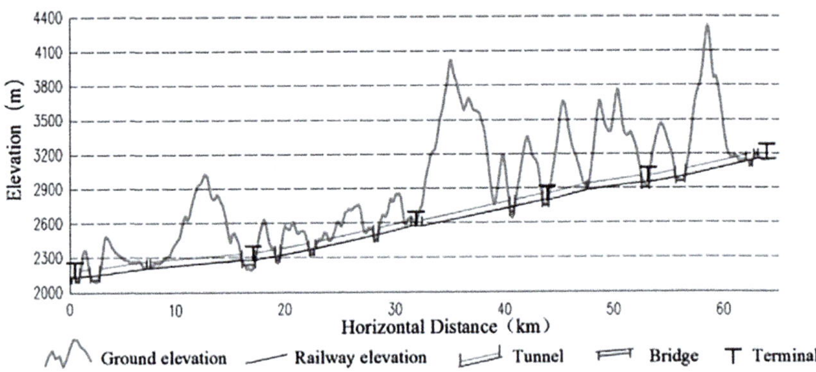

Figure 9.27 Best PSO-generated vertical alignment.

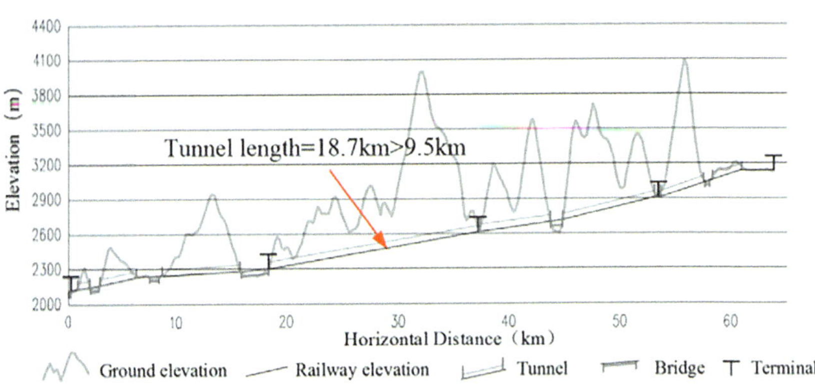

Figure 9.28 GA-generated best vertical alignments.

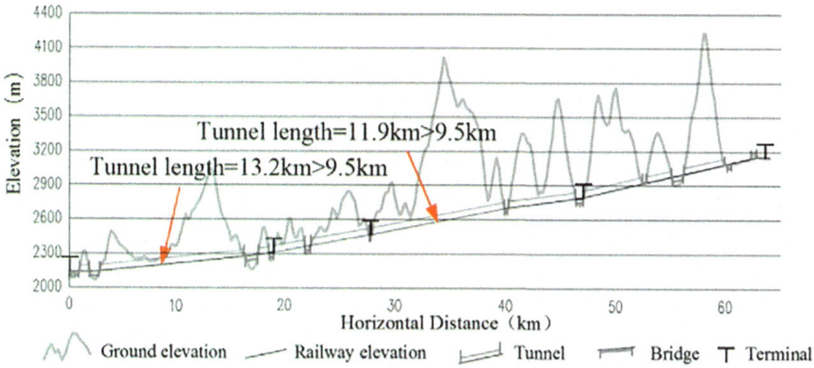

Figure 9.29 Non-stepwise PSO-generated best vertical alignments.

Table 9.9 Check of constraints.

Constraints	**Actual value**	**Allowed value**	**Check**
Minimum circular radius	1600 m	≥1600 m	√
Minimum tangent length between horizontal curves	568.7 m	≥130 m	√
Minimum length of horizontal circular curve	650.7 m	≥130 m	√
Maximum gradient	20‰	≤20‰	√
Minimum length of slope section	600 m	≥350 m	√
Absolute difference between adjacent gradients	9‰	≤15‰	√
Maximum bridge height	62.7 m	≤100 m	√
Maximum tunnel length	8555.9 m	≤19,000 m	√
Terminals overlapping with tunnels	NO	NO	√
Forbidden zones	Bypassed	Bypassed	√
If cross the river, clearance ≥6 m	No crossing	≥6 m	√
Elevation tendency	Increase	Increase	√

length-related costs are reduced. The lengths of tunnels are also reduced by a total of 2949.19 m. In particular, the length of the long tunnels (7000 m ≥ L ≥ 4500 m) is greatly reduced by 17,766.69 m and the number of long tunnels (7000 m ≥ L ≥ 4500 m) is reduced by 3. In addition, the embankment and excavated volume and right-of-way area are also reduced. Thus, the annual comprehensive cost is 10.33% below that of the manual solution.

Comparing the GA and non-stepwise PSO with the stepwise & hybrid PSO shows that, in spite of the similarity of the horizontal

Table 9.10 Our PSO-generated best alignment compared with manual solution and existing methods.

Item	Manual work	Non-stepwise PSO-generated best alignment	GA-generated Best alignment	Stepwise and hybrid PSO-generated best alignment
Length/m	66,400	62,460.65	62,271.11	62,698.88
Right-of-way/m^2	233,173.34	235,155.33	305,426.67	208,100.36
Embankment volume/m^3	300,499	431,833	2,531,550	127,702
Excavated volume/m^3	4,091,418	183,625	803,246	632,086
($H \geq 50$, $L < 500$ m) Bridge/Number-m	–	–	–	1–423.98
($H \geq 50$, $L \geq 500$ m) Bridge/Number-m	–	3–3015.45	2–3516.64	1–1177.12
($H < 50$ m) Bridge/Number-m	13–4998.78	8–3035.28	4–2282.68	10–4560.45
($L \geq 9500$ m) Tunnel/Number-m	–	2–25193.77	1–18700.03	–
($L \geq 7000$ m) Tunnel/Number-m	2–15385.30	–	2–15310.46	2–16905.21
($L \geq 4500$ m) Tunnel/Number-m	6–33352.69	3–16662.15	1–6286.39	3–15586.00
($L \geq 2000$ m) Tunnel/Number-m	2–5306.81	3–8831.32	3–9791.41	6–18676.36
($L < 2000$ m) Tunnel/Number-m	5–3134.72	3–2380.90	1–1064.63	3–3062.76
Total construction costs/million ¥	4349.43	4121.23	4073.83	3876.67
Total construction costs saving/million ¥	–	228.2	275.6	472.76
Annual construction costs ($\Delta = 0.065$)/ million ¥	282.71	267.88	264.80	251.98
Annual operating cost/million ¥	7.24	6.59	6.39	8.02
Annual comprehensive cost/million ¥	289.95	274.47	271.19	260.00
Annual comprehensive cost saving/million ¥	–	15.48	18.76	29.95
Ratio of annual comprehensive cost saving	–	5.34%	6.47%	10.33%

convergence, they generate a comprehensive cost lower than the manual one. However, they are still more expensive than the solution generated by stepwise & hybrid PSO: (1) the GA restricts all the PIs to cutting planes; however, allowing PIs only on cutting planes requires long tunnels and high bridges with high construction cost (e.g., one tunnel exceeding 9.5 km; two bridges whose $H \geq 50$ and $L \geq 500$m); (2) the non-stepwise PSO does not sufficiently account for the undulating terrain line, structures, and code provisions. Consequently, the distribution of VPIs may result in long tunnels and high bridges (e.g., two tunnels exceeding 9.5 km; three bridges whose $H \geq 50$ and $L \geq 500$m).

These results confirm that our proposed approach can improve upon the best alternatives developed by experienced designers and can offer better solutions compared to GA and non-stepwise PSO.

(2) Operators analysis

- Convergence analysis: In order to analyze the influences of GA operators on the optimization ability and convergence performance of the algorithm, we compare the convergence curves of hybrid PSO and basic PSO (see Fig. 9.30 for the convergence trend comparison). We find that the GA operators (1) accelerate the convergence at beginning of the run significantly; (2) provide lasting diversity and momentum at the end of the search to avoid premature convergence.

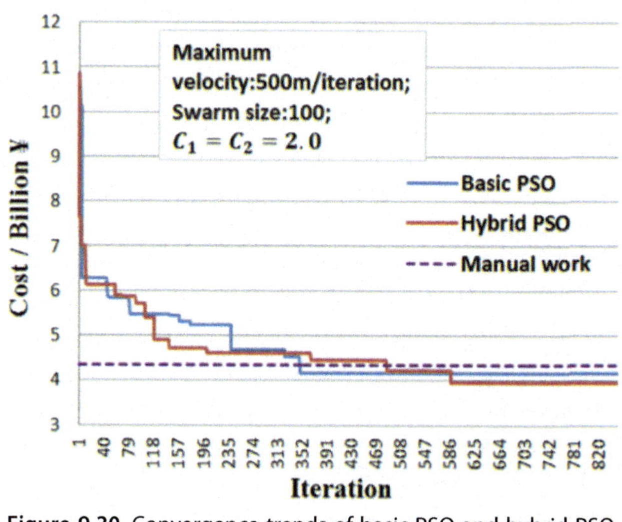

Figure 9.30 Convergence trends of basic PSO and hybrid PSO.

- **Sensitivity analysis** Considering that very small or large values of maximum velocity may decrease the algorithm efficiency, we suggest a maximum velocity between 350 and 1000 m/iteration in this experiment. With other parameters fixed, we run the program 10 times independently under four groups of different maximum velocity values (350, 500, 750 and 1000) for basic PSO and hybrid PSO. The 10 computed results of each group are then sorted in an ascending order. Due to the problem complexity and algorithm randomness, the final comprehensive cost obtained in each run may be different. However, the 10 computed results of each group should exhibit a convergence behavior that fluctuates within a small range. Considering the complexity of mountain alignment optimization, we assess that the fluctuation (explained in Table 9.11) below 5% is acceptable.

It can be clearly seen from Figs. 9.30 and 9.31 and Table 9.11 that for pure PSO: (1) the threshold of maximum velocity value that causes divergence is in the (500, 750) interval; (2) the maximum velocity values of 350 and 500 yield the best results on convergence but do not reach near-optimal results; however, for hybrid PSO, after integrating GA operators, all four groups of experiments converge and

Table 9.11 Objective function values computed with different values of maximum velocity.

	Maximum velocity /(meters/iteration)	350	500	750	1000
Pure PSO	Minimum cost/ (million ¥)	403,046.53	404,964.51	386,881.19	388,369.41
	Maximum cost/ (million ¥)	421,724.05	419,500.90	438,399.99	438,022.29
	Average cost/ (million ¥)	411,740.59	414,362.15	412,308.96	411,135.53
	Fluctuationa	4.63%	3.59%	13.32%	12.78%
Hybrid PSO	Minimum cost/ (million ¥)	384,265.24	382,157.49	383,209.70	383,336.70
	Maximum cost/ (million ¥)	402,842.77	397,210.52	399,973.92	401,544.61
	Average cost/ (million ¥)	391,613.37	389,607.46	391,139.94	390,110.55
	Fluctuation[a]	4.83%	3.94%	4.37%	4.75%

[a]Fluctuation = (Maximum cost - Minimum cost)/Minimum cost.

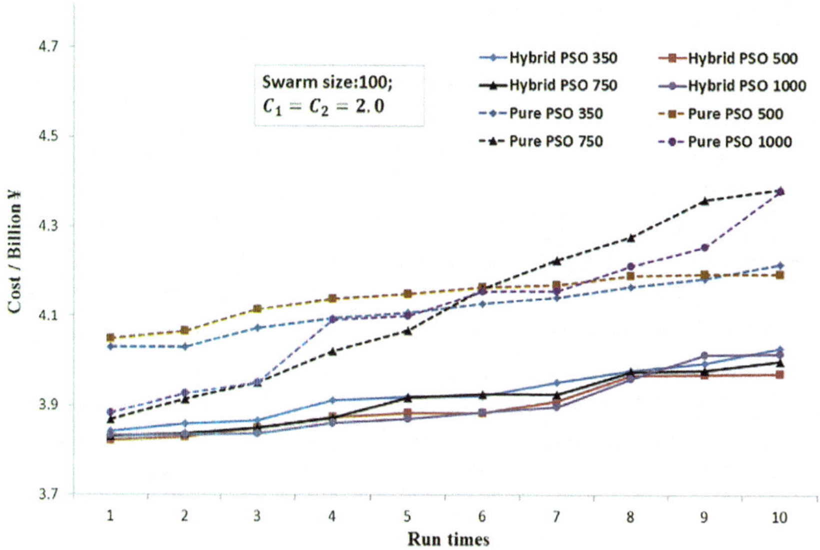

Figure 9.31 Sensitivity analysis of maximum velocity.

the comprehensive costs are reduced compared to the pure PSO. In other words, our hybrid mechanism has successfully avoided getting trapped in local optima and addressed the parameter sensitivity while the convergence and practicability have been greatly enhanced.

(3) Inertia weight analysis

With other parameters fixed, we run the program 10 times independently with two different inertia weight update formulas. It can be observed from Table 9.12 that the PSO with a nonlinear third-order formulation for inertia weight updating has reached lower final comprehensive costs than the linearly decreasing method. It shows that the nonlinear third-order updating strategy can search the solution space near the optimum more thoroughly than the linear descent method and is better suited for our PSO algorithm.

Table 9.12 Objective function value computed with different inertia weight update formulas.

Result	Linear rule	Nonlinear rule
Minimum cost/(million ¥)	396,256.58	384,063.12
Maximum cost/(million ¥)	411,612.31	398,613.90
Average cost/(million ¥)	45,371.47	389,251.94
Fluctuation	3.88%	3.79%

(4) Acceleration coefficients analysis
In this case, with other parameters fixed, we run the program 10 times independently under five groups of different values of accelerator coefficients (C_1:C_2). Similarly to the above section, the results are shown as follows. Fig. 9.32 and Table 9.13 show that C_1:C_2 = 2:2 and 2:1 can guarantee convergence in this case. However, C_1:C_2 = 2:2 leads to the lowest objective function value for our problem.

(5) Swarm size analysis
With other parameters fixed, we compare the convergence history with different swarm size values (50, 75, and 100). The computing times are 86 min (for size of 50), 92 min (for size of 75) and 95 min (for size of 100), respectively. Fig. 9.33 shows that the swarm size of 100 particles yields here the lowest comprehensive cost as well as faster convergence. Moreover, its computing time is acceptable.

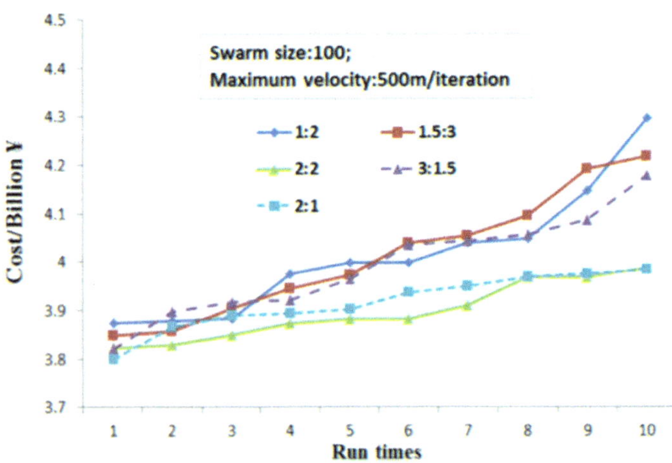

Figure 9.32 Sensitivity analysis of accelerator coefficients.

Table 9.13 Average objective function value computed with different accelerator coefficients.

C_1:C_2	1:2	1.5:3	2:2	3:1.5	2:1
Minimum value	387,447.54	384,946.76	382,234.68	382,127.80	379,909.47
Maximum value	429,678.41	421,692.94	398,863.43	417,819.55	398,364.39
Average value	401,379.37	401,244.85	389,780.47	399,196.53	391,620.11
Fluctuation	10.90%	9.55%	4.35%	9.34%	4.86%

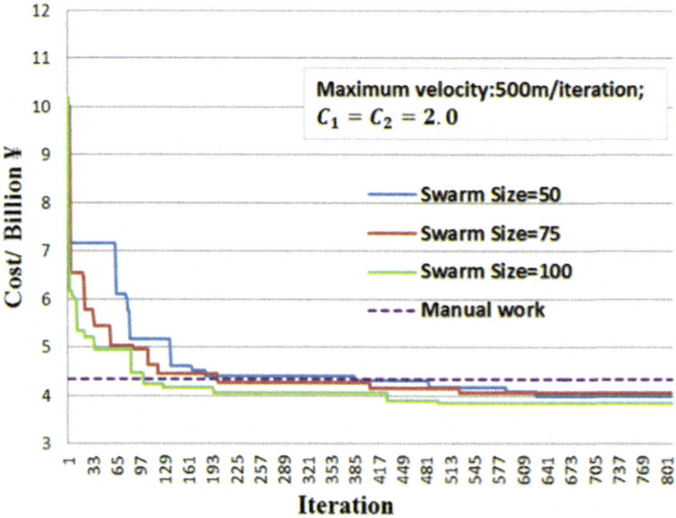

Figure 9.33 Convergence history with different swarm sizes.

References

[1] Z. Zhan, Gradient projection method and its application in railway vertical alignment design, J. Changsha Railway Univ. 4 (1979) 83–88, https://doi.org/10.19713/j.cnki.43-1423/u.1979.04.008.

[2] J.C. Jong, Optimizing Highway Alignment with Genetic Algorithms, University of Maryland, College Park, 1998.

[3] Y. Shafahi, M. Bagherian, A customized particle swarm method to solve highway alignment optimization problem, Comput. Aided Civ. Infrastruct. Eng. 28 (1) (2013) 52–67.

[4] W. Li, H. Pu, P. Schonfeld, et al., Mountain railway alignment optimization with bidirectional distance transform and genetic algorithm, Comput. Aided Civ. Infrastruct. Eng. 32 (8) (2017) 691–709.

[5] J.C. Jong, P. Schonfeld, An evolutionary model for simultaneously optimizing three-dimensional highway alignments, Transp. Res. Part B Methodol. 37 (2) (2003) 107–128.

[6] J. Kennedy, R. Eberhart, Particle swarm optimization, Proc. ICNN'95 - Int. Conf. Neural Networks 4 (1995) 1942–1948.

[7] A. Lipowski, D. Lipowska, Roulette-wheel selection via stochastic acceptance, Phys. Stat. Mech. Appl. 391 (6) (2012) 2193–2196.

CHAPTER TEN

Models and methods for the collaborative optimization of infrastructure structures and railway alignments

Abbreviations and corresponding full texts

API Alignment Point of Intersection
CGIM Comprehensive Geographic Information Model
D&C Divide & Conquer
DEM Digital Elevation Model
DT Distance Transform
HAPI Horizontal Alignment Point of Intersection
HPI Horizontal Point of Intersection
LACP Large-scale Auxiliary Construction Project
MWM Moving Window Method
PI Point of Intersection
PSO Particle Swarm Optimization
SSSP Single-Source Shortest-Path
TVAC Time-Varying Acceleration Coefficients
VAPI Vertical Alignment Point of Intersection
VPI Vertical Point of Intersection

A well-conceived alignment not only fulfills the functional requirements of the railway but also harmonizes with the layout of critical structures such as stations, bridges, tunnels, large-scale auxiliary construction projects, and transport network. Therefore, ideal alignment design goes beyond the geometric configuration in horizontal and vertical alignments (Easa, 2008); it must also integrate infrastructural factors to achieve system-wide, collaborative optimization of infrastructure components.

Based on this perspective, this chapter focuses on three key components of collaborative railway design: station-alignment coordination, Large-scale Auxiliary Construction Projects (LACP)-alignment coordination, and grade-separated crossing optimization. Specifically, the optimization models and methods for these three kinds of collaborative optimization are presented in the following sections.

Intelligent Railway Alignment Design
ISBN: 978-0-443-45923-8
https://doi.org/10.1016/B978-0-443-45923-8.00008-X

10.1 Concurrent optimization of mountain railway alignments and station locations

Determining a method for optimizing the intricate interactions among mountain railway alignment and station locations is a crucial problem. Overcoming the huge elevation difference between the start and end points in complex mountainous areas requires a circuitous horizontal alignment to provide a sufficient length, and the vertical design elevation should be continuously increased or decreased based on the allowed maximum gradient to reach the target elevation. However, for safety, the entire length of a station should be located on a tangent section with a nearly flat slope and outside the tunnel. Because of the flat slope constraint, the maximum design gradient cannot be used within the station. Instead, the alignment must be fairly circuitous to reach the target elevation. All these factors increase the difficulty of finding sufficiently long and flat tangents for station locations. Moreover, numerous bridges and tunnels are required in complex mountains areas, stations are generally avoided within tunnels (partial locations within tunnels are allowed in complex mountainous areas when unavoidable), and station sites should also avoid forbidden areas (e.g., nature protection areas and geological hazards areas). Furthermore, the distances between each pair of adjacent stations are also limited within a range to satisfy the required traffic capacity of the alignment and guarantee the safety of operations.

Considering these concerns, this section mainly focuses on developing a concurrent optimization method for railway alignment and station locations in complex mountainous areas. The optimization model and method are presented below.

10.1.1 Optimization model

Here, the concurrent optimization problem of railway alignment and station locations is to find the most economical railway alignment and station locations between two specified points based on topography, soil conditions, and environmental impacts, while satisfying multiple constraints, such as railway design specifications, forbidden zones, and coupling constraints among railway alignment and station locations. Because the railway alignment and station locations can be represented by a set of points of intersection (PIs) with curve configurations and structures locations, they can be used as the decision variables. The objective function usually includes railway construction costs and operation costs. The detailed formulation of the optimization model as well as the input data and output data of the model are introduced below.

(1) Input data

The input data for the optimization model include geographic information parameters, constraint parameters, and structure parameters.

① Geographic information parameters

Geographic information parameters are essential spatial data for railway location design and include the topography, forbidden areas, soil conditions, ground objects, and unit costs. In this study, we use a comprehensive geographic information model (CGIM) to store and manage these data. The data structure of the CGIM is a regular rectangular lattice, which is similar to a digital elevation model (DEM), as presented in Section 2. All the required geographic information is stored in the cells of the lattice. These geographic information parameters can be obtained through field surveys and economic investigations of the study areas.

② Constraint parameters

Multiple constraints affect the railway location design, such as the railway design specifications, the forbidden zones, the maximum bridge height, the maximum tunnel length, and the coupling constraints among railway alignment and station locations. According to their characteristics, these constraints can be classified into three categories: geometric constraints, location constraints, and structure constraints.

i. Geometric constraints: These include the minimum radius and length of the horizontal circular curves, the minimum length of tangent between two adjacent horizontal curves, the maximum gradient of each section in vertical alignment, the minimum and maximum station spacing between adjacent stations, and the horizontal and vertical alignment constraints within the station.

ii. Location constraints: These are mainly specified to avoid forbidden zones;

iii. Structure constraints: These include constraints on maximum bridge height, maximum tunnel length, and that stations should avoid overlapping with tunnels and bridges.

The constraint parameters are determined according to the Code for Design of Railway Line (2017) [1], the construction technology, and the required traffic capacity.

③ Structure parameters

Structure parameters are used to standardize the construction of stations, bridges, and tunnels in railway location design, including the boundary depth to set a tunnel, the boundary height to set a bridge, and the length and width of the station. These parameters are determined according to the conditions of the study area according to the Code for Design of Railway Station and Terminal [2], geographic environment, and designers' experience.

(2) Decision variables

The railway location can be defined by a set of PIs and structure locations. In this study, we set them as the decision variables and compute them via the search process using the algorithm proposed in Section 3. PIs include both horizontal points of intersection (HPIs) and vertical points of intersection (VPIs). The decision variables are specified with the following data:

① The HPIs data include the coordinates X_i and Y_i, the horizontal curve radius R_i, and the transition curve length L_i of each horizontal intersection point. Therefore, the HPIs data of the railway location are four column vectors: $\boldsymbol{X}$, $\boldsymbol{Y}$, $\boldsymbol{R}$, and $\boldsymbol{L}$.

② The VPIs data include the mileage K_i, the design elevation H_i, and the vertical radius R_{Vi} of each vertical intersection point. Therefore, the VPIs data of the railway location are three column vectors: $\boldsymbol{K}$, $\boldsymbol{H}$, and $\boldsymbol{R_V}$.

③ The data of structure locations include the start and end mileages of every bridge KB_{Si} and KB_{Ei}, the start and end mileages of every tunnel KT_{Si} and KT_{Ei}, and the coordinates of every station center X_{Ci} and Y_{Ci}. Therefore, the data of the structure locations of the railway are six column vectors: $\boldsymbol{KB_S}$, $\boldsymbol{KB_E}$, $\boldsymbol{KT_S}$, $\boldsymbol{KT_E}$, $\boldsymbol{X_C}$, and $\boldsymbol{Y_C}$.

(3) Objective function

The main factor affecting the evaluation of a railway alternative is the railway costs. In this study, we establish an objective function that includes railway alignment-related costs, railway station-related costs, and operational costs:

$$F(\boldsymbol{X}, \boldsymbol{Y}, \boldsymbol{R}, \boldsymbol{L}, \boldsymbol{K}, \boldsymbol{H}, \boldsymbol{R_V}, \boldsymbol{KB_S}, \boldsymbol{KB_E}, \boldsymbol{KT_S}, \boldsymbol{KT_E}, \boldsymbol{X_C}, \boldsymbol{Y_C}) = \Delta\left(C_{\text{alignment}} + C_{\text{station}}\right) + C_{\text{operation}} \quad (10.1)$$

where $\Delta =$ capital recovery factor, which converts the total construction costs to annual construction costs, as shown in Eq. (10.2):

$$\Delta = i\,(1+i)^n \,/\, [(1+i)^n - 1] \tag{10.2}$$

where i is the interest rate (%per period) and n is the number of compounding periods (economic life). The components of Eq. (10.1) include:

$C_{\text{alignment}} =$ railway alignment-related costs (¥), as shown in Eq. (10.3):

$$C_{\text{alignment}} = C_E + C_B + C_T + C_L + C_{RL} + C_{IL} \tag{10.3}$$

where $C_E =$ earthwork cost (¥), $C_B =$ bridge construction cost (¥), $C_T =$ tunnel construction cost (¥), $C_L =$ length dependent construction cost (¥), $C_{RL} =$ right-of-way cost of the alignment (¥), $C_{IL} =$ environmental impact cost of the alignment (¥).

$C_{\text{station}} =$ railway station-related costs (¥), as shown in Eq. (10.4):

$$C_{\text{station}} = C_S + C_{RS} + C_{IS} + C_{FS} \tag{10.4}$$

where $C_S =$ station construction cost (¥), $C_{RS} =$ right-of-way cost of the station (¥), $C_{IS} =$ environmental impact cost of the station (¥), $C_{FS} =$ station facility cost (¥).

$C_{\text{operation}} =$ operational costs (¥), as shown in Eq. (10.5):

$$C_{\text{operation}} = C_G + C_A + C_M \tag{10.5}$$

where $C_G =$ operating cost sensitive to gradient (¥), $C_A =$ operating cost sensitive to the horizontal turning angle (¥), $C_M =$ operating cost sensitive to length (¥).

In this section, we formulate the railway station-related costs functions, which include the station construction cost (C_S), the right-of-way cost of the station (C_{RS}), the environmental impact cost of the station (C_{IS}), and the station facility cost (C_{FS}). The formulations of the railway alignment-related cost functions (C_E, C_B, C_T, C_L, C_{RL}, C_{IL}) and the operational cost functions (C_G, C_A, C_M) included in Eq. (10.1) can be found in Chapter 5 and are not duplicated here.

① Station construction cost (C_S)

The station construction cost consists of three parts: (1) construction cost when the station is set on a subgrade; (2) construction cost when the station is set in a tunnel; and (3) construction cost when the station is set on a bridge.

i. Construction cost when the station is set on a subgrade (C_{SE}), as shown in Eq. (10.6):

$$C_{SE} = \sum_{i=1}^{m} \left[(H_{S0} - H_{Gi}) \times w^2 \times C_f \right] + \sum_{j=1}^{n} \left[\left(H_{Gj} - H_{S0} \right) \times w^2 \times C_c \right] \tag{10.6}$$

where m = number of fill cells within the station (integer), n = number of cut cells within the station (integer), C_f, C_c = unit fill/cut cost (¥/m^3), H_{Gi} = ground elevation of the ith fill cell within the station (m), H_{Gj} = ground elevation of the jth cut cell within the station (m), H_{S0} = design elevation of the station (m), w = cell width (m).

ii. Construction cost when the station is set in a tunnel (C_{ST}), as shown in Eq. (10.7):

$$C_{ST} = L_{S\&Ti} \times C_{Ti} \tag{10.7}$$

where $L_{S\&Ti}$ = overlap length between the station and the ith tunnel (m), C_{Ti} = unit length cost of the ith tunnel (¥/m).

iii. Construction cost when the station is set on a bridge (C_{SB}), as shown in Eq. (10.8):

$$C_{SB} = L_{S\&Bi} \times C_{Bi} \tag{10.8}$$

where $L_{S\&Bi}$ = overlap length between the station and the ith bridge (m),

C_{Bi} = unit length cost of the ith bridge (¥/m).

The construction cost will significantly increase if the station is set in a tunnel or on a bridge. Thus, we introduce two penalty factors ($Penl_b$ and $Penl_t$) to prevent the station from being set in a tunnel or on a bridge. $Penl_b$ is the penalty factor of setting a station on a bridge, and $Penl_t$ is the penalty factor of setting a station in a tunnel. Usually, $Penl_t$ is larger than $Penl_b$ because there are more problems with setting a station in a tunnel than on a bridge, such as ventilation problems and lighting problems, especially in the operation stage.

The sum of station construction costs C_S is Eq. (10.9):

$$C_S = C_{SE} + (1 + \gamma_t \; Penl_t) \; C_{ST} + (1 + \gamma_b Penl_b) \; C_{SB} \tag{10.9}$$

where γ_t = binary variable that is equal to 1 if the station lies in a tunnel and equal to 0 if it lies outside the tunnel,

γ_b = binary variable that is equal to 1 if the station lies on a bridge and equal to 0 if it lies outside the bridge.

② Right-of-way cost of the station (C_{RS}), as shown in Eq. (10.10):

$$C_{RS} = p \times w^2 \times C_R \tag{10.10}$$

where p = number of right-of-way cells within the station (integer), w = cell width (m), C_R = unit right-of-way cost (¥/m^2).

③ Environmental impact cost of the station (C_{IS}), as shown in Eq. (10.11):

$$C_{IS} = q \times w^2 \times C_I \tag{10.11}$$

where q = number of environmentally sensitive cells within the station (integer), w = cell width (m), C_I = unit area environmental impact cost (¥/m^2).

④ Station facility cost (C_{FS})

The station facility cost is a fixed cost and does not vary with the station location. The station facility cost varies only with how many facilities are included in the station, i.e., the numbers of fouling posts, bumper posts, signals, and tracks, which depend on the station type and the Code for Design of Railway Station and Terminal.

(4) Multiple constraints

The railway location design is affected by multiple constraints, such as the rail design specifications, forbidden zones, maximum bridge height, and maximum tunnel length. In Chapter 5, the multiple constraints were classified into geometric constraints, location constraints, and structure constraints according to their characteristics. However, coupling constraints among railway alignment and station locations were not considered. These constraints are crucial in the process of optimizing railway alignment and station locations concurrently.

In this chapter, coupling constraints among railway alignment and station locations are taken into account. Because these constraints have the same characteristics as the constraints in Chapter 5, we also classify them into three categories: geometric constraints, location constraints, and structure constraints.

① Geometric constraints

i. Station spacing constraint between adjacent stations

According to traffic capacity and operation safety demands, railway designers predetermine the maximum and minimum spacing between two adjacent stations. The station distribution should satisfy the station spacing constraint, as shown in Eq. (10.12):

$$D_{\min} \leq D_S \leq D_{\max} \tag{10.12}$$

where D_S = spacing between two adjacent stations (km), $D_{\min}$ = minimum spacing required between two adjacent stations (km), $D_{\max}$ = maximum spacing required between two adjacent stations (km).

ii. Gradient constraint within the station

The station should be set on a flat slope section to prevent trains from sliding. If there is a need to set a station on a local path between 2 cells (e.g., $p_{i,j}$ and $p_{m,n}$), then the gradient between $p_{i,j}$ and $p_{m,n}$ must satisfy Eq. (10.13) to ensure that when setting a flat slope section along the station (see Fig. 10.1), the segment linking these 2 cells still satisfies the gradient constraint:

$$i_g \leq i_{s\,\max} \tag{10.13}$$

where i_g = gradient between two linked cells (‰), as shown in Eq. (10.14):

$$i_g = \left|H_{Gi,j} - H_{Gm,n}\right| / \left[\left(x_{i,j} - x_{m,n}\right)^2 + \left(y_{i,j} - y_{m,n}\right)^2\right]^{0.5} \tag{10.14}$$

where $H_{Gi,j}$ = ground elevation of the cell located at row i, column j in the lattice (m), $H_{Gm,n}$ = ground elevation of the cell

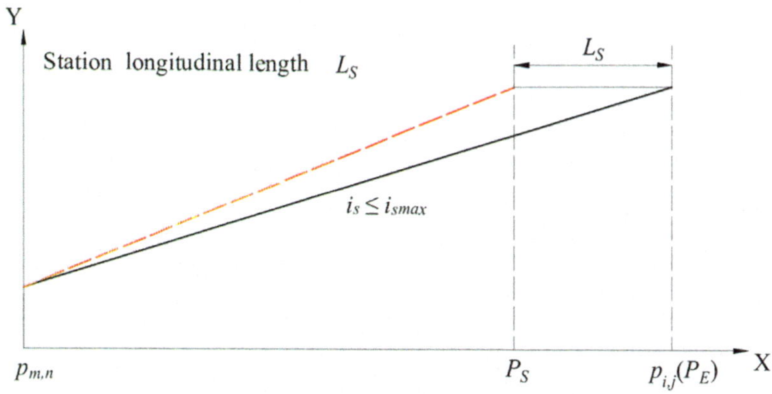

Figure 10.1 Vertical coupling constraint.

located at row m, column n in the lattice (m), $x_{i,j}$, $y_{i,j}$ = coordinates of the cell located at row i, column j in the lattice (m), $x_{m,n}$, $y_{m,n}$ = coordinates of the cell located at row m, column n in the lattice (m).

i_{smax} = maximum gradient when a station must be set on the local path (‰), as shown in Eq. (10.15):

$$i_{smax} = \left[\left[\left[\left(x_{i,j} - x_{m,n}\right)^2 + \left(y_{i,j} - y_{m,n}\right)^2\right]^{0.5} - L_S\right] \times i_{\max}\right] / \left[\left(x_{i,j} - x_{m,n}\right)^2 + \left(y_{i,j} - y_{m,n}\right)^2\right]^{0.5} \quad (10.15)$$

where L_S = station longitudinal length (m),

$i_{\max}$ = maximum gradient according to design specification (‰).

iii. Horizontal alignment constraint within the station

The station should be set on the tangent section. Various pieces of signal equipment are placed within the station, and a driver's vision is hindered if a station is set on a curve, which is a safety hazard. In addition, a curved track also increases the resistance of train starting. Assuming that P_S and P_E are the two ends of a station on the alignment, the horizontal alignment within the station should satisfy Eq. (10.16).

$$AD_P = SD_P = \left[(x_{PS} - x_{PE})^2 + \left(y_{PS} - y_{PE}\right)^2\right]^{0.5} \quad (10.16)$$

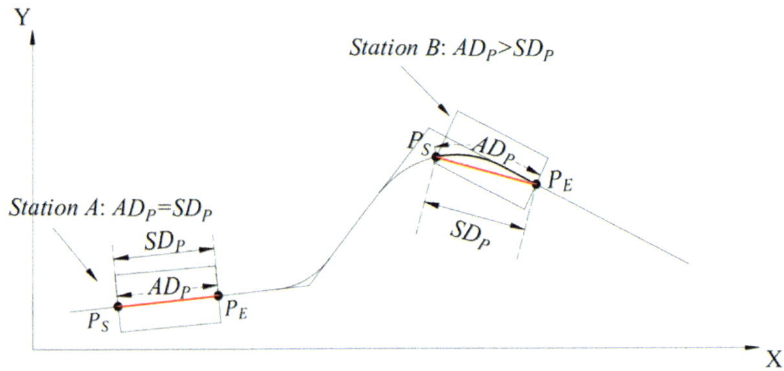

Figure 10.2 Horizontal coupling constraint.

where AD_P = actual distance from P_S to P_E along the alignment (m); x_{PS}, y_{PS} = coordinates of P_S (m); x_{PE}, y_{PE} = coordinates of P_E (m).

If $AD_P > SD_P$, then the station is set on a curve (see station B in Fig. 10.2); if $AD_P = SD_P$, then the station is set on the tangent section (see station A in Fig. 10.2).

② Location constraints

Stations should not be set in forbidden zones, such as cemeteries, historically significant zones, wetlands, geologic hazard areas, and ecological protection areas, in order to protect the environment, preserve cultural heritage, or bypass high-cost zones. In the comprehensive geographic information model, all the cells in forbidden zones are included in the set U_F, all cells within the station are included in the set U_S, and their intersection must be empty to ensure that stations bypass the forbidden zones, as shown in Eq. (10.17):

$$U_F \cap U_S = \emptyset \tag{10.17}$$

③ Structure constraints

i. Station-tunnel constraint

Setting a station in a tunnel will significantly increase the construction investment and difficulty of construction and maintenance. Therefore, the station location should avoid overlapping with tunnels. During the railway location design process, tunnel locations can be determined according to the elevation difference between the ground elevation and the design elevation (Δh). Thus, we use the elevation difference Δh to avoid setting stations in tunnels. We assume that the elevation of a random cell located at row i, column j is $H_{Gi,j}$ and the design elevation of the station is H_{S0}. If $H_{Gi,j} \geq H_{S0}$, then the elevation difference between them should satisfy Eq. (10.18) to allow a station to be set.

$$H_{Gi,j} - H_{S0} < H_T \tag{10.18}$$

where H_T = boundary depth to set a tunnel (m).

ii. Station-bridge constraint

Similarly, setting a station on a bridge also leads to problems. Therefore, the station location should also avoid overlapping with bridges. Bridge locations can also be determined according to the elevation difference (Δh). We assume that the elevation of a random cell located at row i, column j is $H_{Gi,j}$ and the design elevation of the station is H_{S0}. If $H_{Gi,j} < H_{S0}$, then the elevation difference between them should satisfy Eq. (10.19) to set a station.

$$H_{S0} - H_{Gi,j} < H_B \tag{10.19}$$

where H_B = boundary height to set a bridge (m).

(5) Output data

The output data are the values of the decision variables. Assuming that m HPIs and n VPIs are generated after completing the whole search process, then the number of bridges, tunnels, and stations is x, y, and z, respectively. Thus, the output data are the following vectors.

① Horizontal points of intersection (HPIs) data, as shown in Eq. (10.20):

$$\begin{aligned} \boldsymbol{X} &= [X_1, X_2, \ldots, X_m]^{\mathrm{T}} \\ \boldsymbol{Y} &= [Y_1, Y_2, \ldots, Y_m]^{\mathrm{T}} \\ \boldsymbol{R} &= [R_1, R_2, \ldots, R_m]^{\mathrm{T}} \\ \boldsymbol{L} &= [L_1, L_2, \ldots, L_m]^{\mathrm{T}} \end{aligned} \tag{10.20}$$

② Vertical points of intersection (VPIs) data, as shown in Eq. (10.21):

$$\begin{aligned} \boldsymbol{K} &= [K_1, K_2, \ldots, K_n]^{\mathrm{T}} \\ \boldsymbol{H} &= [H_1, H_2, \ldots, H_n]^{\mathrm{T}} \\ \boldsymbol{R_V} &= [R_{V1}, R_{V2}, \ldots, R_{Vn}]^{\mathrm{T}} \end{aligned} \tag{10.21}$$

③ Structure location data, as shown in Eq. (10.22):

$$\begin{aligned} \boldsymbol{KB_S} &= [KB_{S1}, KB_{S2}, \ldots, KB_{Sx}]^{\mathrm{T}} \\ \boldsymbol{KB_E} &= [KB_{E1}, KB_{E2}, \ldots, KB_{Ex}]^{\mathrm{T}} \\ \boldsymbol{KT_S} &= \left[KT_{S1}, KT_{S2}, \ldots, KT_{Sy}\right]^{\mathrm{T}} \\ \boldsymbol{KT_E} &= \left[KT_{E1}, KT_{E2}, \ldots, KT_{Ey}\right]^{\mathrm{T}} \\ \boldsymbol{X_C} &= [X_{C1}, X_{C2}, \ldots, X_{Cz}]^{\mathrm{T}} \\ \boldsymbol{Y_C} &= [Y_{C1}, Y_{C2}, \ldots, Y_{Cz}]^{\mathrm{T}} \end{aligned} \tag{10.22}$$

10.1.1.1 DT-based solution method

The distance transform (DT) algorithm provides a very simple and extremely fast method for the approximation Euclidean distances from every cell of a rectangular lattice to the nearest cell in a target set. The time complexity of the DT is $O(N)$ using the raster scan method, where N is the number of cells in the lattice. In a further study, de-Smith [3] introduced this algorithm to alignment design and considered the slope constraints in the scanning process. Li et al. [4,5] improved the algorithm and proposed a method applicable for railway alignment optimization in mountainous areas. However, the incorporation of station location introduces additional constraints and increases the complexity of the optimization problem.

In this section, we propose a concurrent optimization method for railway alignment and station locations in a complex mountainous area based on the DT, and coupling constraints for railway alignment and station locations are considered.

(1) Checking station spacing
During the search process, station spacing is simultaneously computed. We create a station spacing attribute *DistToSta* for all cells in the lattice to store the distance value from the current cell to the nearest generated station in the previous search process. This attribute is used to determine when to start searching for the station locations.
Assuming that $p_{i,j}$ is the current cell, $p_{m,n}$ is one of the cells in its neighboring mask. The neighboring mask is a basic concept in the DT algorithm, and it is described as a static adjacency matrix in the traditional DT algorithm. Li et al. [4] improved the static adjacency to a dynamic adaptive mask to adapt the complex constraints in mountainous areas, as shown in Fig. 10.3. The dynamic adaptive mask is used in this section.
If *DistToSta* $(p_{m,n}) \geq D_{\max}$, then a backtracking process is conducted along the generated path until the cell $p_{b,d}$ corresponding to *DistToSta* $(p_{b,d}) \leq D_{\min}$ (see Fig. 10.4), and then the station location is searched on the backtracking path.
If *DistToSta* $(p_{m,n}) < D_{\max}$, then a station does not need to be set. The alignment search process is conducted, and the *DistToSta* $(p_{i,j})$ value is renewed by Eq. (10.23) after finding the best link cell. The search process will be described in Section 3.3.

$$DistToSta\left(p_{i,j}\right) = DistToSta\left(p_{m,n}\right) + D_{i,j} \tag{10.23}$$

where $p_{m,n}$ = the best linked cell,
$D_{i,j}$ = distance from cell $p_{i,j}$ to cell $p_{m,n}$ (m).

(2) Optimizing station location
The railway station can be regarded as a rectangular mask with a given length and width by designers. The location of the rectangular mask (railway station) can be determined when the rectangular mask center (station center) and macro-axis direction (the direction of the station) are determined. When a local path is generated, the station's direction, which is same as the local path's direction, can be determined. Thus, the process of searching the station location on a local path can be regarded as a search process for the railway station center.

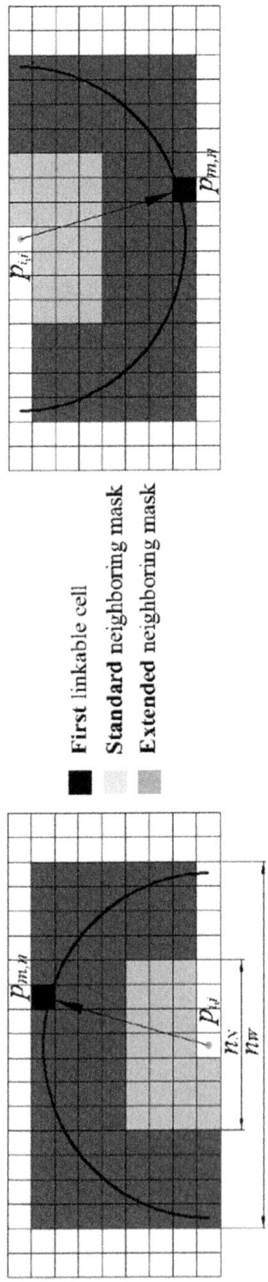

Figure 10.3 Dynamic adaptive neighboring mask.

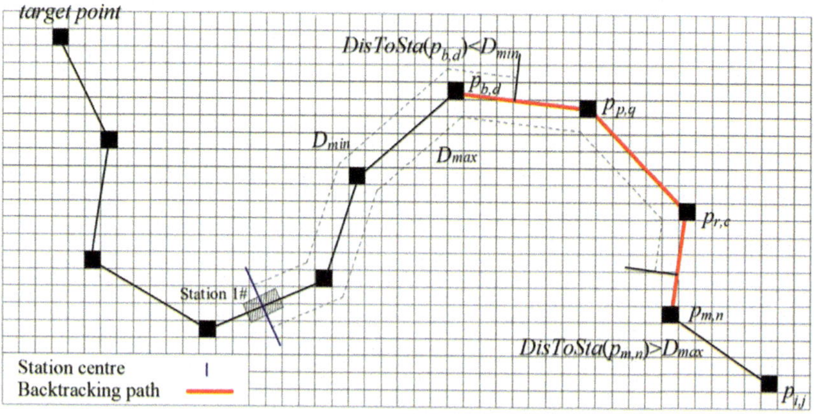

Figure 10.4 Backtracking process.

Many local paths may be located within the backtracking path. As shown in Fig. 10.4, three local paths are observed from $p_{m,n}$ to $p_{b,d}$. The station may be located between $p_{m,n}$ and $p_{r,c}$, between $p_{r,c}$ and $p_{p,q}$, or between $p_{p,q}$ and $p_{b,d}$. We should search the station location along every local path and choose the best one as the optimized station location. This chapter proposes a method of searching for the optimized station locations on the determined local paths. Multiple constraints among railway alignment and station locations are considered during the searching process.

Step 1: Select a local path.

Step 2: Compute the gradient (i_g) of the local path using Eq. (10.13). If the gradient constraint within the station in Section 10.1.1 is satisfied, then go to step 3; otherwise, a station cannot be set on the local path.

Step 3: Determine the moving range of the station center as shown in Fig. 10.5. This step ensures that the station can be set on the local path.

Step 4: Create N_{sc} station centers along the local alignment at equal intervals (see Fig. 10.5a). The distance is set by the designer based on station design standards.

Step 5: Select the ith station center and set $i = 1$, $i = i{+}{+}$.

Step 6: If $i \leq N_{sc}$, then compute the value of $DisToSta$ of the ith station center and go to step 7; otherwise, go to step 13.

Step 7: If $D_{\min} \leq DisToSta \leq D_{\max}$, then set the flat slope along the local path (see Fig. 10.5b), determine the station area according to

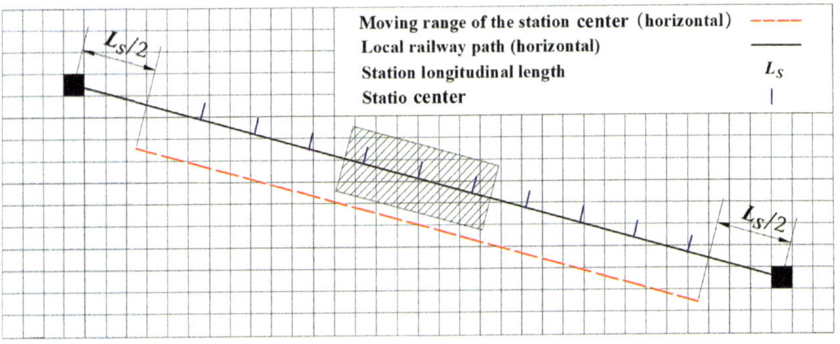

a Horizontal

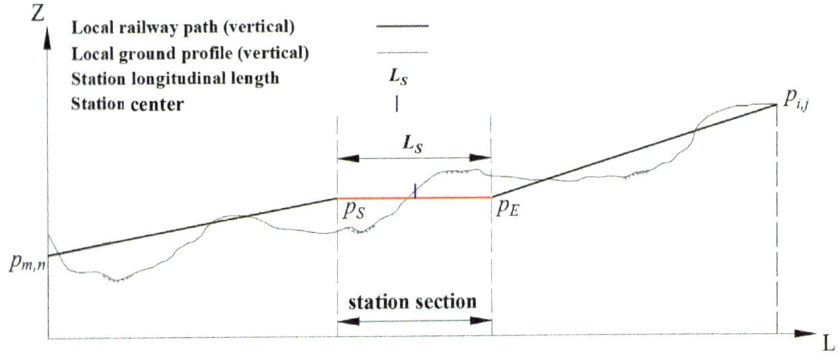

b Vertical

Figure 10.5 Searching station location along the local path.

the given station length and width (see Fig. 10.5a), and then go to step 8; otherwise, return to step 5.

Step 8: If the station satisfies the forbidden area constraint, then go to Step 9; otherwise, go back to step 5.

Step 9: If the station satisfies the station-tunnel constraint, then set $\gamma_t = 0$; otherwise set $\gamma_t = 1$.

Step 10: If the station satisfies the station-bridge constraint, then set $\gamma_b = 0$; otherwise set $\gamma_b = 1$.

Step 11: Check the alignment-related constraints. If all these constraints are satisfied, then go to step 12; otherwise, return to step 5.

Step 12: Compute the local construction costs, and then return to Step 5 until all station centers are scanned.

Step 13: Determine whether a feasible station alternative is found. If there are many feasible alternatives, select the minimum cost

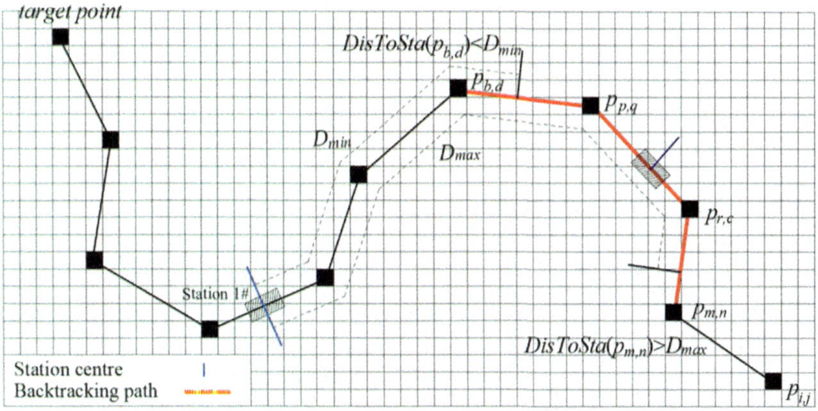

Figure 10.6 Selecting the minimum cost alternative.

alternative. If feasible alternatives are not available, then a station cannot be set on the local path.

After setting a new station, the total costs of the link cell and the value of *DistToSta* must be recomputed. Assuming the new station is set between $p_{p,q}$ and $p_{r,c}$, as shown in Fig. 10.6, then the total cost of $p_{m,n}$ is recomputed with Eq. (10.1), which includes the new station's cost, and the value of *DistToSta* ($p_{m,n}$) is renewed as Eq. (10.24):

$$DistToSta\left(p_{m,n}\right) = \left[\left(x_{r,c} - x_C\right)^2 + \left(y_{r,c,} - y_C\right)^2\right]^{0.5} + D_{m,n} \quad (10.24)$$

where $x_{r,c}$, $y_{r,c}$ = coordinates of the cell located at row r, column c in the lattice (m), x_C, y_C = coordinates of the new station center (m), $D_{m,n}$ = distance from cell $p_{m,n}$ to cell $p_{r,c}$ (m).

(3) Searching for the optimized path

We introduce the DT algorithm to search for the optimized path. Assuming that $p_{i,j}$ is one random cell in the lattice of the CGIM, $C_{total}(p_{i,j})$, $\Delta r(p_{i,j})$ and $\Delta c(p_{i,j})$ are stored in this cell; $C_{total}(p_{i,j})$ is the minimum total cost from $p_{i,j}$ to the target point; and $\Delta r(p_{i,j})$ and $\Delta c(p_{i,j})$ are the incremental row and column movements, respectively, at $p_{i,j}$ in the corresponding minimum cost path.

First, the method sets the start point as the target point and initializes $C_{total}(p_{i,j})$, $\Delta r(p_{i,j})$, and $\Delta c(p_{i,j})$ for all cells, as shown in Eq. (10.25):

$$C_{total}\left(p_{i,j}\right) = \begin{Bmatrix} 0 & p = B \\ \infty & p \neq B \end{Bmatrix}$$

$$\Delta r\left(p_{i,j}\right)=0$$

$$\Delta c\left(p_{i,j}\right)=0 \tag{10.25}$$

where $B=$ the target point.

A two-pass scan of the lattice data is conducted, consisting of a forward scan from the top left to the bottom right, followed by a backward scan from the bottom right to the top left. When 1 cell is scanned, the center of the neighboring mask is placed over it (see Fig. 10.7) and the total local cost $C_{\text{total}}(x,y)$ is added to the value of the current cell. The local total cost $C_{\text{total}}(x,y)$ is the cost between neighboring cells.

The values are refreshed with Eq. (10.26):

$$C_{\text{total}}\left(p_{i,j}\right)=\min\left\{C_{\text{total}}\left(p_{m,n}\right)+C_{\text{total}}(x,y)\right\}$$

$$\Delta r\left(p_{i,j}\right)=a$$

$$\Delta c\left(p_{i,j}\right)=b \tag{10.26}$$

where $C_{\text{total}}(p_{i,j})=$ cost value of cell $p_{i,j}$,

$p_{m,n}=$ one random cell in the neighboring mask (as shown in Fig. 10.7),

$C_{\text{total}}(p_{m,n})=$ cost value of cell $p_{m,n}$,

$C_{\text{total}}(x,y)=$ cost from cell $p_{i,j}$ to cell $p_{m,n}$,

a, $b=$ values of the incremental row and column movements at the current cell (i.e., cell $p_{i,j}$) in the corresponding minimum cost path.

By iterating the two-pass scan process until the values of each cell are static, the start distance image DT_S is generated. The final value of $C_{\text{total}}(p_{i,j})$ in DT_S is the minimum cost from $p_{i,j}$ to the target point. The corresponding shortest path can be generated by $\Delta r(p_{i,j})$ and $\Delta c(p_{i,j})$.

Similarly, the end distance image DT_E is formed by setting the endpoint as the target point.

Finally, various paths are generated by traversing all cells. The paths from $p_{i,j}$ to the start and end points can be found by DT_S and DT_E, respectively. The integrated path via $p_{i,j}$ is generated by linking the above two paths (see Fig. 10.8).

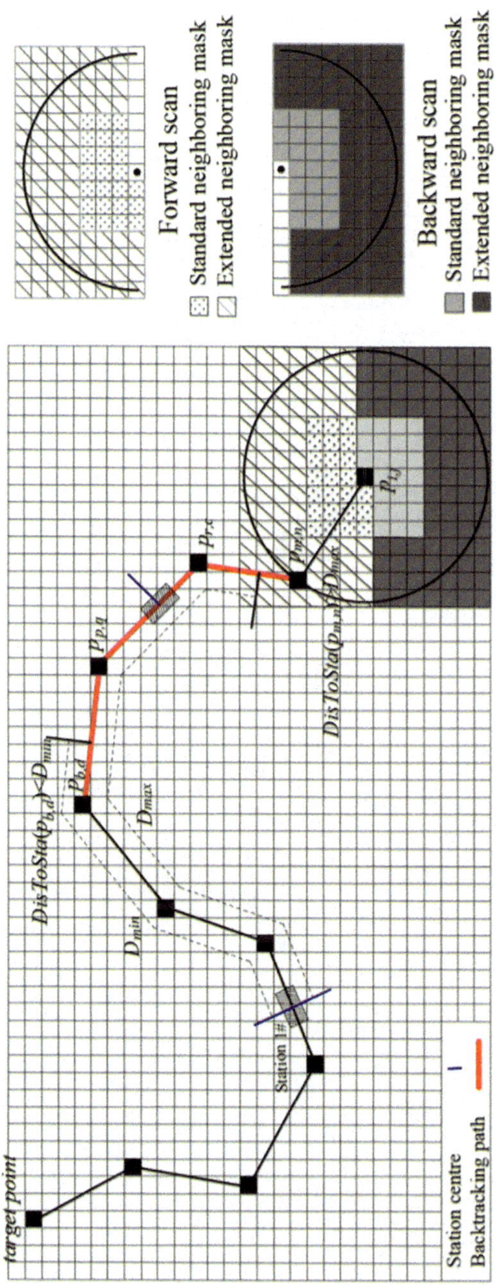

Figure 10.7 Generation of the lowest-cost path.

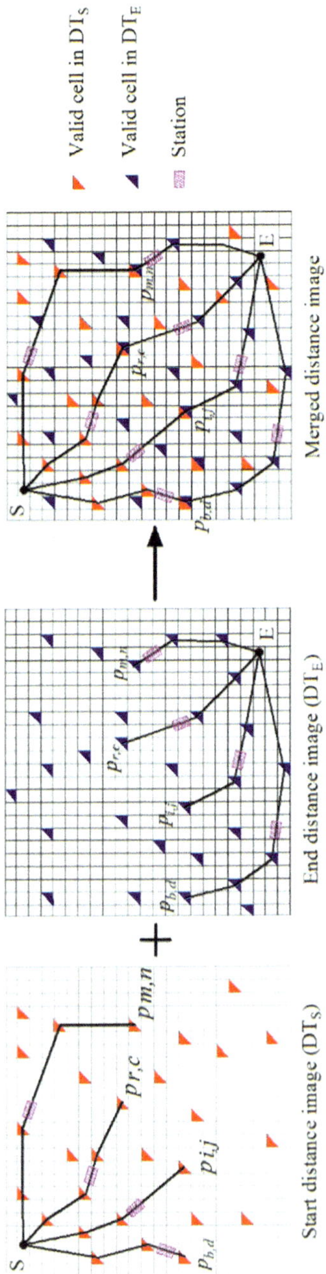

Figure 10.8 Generation of paths and stations.

(4) Fitting process

Because the generated alternatives are piecewise-linear polylines rather than smooth curves, these alternatives must be refined. The fitting process includes initializing and optimizing the generated alternatives. First, we use the successive subdivision method to select control points in the polyline path as the points of intersection to generate the initial refined alternative with the minimum circular curve radii. Next, the mesh adaptive direct search algorithm is employed to optimize the alternative to obtain the final alternative.

10.2 Concurrent optimization of railway alignments and LACPs

An ideal railway alignment design must not only satisfy the fundamental requirements of safety, economy, comfort, and environmental sustainability, but also ensure the smooth execution and on-time delivery of the railway construction project. Large-scale Auxiliary Construction Projects (LACP), as fundamental supporting facilities for railway construction, are closely tied to and mutually influence the alignment design. Their layout must be carefully planned in accordance with the spatial positioning of the alignment and the distribution of major structures such as bridges, tunnels, and stations, while also accounting for terrain, geology, and existing road networks along the alignment. Key facilities such as vertical shafts, inclined shafts, access roads, beam fabrication yards, and spoil disposal areas must be carefully arranged to meet requirements for construction schedule, quality, and investment control.

Given the natural sequencing between alignment and LACP design, where LACPs primarily serve critical decisions during the construction phase, existing practices typically adopt a “serial and decoupled” design approach. In this process, the alignment is determined first, and the LACPs are subsequently arranged based on the alignment plan. This results in the alignment dominating and the LACP design being passively constrained. In relatively simple environments such as plains and gentle hills, where both alignment and LACP have sufficient flexibility, this method can still yield feasible solutions. However, in complex environments, the high proportion of bridges and tunnels in the alignment significantly increases the number of LACPs required, often hampering adherence to the principle of using simple and economical auxiliary structures. The frequent appearance of high-pier long-span bridges and ultra-long, deep-buried tunnels greatly increases the complexity of LACP design. Furthermore, the lack of existing road infrastructure often makes it extremely difficult to connect the alignment to external transport networks, imposing severe limitations on construction logistics. These compounded challenges significantly intensify the difficulty of LACP planning in complex terrain.

If the conventional "alignment-first, LACP-later" serial approach continues to be followed, it may produce alignment schemes that are near-optimized but fail to support feasible LACP configurations, and ultimately may even preclude the alignment solution feasibility. This clearly demonstrates that LACP design can become a decisive factor in alignment decision-making. Traditional serial approaches, which overemphasize the dominance of alignment over LACP and neglect the potential reverse constraints imposed by LACPs, are no longer suitable in complex engineering contexts.

Therefore, under complex conditions, railway alignment and LACP planning must be jointly optimized. The final design should provide both the optimized spatial alignment and an efficient, compatible LACP configuration. The corresponding optimization model and solution method are provided below.

10.2.1 Optimization model

10.2.1.1 Design variables

As shown in Fig. 10.9(a) and (b), from a geometric perspective, a 3-Dimensional (3-D) railway alignment is comprised of tangents and curves. Thus, to specify a 3-D alignment, its geometric parameters of Alignment Points of Intersection (APIs) are generally used as decision variables. According to Ref. [4], those include the coordinates ($\boldsymbol{X_{A\text{-}H}}$, $\boldsymbol{Y_{A\text{-}H}}$) and circular curve radius ($\boldsymbol{R_{A\text{-}H}}$) of every Horizontal API (HAPI) as well as the station point ($\boldsymbol{P_{A\text{-}V}}$) and the design elevation ($\boldsymbol{E_{A\text{-}V}}$) of every Vertical API (VAPI).

Note that, according to the Code for Design of Railway Line (2017) [1], the horizontal transition curve length and vertical curve radius can be obtained after determining the above-mentioned variables and railway design speed. Thus, they are not treated as decision variables here. It is important to note that the codes cited in this paper are all authoritative and mainly used for providing references for values of key parameters in our model. However, these codes do not limit the international application of proposed methods. To apply the model to other countries, only the required input

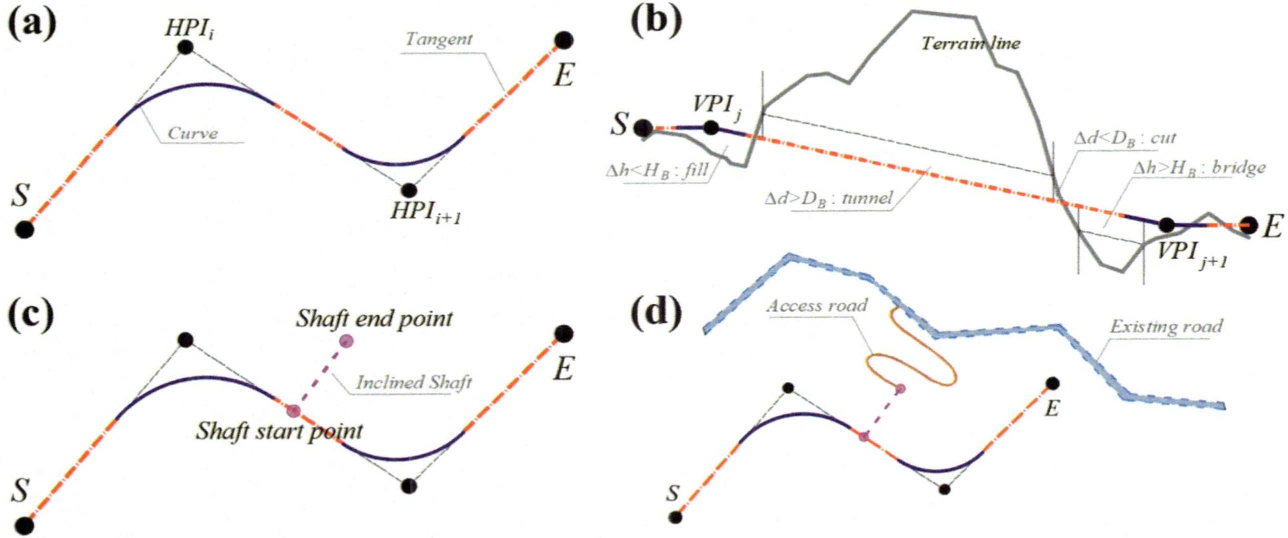

Figure 10.9 Steps in determining design variables for railway alignment-LACPs concurrent optimization:(a) Generate a horizontal railway alignment; (b) determine the vertical alignment and configure tunnels, bridges, and cut/fill sections; (c) locate a tunnel shaft; (d) connect the shaft with existing roads by generating an access road.

parameters need to be changed flexibly, according to other countries' codes or standards.

As explained in Fig. 10.9(b), after obtaining these geometric variables (denoted as a set of $\boldsymbol{D_A}$), a railway alignment can be further determined by configuring bridges, tunnels and earthwork sections on it, according to railway locations, design standards and terrain conditions.

Then, based on tunnel sections, vertical shafts and inclined shafts can be determined, as shown in Fig. 10.9(c). Given that the major difference between these two kinds of service galleries lies in their spatial intersection angles with the horizontal plane (with more details in Section 2.3), for simplicity, shafts are treated as spatial straight lines in this study. Thus, the parameters of their two endpoints are defined as shafts' decision variables (denoted as a set of $\boldsymbol{D_S}$). Specifically, considering that their start points are located exactly on the alignment and can slide along tunnel sections, to reduce model complexity, the location of start points can be described just using the station point ($\boldsymbol{P_{S-S}}$) based on the railway alignment's information. The shafts' end points are exposed on the ground surface and can be parameterized directly by using their 3-D coordinates ($\boldsymbol{X_{S\text{-}E}}$, $\boldsymbol{Y_{S\text{-}E}}$, $\boldsymbol{Z_{S\text{-}E}}$).

Finally, access roads are usually low-grade and built for temporary purpose of railway construction. Similarly to a railway alignment, an access road can be described by its Road Points of Intersection (RPIs). However, the difference is that the horizontal and vertical RPIs can coincide [6]. Therefore, given that access roads are low-grade and to reduce model complexity, the horizontal and vertical RPIs are assumed to coincide in this paper. Hence, the design variables of an access road (denoted as a set of $\boldsymbol{D_R}$) include the 3-D coordinates ($\boldsymbol{X_{R\text{-}P}}$, $\boldsymbol{Y_{R\text{-}P}}$, $\boldsymbol{Z_{R\text{-}P}}$), horizontal and vertical circular curve radii ($\boldsymbol{R_{R\text{-}H}}$ and $\boldsymbol{R_{R\text{-}V}}$) of RPIs. Besides, it is worth noting that access roads generally start from bridge abutments (or piers), tunnel portals or shaft end points to reach an existing road network. Thus, the start point of an access road can be predetermined. However, an access road's end point ($\boldsymbol{X_{R\text{-}E}}$, $\boldsymbol{Y_{R\text{-}E}}$) should be considered as variable along existing roads, as shown in Fig. 10.9(d). In summary, the complete design variables (all in meters) of the studied problem are shown in Eq. (10.27):

$$\begin{aligned}
&\text{design variables}: [\boldsymbol{D_A}, \boldsymbol{D_S}, \boldsymbol{D_R}] \\
&\text{where}: \boldsymbol{D_A} = [\boldsymbol{X_{A-H}}, \boldsymbol{Y_{A-H}}, \boldsymbol{R_{A-H}}, \boldsymbol{P_{A-V}}, \boldsymbol{E_{A-V}}] \\
&\boldsymbol{D_S} = [\boldsymbol{P_{S-S}}, \boldsymbol{X_{S-E}}, \boldsymbol{Y_{S-E}}, \boldsymbol{Z_{S-E}}] \\
&\boldsymbol{D_R} = [\boldsymbol{X_{R-P}}, \boldsymbol{Y_{R-P}}, \boldsymbol{Z_{R-P}}, \boldsymbol{R_{R-H}}, \boldsymbol{R_{R-V}}, \boldsymbol{X_{R-E}}, \boldsymbol{Y_{R-E}}]
\end{aligned} \tag{10.27}$$

10.2.1.2 Objective function

The construction cost (F_{cost}) is important for every railway project, especially for with complex mountain railways with highly-expensive structures. In this field, the railway alignment construction cost (C_A) has been frequently used as an objective function. In the study presented here, the construction costs for shafts (C_S) and access roads (C_R) have also been included, as shown in Eqs. (10.28) to (10.31):

$$F_{\text{cost}}(\boldsymbol{D_A}, \boldsymbol{D_S}, \boldsymbol{D_R}) = C_A + C_S + C_R \tag{10.28}$$

$$C_A = C_{AE} + C_{AB} + C_{AT} + C_{AR} + C_{AL} \tag{10.29}$$

$$C_S = \sum_{i=1}^{I} (U_{Si} \cdot L_{Si}) \tag{10.30}$$

$$C_R = \sum_{j=1}^{J} \left(C_{REj} + C_{RBj} + U_{Rj} \cdot L_{Rj}\right) \tag{10.31}$$

where C_{AE}, C_{AB}, C_{AT}, C_{AR} and C_{AL} are alignment construction costs (¥) for earthwork, bridges, tunnels, right-of-way areas and length-dependent components (e.g., tracks, power and traction power supply systems), respectively. They can be computed according to Chapter 5.4; I is the total number of shafts; U_{Si} and L_{Si} are the unit cost (¥/m) and length (m) of the ith shaft, respectively; J is the total number of access roads; C_{REj} and C_{RBj} are earthwork and short bridge (length <100m) costs (¥) of the jth access road, respectively. They can be similarly computed as those in Eq. (10.3); U_{Rj} and L_{Rj} are the unit length-dependent cost (¥/m) and length (m) of the jth access road, respectively.

10.2.1.3 Constraints

(1) Geometric constraints

Multiple kinds of geometric constraints are imposed when designing railway alignments and access roads. For both alignment types, the constraints on minimum radius as well as length of horizontal circular curves, and on maximum gradient as well as minimum length of sloped sections should be satisfied. Also, railway alignments should satisfy the minimum length of tangents between adjacent horizontal curves and the maximum difference between adjacent gradients. Furthermore, the access roads should also satisfy the maximum length of sloped sections and the minimum length of vertical curves.

(2) Location constraints

Railway alignments and LACPs should bypass forbidden zones and also satisfy clearance constraints when crossing existing rivers or roads.

Then, the distance between two adjacent shafts (or between the portal and its adjacent shaft) should be within a specific min-max interval for purposes of construction and ventilation. Furthermore, a vertical shaft must be perpendicular to the horizontal plane, as depicted in Fig. 10.10.

For an inclined shaft, its horizontal intersection angle with the railway alignment (e.g., α in Fig. 10.11(a)) must exceed a minimum threshold to avoid impacting the subsequent tunnel construction when excavating the inclined shaft. Moreover, to facilitate the transportation of rocks and soils when excavating tunnels, an inclined shaft's vertical intersection angle with the horizontal plane (e.g., β in Fig. 10.11(b)) must also be within a specific min-max range.

(3) Structural constraints

The maximum tunnel length and maximum bridge height are usually constrained for main alignments.

Access roads are mostly comprised of earthwork sections. When designing access roads, tunnels are generally forbidden and bridge

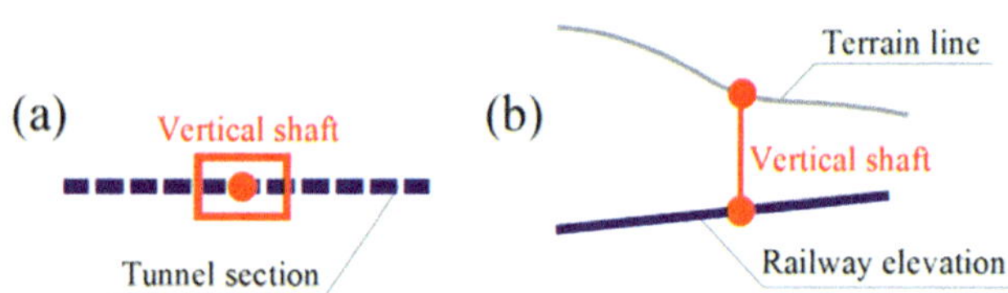

Figure 10.10 Horizontal (a) and vertical (b) views of a vertical shaft.

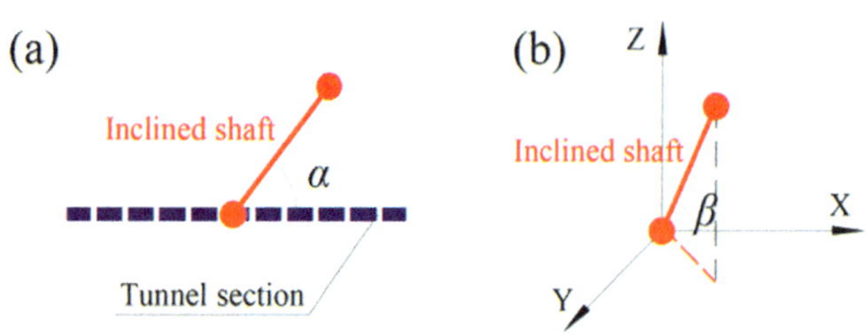

Figure 10.11 Horizontal (a) and 3-D (b) views of an inclined shaft.

lengths are strictly limited, according to the Technical Standard of Low Volume Rural Highway Engineering (China Ministry of Transport, 2019).

Besides, the maximum length of a shaft should also be constrained to save cost and construction time on LACPs. It is noteworthy that the maximum allowable lengths for vertical shafts and inclined shafts are often different.

10.2.1.4 Construction duration constraints

The above design constraints are specified separately for alignments, shafts and access roads. Moreover, when optimizing railway alignments and LACPs concurrently, the duration becomes a deciding but complex factor for construction feasibility analysis of the whole system.

Generally, the total construction duration of a railway project is controlled by the most important and complex structure along the alignment, such as the longest tunnel or the largest bridge. Thus, in this study, Eq. (10.32) is proposed for computing construction duration (T_{total}). During optimization, T_{total} cannot exceed its maximum allowable value ($T_{\max}$):

$$T_{\text{total}}(\boldsymbol{D_A}, \boldsymbol{D_S}, \boldsymbol{D_R}) = \max\{T_{\text{bri}-\max}, T_{\text{tun}-\max}\} / \gamma + T_{\text{check}} \leq T_{\max} \tag{10.32}$$

where $T_{\text{bri-max}}$ and $T_{\text{tun-max}}$ are the maximum construction time among all the bridges and tunnels, respectively. According to the Code for Design of the Railway Engineering Construction Organization [7], $\gamma = 0.85$–0.90 is an empirical coefficient and $T_{\text{check}} = 3$–5 months is additional time needed for, e.g., laying tracks and installing communication equipment.

Then, $T_{\text{bri-max}}$ and $T_{\text{tun-max}}$ are computed as Eqs. (10.33) and (10.34):

$$T_{\text{bri}-\max} = \max_{n=1}^{N}\{T_{\text{bri}-n}\} \tag{10.33}$$

$$T_{\text{tun}-\max} = \max_{m=1}^{M}\{T_{\text{tun}-m}\} \tag{10.34}$$

where N and M are the total numbers of bridges and tunnels, respectively; $T_{\text{bri-}n}$ and $T_{\text{tun-}m}$ are the construction time for the nth bridge and the mth tunnel, respectively. Note that based on the above referred code, only long tunnels (length $>$ 5 km) and large bridges (height $>$50m and length $>$1,000m) are considered for duration analysis in this study. Other bridges' or tunnels' construction times can be directly obtained from that code.

More specifically, for large bridges defined above, $T_{\text{bri-}n}$ is expressed as Eq. (10.35):

$$T_{\text{bri}-n} = T_{B-R}^{n} + \left(T_{U}^{n} + T_{P}^{n} + T_{B}^{n}\right) + T_{B-E}^{n} \tag{10.35}$$

where $T_{B\text{-}R}$ is the access road construction time to connect the bridge with existing roads, which is directly related to the length of an access road; T_U, T_P and T_B are the construction time for a bridge's foundations, piers and beams, respectively. They can be readily obtained by referring to the Code for Design on Railway Bridge and Culvert [8]; $T_{B\text{-}E} = 0.28$–0.35 month is the additional time to connect a bridge's two abutments with earthwork sections. It is important to note that Eq. (10.34) is not applicable for specially-structured bridges, such as super-large suspension bridges. The construction time of such bridges should be specifically estimated, usually independently from alignment design.

Compared with $T_{bri\text{-}n}$, $T_{tun\text{-}m}$ is more complex. First, unlike bridges, a long tunnel section must be first divided into several subsections according to the locations along it of shafts (whose dynamic searching process is explained in Section 3.3). As shown in Fig. 10.12, the midpoints between every pair of adjacent shafts, or between a portal and its adjacent shaft (at the tunnel's two endpoints), are deemed demarcation points.

Then, it is usually assumed at the alignment design stage that manpower and material resources for building the tunnel are sufficient. This assumption is reasonable for simplifying the complex problem because, for real-world alignment optimization at early railway design stages, it is very difficult to obtain a detailed construction duration due to limited resources and time. In-depth analyses regarding construction risks, scheduling and engineering managements for duration assessment are generally conducted after the alignment design process. Moreover, the purpose of duration computation in alignment designs is not to optimize the total construction time but to provide references on controlling locations of complex tunnels, bridges and LACPs. Thus, construction duration is treated as a constraint

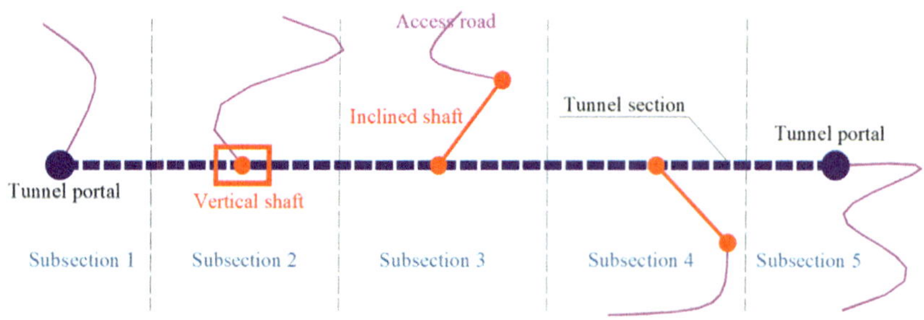

Figure 10.12 Tunnel division for computing construction time.

and estimated based on the above assumption. Note that this computed duration will also be further modified in Eq. (10.32) by γ considering the uncertainties with respect to, e.g., manpower and material resources.

Afterward, $T_{\text{tun-}m}$ is expressed as:

$$T_{\text{tun}-\text{m}} = \max_{s=1}^{S}\{T_{\text{tun}-m-s}\} + T_{T-A} + T_{T-E} \tag{10.36}$$

where S is the number of subsections; $T_{\text{tun-}m\text{-}s}$ is the excavation time for the sth subsection, whose computation is expressed in Eq. (10.36); $T_{T\text{-}A} = 1$–3 months is the additional time for, e.g., building lining and portals [7]; $T_{T\text{-}E} = 0.32$–0.4 month is the time required to connect a tunnel's two portals with earthwork sections, as shown in Eq. (10.37):

$$T_{\text{tun}-m-s} = T_{T-R} + T_{T-S} + T_{T-H} = T_{T-R} + L_S/l_S + L_H/l_H \tag{10.37}$$

According to the construction sequence, $T_{\text{tun-}m\text{-}s}$ contains time components for constructing access roads ($T_{T\text{-}R}$), excavating shafts ($T_{T\text{-}S}$) and excavating tunnel subsections ($T_{T\text{-}H}$). Since the excavation rates (i.e., l_S and l_H in meters/month) are generally assumed to be fixed when the surrounding rocks' grade and excavation technology are known, $T_{T\text{-}S}$ and $T_{T\text{-}H}$ can be estimated when the shaft length (L_S) and tunnel subsection length (L_H) are known.

Finally, by denoting the geometric, location, structural and construction duration constraints as four sets of $\boldsymbol{C_G}$, $\boldsymbol{C_L}$, $\boldsymbol{C_U}$ and $\boldsymbol{C_C}$, the completed model formulation can be simplified as Eq. (10.38):

$$\begin{aligned}
&\text{Minimize}: F_{\text{cost}}(\boldsymbol{D_A}, \boldsymbol{D_S}, \boldsymbol{D_R}) = C_A + C_S + C_R,\\
&\text{subject to}: \boldsymbol{C_G}(\boldsymbol{D_A}, \boldsymbol{D_S}, \boldsymbol{D_R}),\\
&\boldsymbol{C_L}(\boldsymbol{D_A}, \boldsymbol{D_S}, \boldsymbol{D_R}),\\
&\boldsymbol{C_U}(\boldsymbol{D_A}, \boldsymbol{D_S}, \boldsymbol{D_R}),\\
&\boldsymbol{C_C}(\boldsymbol{D_A}, \boldsymbol{D_S}, \boldsymbol{D_R}).
\end{aligned} \tag{10.38}$$

10.2.2 Four-step hierarchical solution method

10.2.2.1 Overall strategy

In the real-world, the railway construction sequence is to build LACPs before the main alignment. However, during a design process, as shown in Fig. 10.13, the railway alignment must be first determined based on environmental inputs, such as topography, geology and existing road network.

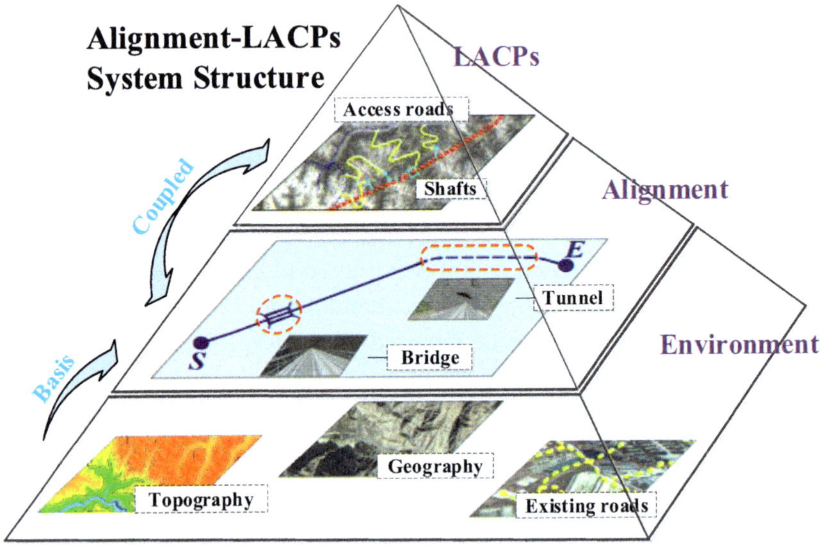

Figure 10.13 Pyramid structure of alignment-LACPs system.

Afterward, LACPs can be configured along the alignment. Lastly, the alignment and LACPs are evaluated and adjusted as a coupled system.

Therefore, in this study, an "alignment-shaft-road-integration" four-step procedure is proposed for solving the developed optimization model. As shown in Fig. 10.14, a hybrid PSO - Divide & Conquer - Dijkstra method is presented to search for alignments, shafts and access roads, respectively. The overall method is embedded in an improved PSO. More details are provided in the following sections.

10.2.2.2 Alignment generation

A PSO-based algorithm proposed in Chapter 9.3 is adopted here for searching railway alignments. In this PSO, every 3-D alignment is abstracted as a 3-D particle, whose spatial locations are characterized by alignment design variables. More specifically, a horizontal-vertical-integration strategy has been proposed for the PSO by first defining the horizontal alignment particle using HPIs' information and then, similarly, obtaining the vertical alignment particle using VPIs' information. The algorithm is implemented through two steps, i.e., initialization and iteration (Fig. 10.14).

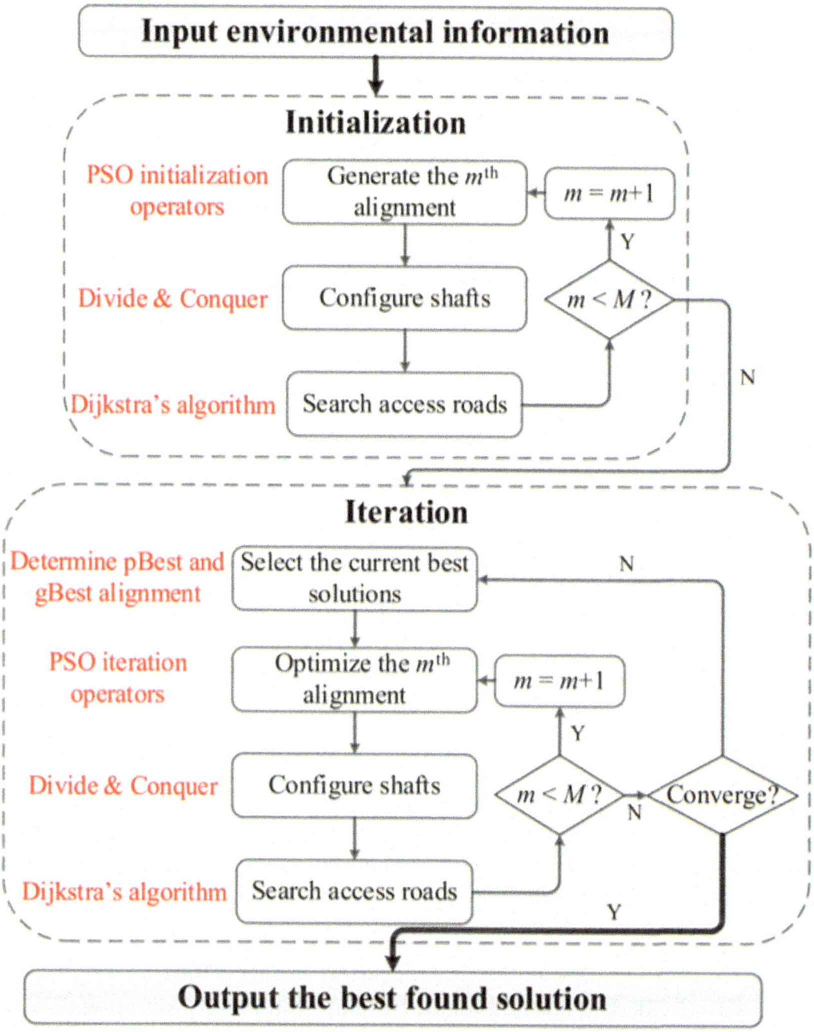

Figure 10.14 Main flow of the solution method, where M is the particle swarm size.

10.2.2.3 Shafts configuration

When an alignment alternative is designed, its tunnel sections are determined. Corresponding shafts can then be configured. It is generally assumed that, for mountain tunnels shorter than 5 km, construction can be directly conducted from their two portals and therefore no shafts are needed. For longer tunnels, a Divide & Conquer (D&C) procedure is proposed.

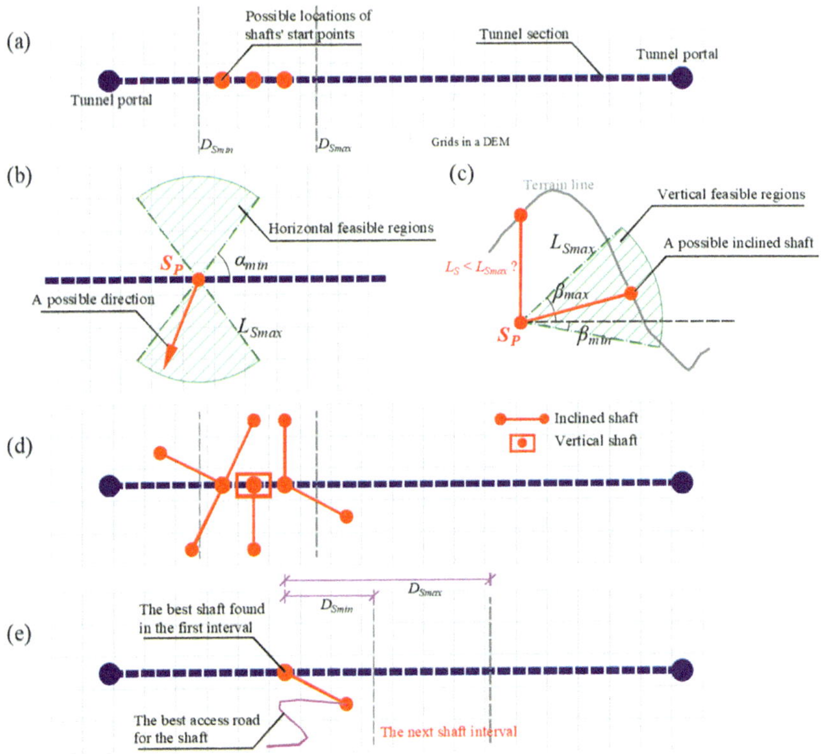

Figure 10.15 The D&C process for configuring shafts: (a) Divide the tunnel section according to the distance interval of [D_{Smin}, D_{Smax}]; (b) determine horizontal feasible regions for inclined shafts; (c) interpolate the terrain line using a DEM for each horizontal direction and generate vertical feasible regions; (d) record all possible shaft alternatives; (e) obtain the next shaft interval.

First, using a grid-based Digital Elevation Model (DEM), a long tunnel can be first divided by considering the distance interval constraint of [D_{Smin}, D_{Smax}] for shafts, as shown in Fig. 10.15(a). The grids within the spacing of [D_{Smin}, D_{Smax}] can be deemed as potential horizontal locations for shafts on the railway alignment.

Then, for every such horizontal position, fan-shaped feasible regions for inclined shafts can be obtained considering the constraints of horizontal intersection angle (α_{min}) and maximum shaft length (L_{Smax}), as shown in Fig. 10.15(b). Thus, all possible directions for inclined shafts within these horizontal regions can be determined.

Afterward, for every feasible horizontal direction, its terrain line can be interpolated using the DEM, as shown in Fig. 10.15(c). If the elevation difference (L_S) between the ground and the railway at point S_P is lower than L_{Smin}, a vertical shaft can be set at S_P. For inclined shafts, considering the constraints of vertical allowable angles (i.e., β_{min} and β_{max}) and L_{Smax}, a vertical feasible region can be generated. Therefore, every ground point within this region can be connected with S_P to form possible inclined shafts.

Ultimately, as in Fig. 10.15(d), all possible shafts within [D_{Smin}, D_{Smax}] will be recorded and their construction cost can be computed. Note that during the above process the sampling distance between DEM grids can be enlarged to improve efficiency but, according to experience, the maximum threshold is $(D_{Smin} - D_{Smax})/10$ (while the minimum sampling distance between grids is the grid width). It is also important to note that construction cost is not sufficient by itself to determine the best shaft among these alternatives. During an actual design, many shafts may be short and easy to construct. However, it may be very difficult to construct access roads for these shafts to link their endpoints with existing roads. Thus, shafts should be evaluated jointly with their corresponding access roads to determine which shaft-road combinations are the best. This will be elaborated in the next section.

After finding the best location of the shaft on the alignment, it will be used as a new start point to determine the next distance interval to search for the next shaft (Fig. 10.15(e)), until the other tunnel portal is reached.

10.2.2.4 Access road search

To connect existing roads, an access road can start from three kinds of key points, namely tunnel portals, shafts' endpoints and bridge abutments, which can be pre-determined in previous sections and then fixed in the landscape. However, the end point of an access road is variable along exiting roads. Thus, this is a typical Single-Source Shortest-Path (SSSP) problem. To this end, a customized Dijkstra algorithm is proposed.

Dijkstra's algorithm [9] is a well-known method for solving SSSP problems and has been widely-applied to studies such as ecological connectivity [10] and transportation networks [11].

To implement Dijkstra's method, a weighted directed graph with nodes and links should be first constructed. Then, to find the shortest paths from the original node to other nodes, in every step only the closest node to the original one is searched and added into original node's neighborhood until none remains. In this work, Dijkstra's original algorithm is improved to search for access roads with two procedures, i.e., grid preprocessing and path generation.

Step 1: Gird preprocessing

Regarding access road design, there exist no such links or nodes in the study area. Thus, the first step for Dijkstra's implementation is to construct a specific weighted directed graph for the present problem.

Based on the DEM, the study area is first discretized into regular grids, as mentioned in Section 3.3. Then, a Moving Window Method (MWM), which is usually adopted for geomorphic analysis, is developed to scan every grid in the DEM. As shown in Fig. 10.16, unlike a traditional N×N square window, a circular-shaped one is designed which considers the constraint on minimum sloped section length ($L_{P\min}$) for access roads.

During the scan, every grid located on the horizontal boundary of the window is linked with the central grid. Other grids within the window are not considered since they violate the $L_{P\min}$ constraint. Then, the terrain line of every linkage is interpolated. The maximum gradient constraint and structural constraints can be checked for the linkage. If these constraints are satisfied, the access road cost for this local linkage can be computed and stored in the corresponding boundary grid. By checking all grids along the window's boundary, a set containing all feasible boundary grids can be found to form a neighborhood of the central grid. Hence, after scanning the entire study area, a weighted directed graph can be generated by assigning such a neighborhood to every grid.

It is important to note that the above process is actually a one-time workaround that should be done before the PSO initialization process (Fig. 10.14). That is, after the DEM is input, this grid preprocessing can be executed before detailed alignment optimization.

Step 2: Path generation

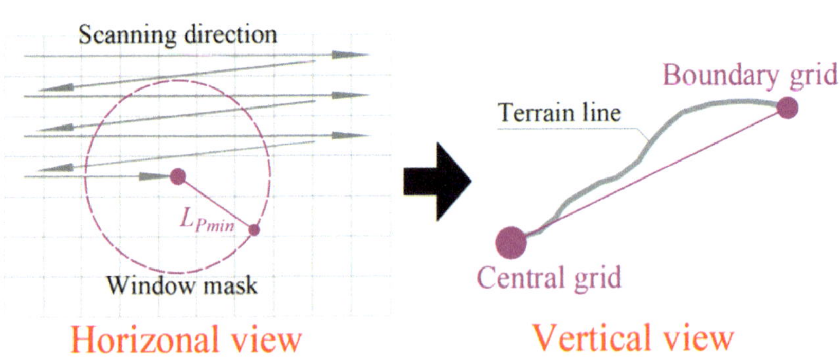

Figure 10.16 The MWM for scanning the study area.

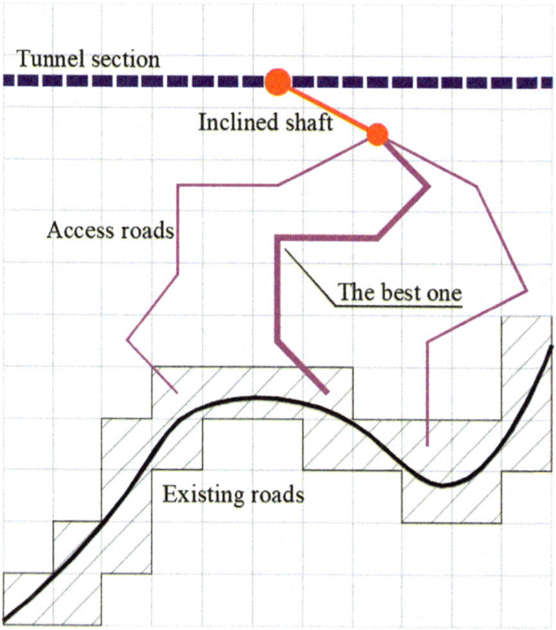

Figure 10.17 Searching access roads with Dijkstra's algorithm.

Based on the preprocessed study area, possible start points of an access road, i.e., tunnel portals, shafts' endpoints and bridge abutments, can be located by using their grid indexes. Moreover, the grids traversed by existing roads should also be determined as possible end points for access roads. Then, as shown in Fig. 10.17, from a specified start point, the search is executed using Dijkstra's algorithm until the end points are reached. It is worth noting that during the search, the "distance" between two grids is characterized by their link cost. Also, other related constraints introduced in Section 2.3 should be checked. Finally, among several paths that connect the start point and a set of end points, the one with the lowest access road cost is selected as the best one for subsequent evaluations.

During the search, between two possible start points of access roads (e.g., between two adjacent shafts' endpoints or between one bridge abutment and a tunnel portal), if their 3-D distance is under 100m and their elevation difference is under 25m, it is deemed that the two start points are considered close enough to be connected by a short linking segment within the tunnel and share one access road and, thus, the Dijkstra search is only executed once.

After the above processes, every shaft generated in Section 3.3 is configured with an access road. Then, their shaft cost and access road cost are added. The best shaft with the lowest shaft and road costs is recorded together with its access road. Other shaft alternatives are eliminated. Note that after determining the best access road, this road is included into the existing road network for the next Dijkstra search.

10.2.2.5 Integration and iteration

So far, the railway alignment, shafts and access roads are all generated. Thus, the alignment-LACPs system can be integrated. Earlier, the design constraints separately for railway alignments, shafts and access roads have been handled using different methods. In this part, the construction duration must be computed and checked for the entire alignment-LACPs system. If the duration constraint is satisfied, the total construction cost for railway alignments and LACPs can be computed. Therefore, the best solution with the lowest construction cost can be determined and the PSO evolution can be implemented until convergence.

Basically, the original PSO functions are as Eqs. (10.39) and (10.40):

$$\boldsymbol{V}_n(t+1) = \omega(t)\cdot\boldsymbol{V}_n(t) + C_g\cdot r_g\cdot[\boldsymbol{gBest}(t) - \boldsymbol{P}_n(t)] + C_p\cdot r_p\cdot[\boldsymbol{pBest}_n(t) - \boldsymbol{P}_n(t)] \quad (10.39)$$

$$\boldsymbol{P}_n(t+1) = \boldsymbol{P}_n(t) + \boldsymbol{V}_n(t+1) \quad (10.40)$$

where $\boldsymbol{V}_n$ and $\boldsymbol{P}_n$ are the nth ($1 \leq n \leq N$, N is the swarm size) particle's velocity and position vectors, respectively. They are characterized by design variables of the problem; $1 \leq t \leq T$ is the current iteration number and T is the iteration threshold; $0 \leq r_1, r_2 \leq 1$ are random values; $\boldsymbol{gBest}(t)$ denotes the global best one among all the generated particles up to the tth iteration; $\boldsymbol{pBest}(t)$ is a group of personal best solutions recorded by every individual in the swarm up to the tth iteration; ω denotes an inertia weight; C_p and C_g are cognitive and social acceleration coefficients, respectively. ω, C_p and C_g are the major parameters for PSO algorithm.

The value of ω is gradually reduced during successive iterations in order to improve the *pBest* exploration and *gBest* exploitation of PSO.

C_p and C_g have different significances: C_p guides the particles to "study" from their own history while C_g provides momentum for every particle to get closer to the current swarm optimum. Previously, C_p and C_g are both set at a simple empirical value of 2. In theory, with a higher C_p value, the solution space can be searched more thoroughly but that may reduce search efficiency. In contrast, a relatively high value of C_g can lead the PSO to converge faster but may also get the search trapped in a local optimum. Moreover, for the present problem, it is worth noting that LACPs are accessories for alignments and are not directly involved in the PSO iteration. In other words, LACPs should be configured uniquely for every alignment without reference to other alignments' information (i.e., swarm communications). In this condition, it is vital for every alignment particle to "study" its own search history thoroughly in order to find its *pBest* alignment accompanied with specific LACPs. Therefore, the motivation of this section is to improve the *pBest* search ability of PSO while considering the global convergence on *gBest*.

Based on the above discussions, a Time-Varying Acceleration Coefficients (TVAC) concept proposed in Ref. [12] is adopted here. According to this influential TVAC method, C_p is gradually reduced during a search history from 2.5 to 0.5 whereas C_g is increased from 0.5 to 2.5. Thus, by first combining a large C_p and a small C_g, the search space can be detected by particles in greater detail. Meanwhile, by assigning a small value to C_p and a large value to C_g when approaching the end of search, the global optimum is more likely to be found. Initially, a simple linearly decreasing TVAC mechanism is frequently used to update the two acceleration coefficients. In this study, novel polynomial functions are proposed as TVAC formulas Eqs. (10.41) and (10.42):

$$C_g(t) = \left(9 \cdot t^3\right) / \left(7 \cdot T^3\right) + (5 \cdot t) / (7 \cdot T) + 0.5 \tag{10.41}$$

$$C_p(t) = 2.5 - \left(9 \cdot t^3\right) / \left(7 \cdot T^3\right) - (5 \cdot t) / (7 \cdot T) \tag{10.42}$$

More explicitly, this TVAC method is illustrated in Fig. 10.18. Using the four key points' coordinates marked on each curve, Eqs. (10.41) and (10.42) can be easily derived. The value of C_p (or C_g) decreases (or increases) slowly at the beginning of the search but quickly at the end. Compared to traditional linear TVAC, this method, to a higher degree, favors a *pBest* solution search at the beginning of the optimization and facilitates global search later. This increases the likelihood of finding better alignment-LACPs combinations for this particular problem. Moreover, it is worth noting that this polynomial TVAC is not limited to the present problem but can be used for general PSO applications.

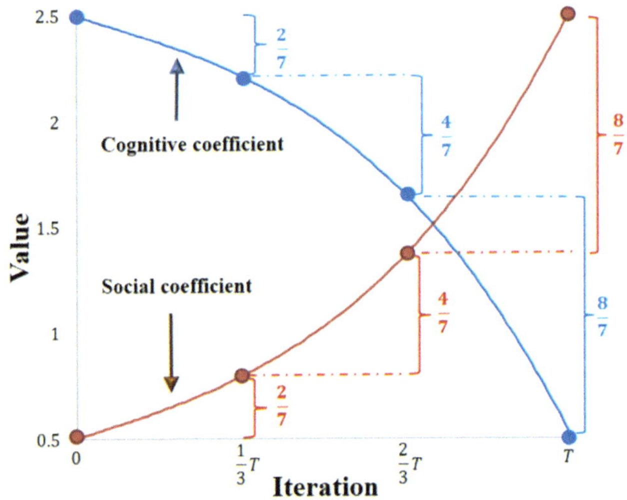

Figure 10.18 Graphical representation of the proposed TVAC.

10.3 Optimization of grade-separated road and railway crossings

During railway construction, numerous intersections with densely distributed existing road networks in populated areas are inevitable. To prevent a newly constructed railway from obstructing existing roads, while also ensuring the safety and capacity of road transport, certain segments of the existing roads must be realigned, and grade-separated crossings must be established at intersections to achieve vertical separation between highways and railways. The layout of such grade-separated crossings is constrained by various factors at the intersection point, including the crossing angle, required clearance height, and geometric design parameters. Given limited time and resources, engineers often rely solely on personal experience to compare a few feasible schemes that meet the constraints and minimize investment costs. However, this subjective approach may overlook promising solutions. Therefore, there is an urgent need to develop an intelligent design method for road–rail grade-separated crossings that accounts for constraints related to road construction, network connectivity, and investment efficiency.

The optimization of road–rail grade-separated crossings falls under the broader framework of alignment optimization. Compared to conventional alignment optimization between two fixed endpoints, this problem introduces additional complexity due to the need to consider road network intersections and connectivity to existing roads. For a grade-separated crossing, its alignment's start and end points, i.e., the two connection points with existing roads, are not predetermined. Furthermore, constraints at the intersection with the new railway, including crossing constraints, connectivity conditions, and geometric requirements, must be addressed during the optimization process.

To tackle these challenges, the construction cost of a grade-separated crossing is formulated as a function of a specific distance between the start and end points of the crossing road, thus transforming the alignment design into a shortest-path search problem. Considering that both endpoints of the grade-separated crossing are not yet fixed, a dedicated optimization model and solution method are proposed to optimize the spatial alignment of road–rail grade-separated crossings along the railway corridor.

10.3.1 Optimization model

(1) Comprehensive Geographic Information Model

Defining the study area's boundary is a prerequisite for organizing geographic information. As shown in Fig. 10.19, an elliptical area is constructed along the existing road centered around the intersection point III between the existing road and the newly planned railway, with a longitudinal length of X meters (here, X = 500 m). The two focal points of the ellipse, A and B, are identified, and the eccentricity is defined as $\gamma = imax/iD$, where iD is the average slope of the ground elevation and imax is the maximum allowable road slope. The study area is defined as the circumscribed rectangle of the ellipse and is subdivided into square grid cells with a width of w. Geographic information within the study area—including terrain (elevation), geology (rock and soil characteristics), unit construction cost, and existing structures (e.g.,

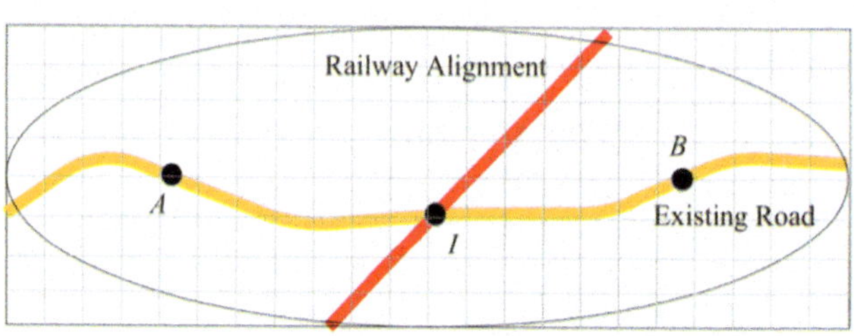

Figure 10.19 Study area.

factories, roads, railways, residential areas)—is collected to construct a Comprehensive Geographic Information Model (CGIM).

(2) Decision Variables

The optimization of grade-separated road–rail crossings is a type of alignment optimization problem. A 3D alignment can be described using a series of intersection points (PIs) between the start and end points, which include horizontal intersection points (HPIs) and vertical intersection points (VPIs). Additionally, bridges and underpasses are represented by MBUs, and connection points with existing roads and the railway crossing point are represented by VCCs. Together, the decision variables HPIs, VPIs, MBUs, VCCs are defined as in Eqs. (10.43)–(10.46):

$$\mathrm{HIP} = [X_H, Y_H, R_H, L_{H_0}, L_{H_1}] \tag{10.43}$$

$$\mathrm{VIP} = [K_V, H_V, R_V] \tag{10.44}$$

$$\mathrm{MBU} = [KB_S, KB_E, KU_S, KU_E] \tag{10.45}$$

$$\mathrm{VCC} = [I, S, E] \tag{10.46}$$

where X_H, Y_H, R_H, l_{H_0} and l_{H_1} represent the coordinates of horizontal intersection points, curve radii, and lengths of the preceding and succeeding transition curves, respectively; K_V, H_V and R_V represent the mileage, elevation, and vertical curve radius of a vertical PI; KB_S, KB_E and KU_S, KU_E are the start and end mileage of bridges and tunnels; I is the crossing point with the railway, and A and B are the start and end connection points with existing roads.

(3) Objective Function

The objective of the road–rail grade-separated alignment optimization model along the railway corridor is to minimize total construction costs. These include the cost of earthwork (C_E), pavement (C_P), land acquisition (C_R), bridge construction (C_B), and tunnel construction (C_U), represented as Eq. (10.47):

$$F(HPI, \mathrm{VPI}, \mathrm{MBU}, \mathrm{VCC}) = \min(C_E + C_P + C_R + C_B + C_U) \tag{10.47}$$

(4) Constraints

Grade-separated crossings between highways and railways must satisfy four types of constraints: geometric alignment constraints (CGeometry), crossing constraints (CCrosse), connection constraints (CConnect), and structural constraints according to the Highway Route Design Code (JTG D20–2017) (CStructural).

1. Geometric Alignment Constraints

Horizontal Geometry: Minimum curve radius; Maximum curve radius; Minimum tangent length; Minimum deflection angle.

Vertical Geometry: Minimum gradient segment length; Maximum gradient segment length; Minimum vertical curve radius; Maximum allowable gradient.

2. Crossing Constraints

The intersection angle A_C between the road and railway must not be less than the minimum allowed value $A_{\min}$, as shown in Eq. (10.48):

$$A_C \geq A_{\min} \tag{10.48}$$

The clearance height (difference between railway design elevation Hrail and road design elevation Hroad) must satisfy the requirement, as shown in Eq. (10.49):

$$\Delta h \leq |H_{\text{road}} - H_{\text{rail}}| \tag{10.49}$$

The alignment at the crossing must remain straight to ensure safe driving visibility.

3. Connection Constraints

At connection points between the reconstructed and existing roads, the azimuth angles must be consistent, as shown in Eq. (10.50):

$$\alpha_{\text{exist}} \leq \alpha_{\text{cross}} \tag{10.50}$$

4. Structural Constraints

Prohibited Zones: Let set UF represent forbidden grid cells and UC represent grid cells occupied by the alignment. To ensure avoidance of prohibited zones, as shown in Eq. (10.51):

$$U_F \cap U_C = \varnothing \tag{10.51}$$

Structural Requirements: The height and length of bridges and the aperture of tunnels must comply with the specifications of the Highway Route Design Code (JTG D20–2017).

In summary, the optimization model for road–rail grade-separated crossing alignments along the railway corridor can be expressed as Eq. (10.52):

$$F(\mathrm{HPI}, \mathrm{VPI}, \mathrm{MBU}, \mathrm{VCC})$$
$$\text{s.t.}\ \ C_{\mathrm{Geometry}}, C_{\mathrm{Crosse}}, C_{\mathrm{Connect}}, C_{\mathrm{Structural}} \tag{10.52}$$

10.3.2 Solution method

(1) Hierarchical Optimization Strategy

Since both the intersection point with the new railway (I) and the connection points with the existing roads (S and E) are variable (as shown in Fig. 10.20), the optimized alignment is highly dependent on terrain, geology, constraints, and objectives. The key issue is how to decompose the optimization process and mitigate the coupled interactions between these elements. The overall path from S → I → E is divided into two segments by point I: S → I and I → E. Thus, the problem is transformed into two main subproblems: 1) path splicing at the variable intersection point I, and 2) path search between varying endpoints (S & I, I & E). Effectively splicing the S → I and I → E paths at point I under crossing constraints and developing a robust path search method between a fixed and a variable endpoint is crucial.

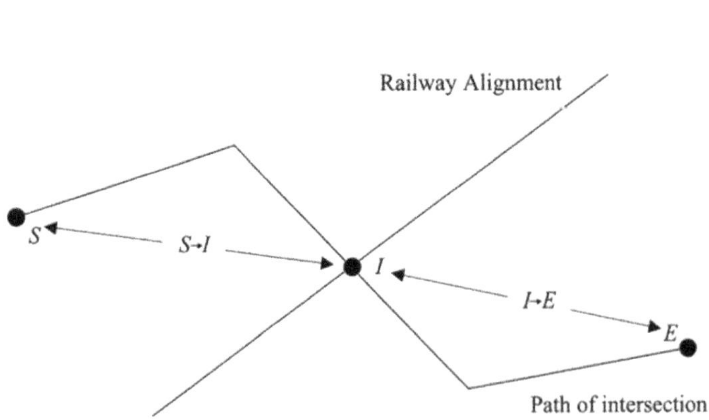

Figure 10.20 Path decomposition at intersection point I.

1. Define the cells adjacent to the new railway as two sets: the left-side cell set L = [L_1, L_2, …, L_p] and the right-side cell set R = [R_1, R_2, …, R_q]. The cells where the existing road intersects the left side of the new railway define the start point set S = [S_1, S_2, …, S_m], and similarly, the endpoint set is E = [E_1, E_2, …, E_n]. As shown in Fig. 10.21, the S → I and I → E segments are further decomposed into S → L, L → R, and R → E. The L → R crossing segment is handled with specific crossing constraints to complete the splicing at point I.
2. The Distance Transform (DT) algorithm is capable of generating optimized paths between any 2 cells. Therefore, each cell in L is taken as a target cell, and each cell in S is used as a start cell to search multiple S → L paths using the DT algorithm. Similarly, feasible paths from R to E are obtained.

(2) Path Search Between a Fixed and Variable Endpoint

For each fixed cell in L (or R), paths are searched from variable endpoint cells in S (or E). Taking a cell Li ∈ L as a fixed endpoint and S = [S_1, S_2, …, S_m] as variable start points, the path search process using the DT algorithm under the constraints involves the following steps:

Step 1: Preprocess the study area on the left side of the new railway. Cells in the area must satisfy elevation accessibility. For any cell (R_x, R_y, R_z), the elevation difference to the fixed endpoint $Li(LR_x, LR_y, LR_z)$ must be overcome using the maximum allowable slope i_{max} and circuity coefficient γ, as shown in Eq. (10.53):

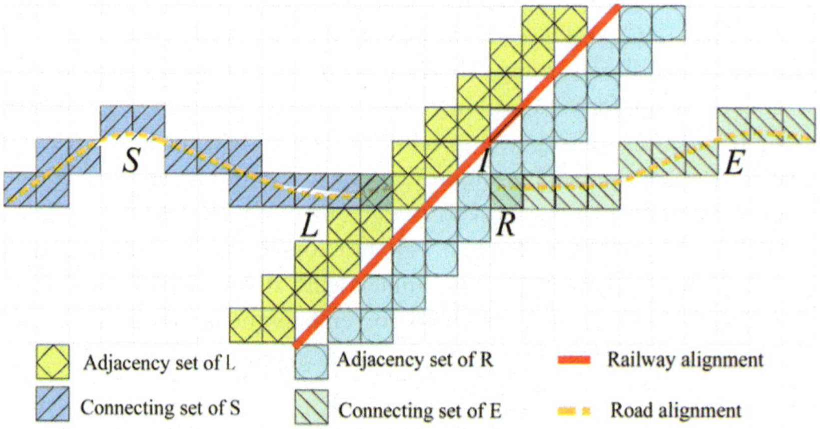

Figure 10.21 Fixed and varied points of four sets.

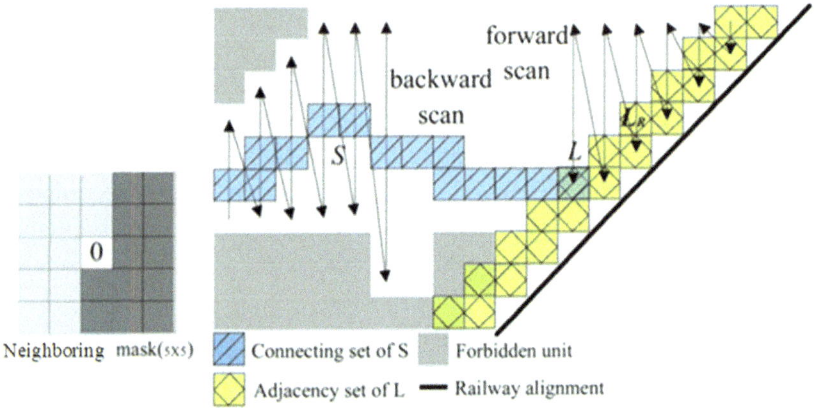

Figure 10.22 Neighborhood mask and bi-directional scan in DT algorithm.

$$(L_{R_Z} - R_Z) / \left[\gamma \times \sqrt{(L_{R_X} - R_x)^2 + \left(L_{R_Y} - R_y\right)^2}\right] \leq i_{\max} \tag{10.53}$$

Cells that fail this condition are marked as forbidden and excluded in the following bi-directional scanning.

Step 2: Initialize the generalized distance value V_{DT} of all cells in the left region, where in Eq. (10.54):

$$\begin{cases} V_{DT}(L_R) = 0 \\ V_{DT}(D) = \infty \end{cases} \tag{10.54}$$

This generalized distance incorporates total cost (e.g., $C_E + C_P + C_R + C_B + C_U$) instead of just Euclidean distance.

Step 3: As shown in Fig. 10.22, apply a bi-directional scanning procedure. Begin forward scanning (bottom to top, left to right), followed by backward scanning (top to bottom, right to left). For each scanned cell G, place the neighborhood mask over it and evaluate each neighboring cell P. For the local path segment G → P, verify compliance with constraints from Section 1.4: horizontal and vertical geometric rules, bridge height and length, tunnel depth and geological conditions, and slope/direction requirements for embankments. If the local path satisfies Eq. (10.55):

$$V_{DT}(G) + d(P \rightarrow G) < V_{DT}(P) \tag{10.55}$$

then update $V_{DT}(P)=V_{DT}(G) + d(P \rightarrow G)$ and record the row/column offset (ΔR, ΔC) for path tracing

Step 4: Using the generated distance map, identify the cell Si ∈ S with the smallest V_{DT} value. This cell becomes the path endpoint, and the shortest path from S_i to L_i is traced back using (ΔR, ΔC). Iterating through all L cells yields all optimized non-crossing paths from S to L. The same process applies for generating all R → E paths.

This chapter addresses the limitation of traditional railway alignment design that optimizes alignment in isolation while ignoring its coupling with critical infrastructure, proposing three core collaborative optimization frameworks for railway alignment and key infrastructure (stations, Large-scale Auxiliary Construction Projects (LACPs), road-rail grade-separated crossings), each with targeted optimization models and tailored solution methods. For concurrent optimization of mountain railway alignments and station locations, the model takes HPIs/VPIs and structure locations as decision variables to minimize total costs, with constraints including station spacing and gradient requirements, and uses an improved DT algorithm to search for optimized alignments and stations. For concurrent optimization of alignments and LACPs, the model minimizes construction costs of alignments, shafts, and access roads, with constraints covering geometry, location, structure, and construction duration, and adopts a four-step method combining PSO, Divide & Conquer, customized Dijkstra, and integrated iteration. For optimization of road-rail grade-separated crossings, the model based on CGIM minimizes total costs, with constraints on geometry, crossing, connection, and forbidden zones, and uses a hierarchical strategy with DT to search for optimized paths by decomposing the route and defining cell sets. Overall, the chapter builds a systematic collaborative optimization paradigm for railway alignments and critical infrastructure, providing practical tools for efficient railway design.

References

[1] Ministry of Railways of the Peoples Republic of China, Code for Design of Railway Line, China Planning Press, Beijing, 2017.

[2] Ministry of Railways of the Peoples Republic of China, Code for Design of Railway Station and Terminal, China Planning Press, Beijing, 2006.

[3] M.J. de Smith, Determination of gradient and curvature constrained optimal paths, Comput. Aided Civ. Infrastruct. Eng. 21 (1) (2006) 24–38.

[4] W. Li, H. Pu, P. Schonfeld, et al., Methodology for optimizing constrained 3-dimensional railway alignments in mountainous terrain, Transport. Res. C Emerg. Technol. 68 (2016) 549–565.
[5] W. Li, H. Pu, P. Schonfeld, et al., Mountain railway alignment optimization with bidirectional distance transform and genetic algorithm, Comput. Aided Civ. Infrastruct. Eng. 32 (2017) 8 691–709.
[6] J.C. Jong, P. Schonfeld, An evolutionary model for simultaneously optimizing three-dimensional highway alignments, Transp. Res. Part B Methodol. 37 (2) (2003) 107–128.
[7] China Railway Corporation, Code for Design of Railway Engineering Construction Organization, Q/CR 9004-2018, 2018.
[8] China Ministry of Railways, Code for Design on Railway Bridge and Culvert, 2017. TB 10002-2017.
[9] E.W. Dijkstra, A note on two problems in connexion with graphs, Numer. Math. 1 (1959) 1 269–271.
[10] T.R. Etherington, Least-cost modelling and landscape ecology: Concepts, applications, and opportunities, Curr. Landscape Ecol. Rep. 1 (1) (2016) 40–53.
[11] A. Pradhan, G. Mahinthakumar, Finding all-pairs shortest path for a large-scale transportation network using parallel Floyd-Warshall and parallel Dijkstra algorithms, J. Comput. Civ. Eng. 27 (2013) 3 263–273.
[12] A. Ratnaweera, S.K. Halgamuge, H.C. Watson, Self-organizing hierarchical particle swarm optimizer with time-varying acceleration coefficients, IEEE Trans. Evol. Comput. 8 (3) (2004) 240–255.

CHAPTER ELEVEN

Intelligent recreation optimization of existing railways

Abbreviations and corresponding full texts

ALMM Augmented Lagrange Multiplier Method
DV Deviation Value
GA Genetic Algorithm
HA Horizontal Alignment
LM Levenberg–Marquardt
MADS Mesh Adaptive Direct Search
OLSM Orthogonal Least Squares Method
PI Points of Intersection
VA Vertical Alignment

11.1 Research overview

After long-term operation of existing railways, their actual alignments inevitably shift due to continuous interaction between wheels and rails. This not only affects daily normal operations but may even lead to major safety accidents in severe cases. Therefore, it is necessary to conduct regular recreation design for existing railways to maintain alignment smoothness and ensure operational safety. When existing railways require renovation for higher train speeds, the alignment of certain sections may fail to satisfy the upgraded design standards or actual engineering requirements. These sections must also undergo recreation design to satisfy the enhanced alignment design criteria for the renovated alignments. Currently, the recreation design of existing railways still relies mainly on manual experience. However, due to limited time and resources, manual design is not only inefficient but also prone to overlooking many potential feasible alignment alternatives. Hence, research on intelligent optimization design for the recreation of existing railways is needed to improve the efficiency and quality of such work.

To minimize interference with the operation of existing railways, the recreation of existing railways must be based on the actual post-operation positions of the existing railways. By exploring the relation between the

Intelligent Railway Alignment Design
ISBN: 978-0-443-45923-8
https://doi.org/10.1016/B978-0-443-45923-8.00007-8

alignment element attribution of the measured points on the existing track center alignment and the geometric parameter configuration of the recreation alignment, a recreation alignment with the minimum horizontal and vertical deviation (d_i) can be obtained. For the recreation of existing railways, it is necessary to consider the complex coupling relations between existing and recreated alignments. The goal is to generate a recreated alignment that satisfies the recreation technical standards with the minimal engineering efforts.

The primary task of existing alignment recreation design is to determine the spatial alignment with minimal deviation that satisfies various constraints, based on the complex coupling relation between the coordinates of measured points on the center alignment of the actually operating alignment and the recreated alignment (as shown in Fig. 11.1). The range of alignment element attribution for the measured points provides a data foundation for the recreated geometric alignment, while the geometric parameter configuration of the recreated alignment will, in turn, readjust the attribution range of the measured points. These two elements interact and influence each other, Based on this relationship, a cyclic iterative adjustment mechanism can be constructed such that they reach a stable state in space. Therefore, in the existing literature, most intelligent recreation

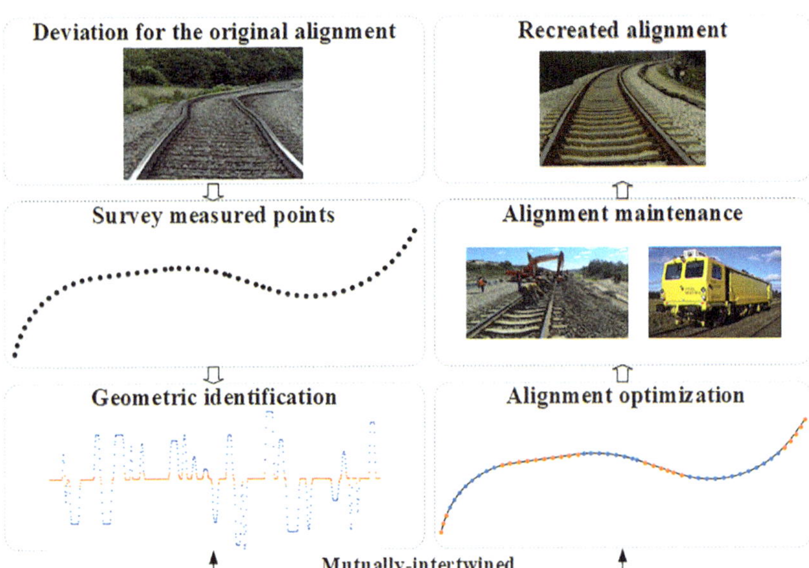

Figure 11.1 Main process of alignment recreation.

methods for existing railways are divided into two sub-processes: Identification of alignment element attribution for measured points and optimization of alignment geometric parameters. Exploring the coupling relation between these two has become a key focus in alignment recreation research.

11.1.1 Identification of alignment element attribution for measured points

In existing studies, the attribution of measured points to alignment elements is primarily determined based on the geometric characteristics of the measured points' alignment. For example:

① Use the discrete spline approximation method to construct the spline function [1]. Compute the geometric feature values of the measurement points through spline interpolation and determine the aligned feature points. This method involves a relatively complex computation process and lacks sufficient robustness. Moreover, it requires manual input of thresholds for identification, resulting in low automation.

② The curvature of the curve at the measured points is calculated by using the geometric principle of determining a circle from multiple measured points [2]. The curvature threshold in this method also needs to be manually specified, and curvature is highly sensitive to measured point data. When there are noise points [3] or insignificant changes in curvature, the identification effect is poor.

③ The azimuth at measured points [4] and their variation characteristics are used as the basis for segmentation. The azimuth and its rate of change in this method depend on the distance and angle between adjacent measured points, as well as segmentation thresholds [5]. When the quality of measured point data is poor, the identification results are unsatisfactory.

11.1.2 Optimization of alignment geometric parameters

In terms of alignment geometric parameter optimization, numerous scholars have explored the fitting and related parameter optimization of straight alignment, circular curve, and transition curve elements using various analytical or heuristic algorithms. The mainstream alignment element optimization methods include:

① Straight alignment fitting: Least Squares Method, Integer Programming;

② Circular curve fitting: Monte Carlo Method, Nelder-Mead Simplex Method, Lagrange Multiplier Method;
③ Transition curve fitting: Cubic Spline Algorithm, Levenberg–Marquardt Algorithm;
④ Heuristic algorithms for optimizing alignment parameters: Powell Algorithm, Particle Swarm Optimization Algorithm, Ant Colony Algorithm.

11.1.3 Interactive influence between the two sub-processes

In most of the aforementioned studies, the identification of alignment element attribution for measured points and the optimization of alignment geometric parameters have been treated as two independent sub-processes. In reality, these two sub-processes are intertwined and mutually influential during the recreation of existing railways [6]. The results of alignment element attribution identification for measured points affect the initial input data for the optimization of alignment geometric parameters; conversely, the results of alignment geometric parameter optimization can also lead to changes in those of alignment element attribution identification. Therefore, to explore the complex coupling relation between these two subprocesses, researchers have proposed some iterative adjustment solution methods to improve the quality of alignment recreation.

The digital and intelligent alignment design group of the Central South University proposed a heuristic recreation method called "swing iteration" [7]. During the iteration process, the ownership of measured points is continuously updated based on the alignment parameters optimized in the previous iteration, and the alignment geometric parameters are re-optimized until the convergence condition is satisfied, thus completing the reconstruction of the horizontal and vertical profiles of the existing alignment. Furthermore, the team developed the aforementioned local "swing iteration" recreation method into an overall swing iteration recreation method for the entire alignment to enhance the recreation effect. In each iterative optimization process, a hybrid algorithm combining the Mesh Adaptive Direct Search algorithm (MADS) with the Augmented Lagrange Multiplier Method (ALMM) [8] or Genetic Algorithm (GA) [9] is applied to optimize the key geometric parameters of the recreated alignment. This chapter will focus on introducing the implementation principle and process of the "swing iteration" method.

11.2 Optimization model for horizontal and vertical alignment recreation of existing railways

The purpose of existing alignment recreation is to fit and restore the 3D spatial alignment of the existing alignment based on field measured data, while satisfying specification requirements and control points of major structures near the alignment, so as to minimize the deviation from each measured point to the recreated alignment. Therefore, the design of an optimization model is the foundation for solving this problem. This section will present the model from aspects such as the compositional characteristics of horizontal and vertical alignments, design variables, objective functions, and constraint conditions.

11.2.1 Optimization model for horizontal alignment recreation

(1) Design variable

In the process of recreating an existing horizontal alignment, the key geometric parameters of the intersection points of the recreated horizontal alignment (**HA**) are usually taken as design variables. These mainly include the coordinates of the intersection points (**X**, **Y**), curve radius (**R**), length of the preceding transition curve ($\mathbf{L^p}$), and length of the following transition curve ($\mathbf{L^s}$), as shown in Fig. 11.2. Therefore, the design variables of the recreated horizontal alignment can be expressed as Eqs. (11.1) and (11.2).

$$\mathbf{HA} = \{\mathbf{X}, \mathbf{Y}, \mathbf{R}, \mathbf{L^p}, \mathbf{L^s}\} \tag{11.1}$$

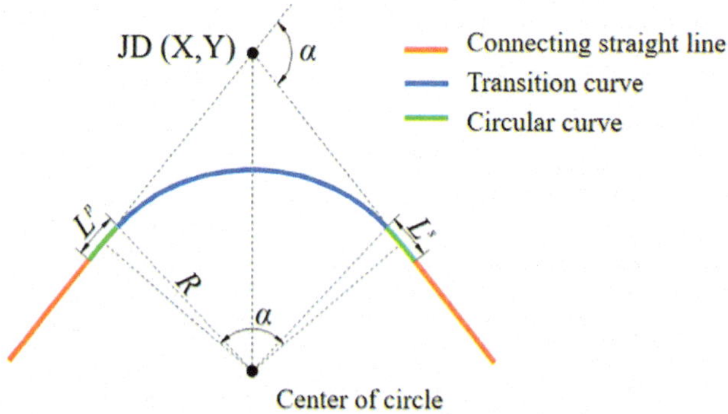

Figure 11.2 Horizontal intersection.

$$\begin{cases} \mathbf{X} = \{X_1, X_2, ..., X_m\} \\ \mathbf{Y} = \{Y_1, Y_2, ..., Y_m\} \\ \mathbf{R} = \{R_1, R_2, ..., R_m\} \\ \mathbf{L^p} = \{L_1^p, L_2^p, ..., L_m^p\} \\ \mathbf{L^s} = \{L_1^s, L_2^s, ..., L_m^s\} \end{cases} \tag{11.2}$$

where m is the number of intersection points of the recreated horizontal alignment.

(2) Objective function

As shown in Fig. 11.3, during the recreation of a horizontal alignment, the sum of squares of the normal offsets (i.e., deviations) from the measured points on the track center alignment to the recreated alignment is typically used as the criterion for evaluating the quality of horizontal alignment recreation. The smaller this value, the smaller the deviation of the spatial alignment of the existing alignment after recreation, and the better the recreation effect.

The set of center stake points obtained via field measurement for reconstructing the alignment is defined here as **M**, as Eq. (11.3).

$$\mathbf{M} = \{P_i(x_i, y_i), i = 1, ..., k\} \tag{11.3}$$

where is the horizontal coordinates of the field-measured point P_i; The objective function (F_{HR}) of the optimization problem can then be described as Eq. (11.4).

$$F_{HR}(\mathbf{HA}, \mathbf{M}) = \min \sum_{i=1}^{k} d_i^2 = \min \sum_{i=1}^{k} f_d^2(\mathbf{HA}, P_i) \tag{11.4}$$

where d_i is the distance from the measured point P to its projected point on the recreated existing alignment, i.e., the deviation; generally, in

Figure 11.3 Horizontal offset of measured points.

accordance with the alignment's direction of advance, if the measured point is located to the right of the alignment's direction of advance, d_i is positive; otherwise, d_i is negative; f_d is the function for computing the deviation of the measured points.

The deviation of the horizontal alignment refers to the distance from the measured points to the fitted alignment, and its computational method can be analyzed primarily based on the following three kinds of alignment element types: Straight alignments, circular curves, and transition curves.

① **Straight alignment element**

Let the equation of a straight alignment element be $Ax + By + c = 0$. The deviation of the measured point $P_s(x_s, y_s)$ from the straight alignment element is using Eq. (11.5).

$$d = \left| \frac{Ax_s + By_s + C}{\sqrt{A^2 + B^2}} \right| \tag{11.5}$$

② **Circular curve element**

Assume the expression of an arbitrary circular curve segment is $(x - x_H)^2 + (y - y_H)^2 = R_H^2$. When the measured point $P_s(x_s, y_s)$ lies within the range of the horizontal circular curve element of the alignment, the deviation d is using Eq. (11.6).

$$d = \left| \sqrt{(x_s - x_H)^2 + (y_s - y_H)^2} \right| - R_H \tag{11.6}$$

where (x_H, y_H) are coordinates of the center of the intersection circle; R_H is radius of the circular curve.

③ **Transition curve element**

For transition curve elements, there is no direct function to compute the deviation. We first need to find the foot of the perpendicular from the measured point $P_s(x_s, y_s)$ to the alignment; the distance from the measured point to this foot of the perpendicular is the deviation. The tangent direction at any point $P_s'(x_s', y_s')$ on the transition curve is B, and the straight-alignment direction from $P_s'(x_s', y_s')$ to the measured point $P_s(x_s, y_s)$ is A. The angle formed by these two directions is γ, as shown in Fig. 11.4. The angle between the straight alignment $\overrightarrow{P_sP_s'}$ and the tangent direction at $P_s'(x_s', y_s')$ is also γ. If γ is equal to or very close to 90 degrees, then

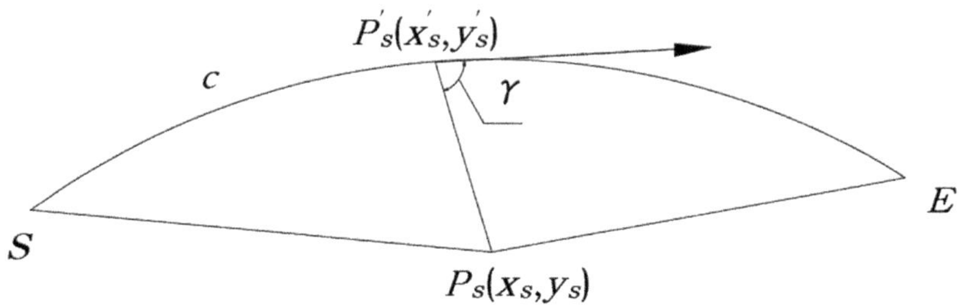

Figure 11.4 Diagram of calculation for measured points on transition curve elements.

γ is the foot of the perpendicular we must find, and solving for it is the key step.

First, construct a function where the independent variable is the length l from a point to the start of the transition curve, and the dependent variable is γ. At the start of the transition curve, $l = 0$ and $\gamma < 90$ degrees; at the end, $l = l_0$ and $\gamma > 90$ degrees. It is also possible that $\gamma > 90$ degrees at the start and $\gamma < 90$ degrees at the end. This function is monotonic, so the bisection iteration method can be used to solve for the l corresponding to $\gamma = 90$ degrees, and then the coordinates of the foot of the perpendicular can be computed. The iteration process is as follows.:

1) Define two variables a and b, which are the lower and upper bounds of the search interval, respectively. Let $a = 0$ and $b = l_0$.
2) Let variable c be the midpoint of the interval, i.e., $c = (a + b)/2$.
3) Compute $f(a)$, $f(b)$ and $f(c)$.
4) If ($f(a)$-90 degrees) ($f(c)$-90 degrees) > 0, this indicates that both $f(a)$ and $f(c)$ are acute or obtuse angles, and there is no right angle in this range. The search should be conducted in the interval [c, b], so update the lower bound of the interval to $a = c$; then proceed to step 2.
5) Otherwise, if ($f(a)$-90 degrees) ($f(c)$-90 degrees) < 0, it indicates that one of $f(a)$ and $f(c)$ is an acute angle and the other is an obtuse angle, so a right angle can be found in the interval [a, c]. Update the upper bound of the interval to $b = c$; then return to step 2.
6) Repeat the above process until the interval [a, b] becomes sufficiently small. At this point, $f(c)$ is very close to 90 degrees, and this point is the foot of the perpendicular.

After obtaining the position of the foot of the perpendicular $P_s'(x_s', y_s')$ via the Bisection Iteration Method, c represents the length from the foot of the perpendicular $P_s'(x_s', y_s')$ to the start point of the transition curve. Based on c, the mileage L of the foot of the perpendicular $P_s'(x_s', y_s')$ can be computed. The computation for the following transition curve segment is similar, as Eq. (11.7).

$$L=\begin{cases} L_{ZH}+c(\text{The measured point is on the preceding transition curve}) \\ L_{HZ}-c(\text{The measured point is on the following transition curve}) \end{cases} \tag{11.7}$$

(3) Constraint conditions

Similarly to new railway design, the recreation of existing railways must satisfy basic alignment specification constraints such as curve radius, length of straight segments between curves, transition curve length, and circular curve length. Additionally, it is necessary to ensure that the deviation at certain measured points does not exceed the allowable value, as shown in Eq. (11.8).

$$|f_d(HA, P_i)| - d_{\max} \le 0 \tag{11.8}$$

If all horizontal constraints are denoted as $c_g(\mathbf{HA})$, the optimization model for horizontal alignment recreation can be expressed as Eq. (11.9).

$$\begin{aligned} &\mathrm{F}_{HR}(\mathbf{HA}, \mathbf{M}) = \min\sum_{i=1}^{k} d_i^2 = \min\sum_{i=1}^{k} f_d^2(\mathbf{HA}, P_i) \\ &S.\text{t.}c_g(\mathbf{HA}) \end{aligned} \tag{11.9}$$

11.2.2 Optimization model for vertical alignment recreation

(1) Design variables

In the process of recreating an existing vertical alignment, the design variables typically include the mileage (**K**), elevation (**H**) of the vertical point of intersection, and vertical curve radius ($\mathbf{R}^{\mathbf{v}}$) of the recreated vertical alignment (**VA**), as shown in Fig. 11.5. Therefore, the design variables of the recreated vertical alignment can be expressed as Eqs. (11.10) and (11.11).

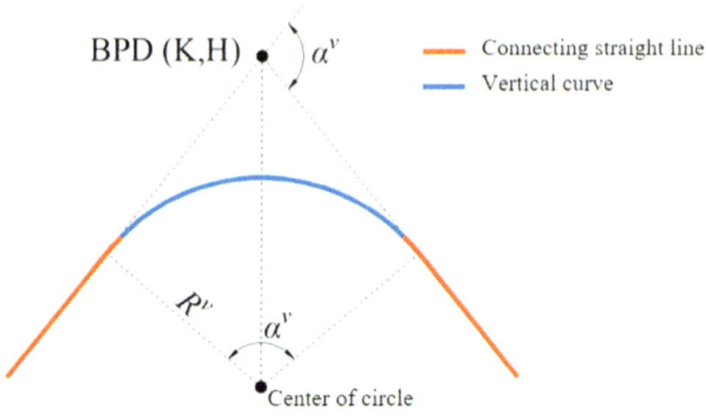

Figure 11.5 Vertical point of intersection.

$$\mathbf{VA} = \{\mathbf{K}, \mathbf{H}, \mathbf{R^v}\} \tag{11.10}$$

$$\begin{cases} \mathbf{K} = \{K_1, K_2, ..., K_n\} \\ \mathbf{H} = \{H_1, H_2, ..., H_n\} \\ \mathbf{R^v} = \{R_1^v, R_2^v, ..., R_n^v\} \end{cases} \tag{11.11}$$

where n is the number of vertical points of intersection in the recreated vertical alignment. As shown in Fig. 11.5.

(2) Objective function

In the process of recreating a vertical alignment, the sum of squares of the lifting or lowering amounts (as shown in Fig. 11.6) from the measured points on the existing rail surface to the recreated vertical alignment is typically used as the criterion for evaluating the quality of vertical alignment recreation.

Therefore, the objective function (F_{VR}) for vertical alignment recreation is computed as Eq. (11.12).

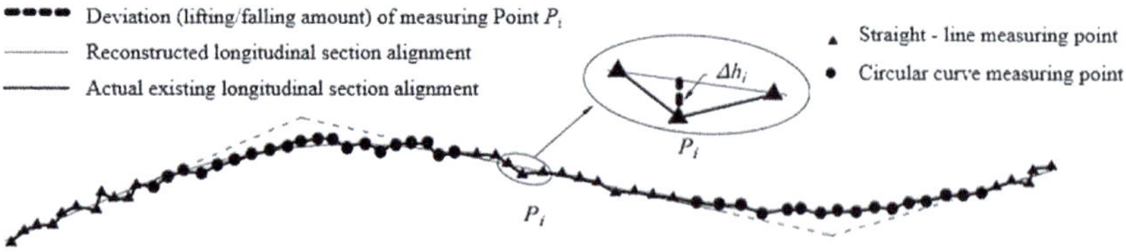

Figure 11.6 Vertical lifting and lowering of the measurement point.

$$F_{VR}(\mathbf{VA},\mathbf{M})=\sum_{i=1}^{k}\Delta h_i^2=\sum_{i=1}^{k}\left[f_h^2(\mathbf{VA},P_i)\right] \tag{11.12}$$

where Δh_i is the lifting or lowering amount (in mm) of the measured point P_i. Generally, in terms of elevation direction, a P_i located below the **VA** indicates that lifting is required, and Δh_i is positive; otherwise, it indicates lowering, and negative Δh_i. f_h is the function for computing the lifting or lowering amount of P_i when it is on different types of alignment elements, which can be divided into two categories: Straight alignments and vertical curves.

① **Straight alignment element**

Letting the equation of a vertical straight alignment element be $h=ks+b$. The lifting or lowering amount of the measured point $P_i(s_i, h_i)$ on the straight alignment element is using Eq. (11.13).

$$\Delta h_i=(k\cdot s_i+b)-h_i \tag{11.13}$$

② **Vertical curve element**

Assume the expression of an arbitrary vertical curve segment is $(s_i-s_O)^2+(h_i-h_O)^2=R_V^2$; if the measured point $P_i(s_i, h_i)$ lies within the range of the vertical curve element, the lifting or lowering amount Δh_i is using Eq. (11.14).

$$\Delta h_i=h_O\pm\sqrt{R_V^2-(s_i-s_O)^2}-h_i \tag{11.14}$$

where (s_O, h_O) are mileage and elevation of the center of the vertical curve, respectively; R_V is radius of the vertical curve; $\pm$ takes the value of "−" for concave vertical curves (crests) and "+" for convex vertical curves (sags).

(3) Constraint conditions

In accordance with the *Code for Design of Railway alignment* [10], a recreated vertical alignment must satisfy the following geometric constraints ($c_g(\mathbf{VA})$):

① The constraints on the length of the grade section between adjacent vertical points of intersection are the same as mentioned above for vertical alignment constraints.

② The equation of a straight grade segment is $h=ks+b$, where k is the grade, which must be below the maximum allowable upward grade $i_{U\max}$ and above the maximum allowable downward grade $-i_{D\max}$, i.e., as Eq. (11.15).

$$-i_{D\max} \le k \le i_{U\max} \tag{11.15}$$

③ The constraints on the algebraic difference of grades Δg_i between adjacent grade segments are the same as those specified in Chapter 4 for vertical alignment constraints.

④ The vertical curve radius R_{Vj} must satisfy the limit values $\mathrm{R}_{V\min}$ and $\mathrm{R}_{V\max}$ specified in the alignment design code, as Eq. (11.16).

$$R_{V\min} \le R_{Vj} \le R_{V\max} \quad j = 1, 2, 3, \ldots, m \tag{11.16}$$

⑤ The vertical curve segments of the vertical alignment should not overlap with transition curves, turnouts, or other equipment. This constraint can be described as the disjointness of the two sets of vertical curves and transition curves, Eq. (11.17).

$$\begin{aligned} U_H &= \{[H_{Si}, H_{Ei}]), i = 1...m\} \\ U_V &= \{[V_{Si}, V_{Ei}], i = 1...n\} \\ U_H &\cap U_V = \varnothing \end{aligned} \tag{11.17}$$

where U_H is the set of mileages for transition curves; m and n are the numbers of transition curves and vertical curves, respectively; $[H_{Si}, H_{Ei}]$ is the interval formed by the start and end mileages of the i-th transition curve; U_V is the set of mileages for vertical curves; and $[V_{Si}, V_{Ei}]$ is the interval formed by the start and end mileages of the i-th vertical curve.

⑥ To satisfy the requirements of alignment smoothness and operational safety and stability, the vertical curve elements at each vertical point of intersection must be tangent to the adjacent straight grade segments.

In addition, the vertical lifting or lowering amount of each measured point should satisfy its maximum thresholds for lifting and lowering (*ht max, hl max*), as Eq. (11.18).

$$h^l_{\max} \le f_h(\mathbf{VA},\ P_i) \le h^t_{\max}, \tag{11.18}$$

Therefore, the optimization model for vertical alignment recreation can be expressed as Eq. (11.19).

$$\begin{aligned} &\min: F_{VR}(\mathbf{VA}, \mathbf{M}) = \sum_{i=1}^{n} \left[f_h^2(\mathbf{VA}, P_i) \right] \\ &S.t.\ c_g(\mathbf{VA}) \end{aligned} \tag{11.19}$$

11.3 Swing iteration method for recreating existing 3D railway alignments

The recreation optimization of existing railways is generally divided into two sub-processes: Identification of alignment element attribution for measured points and optimization of alignment geometric parameters. Most recreation methods, after completing the initial identification of the geometric attribution of measured points, fit alignment elements in sequence and then combine them to form a complete recreated alignment. Such one-time identification-restoration methods lack a callback mechanism for the initial alignment element attributes, resulting in poor recreation effects. In response, our team proposed the idea of point-alignment consistency and conducted iterative recreation based on this idea, enabling repeated adjustments of the alignment element attribution of measured points until a stable state is reached, thereby improving the recreation precision.

11.3.1 Identification of alignment element attribution for measured points

(1) Horizontal alignment

Existing methods typically use curvature diagrams to identify the alignment elements of the horizontal alignment of existing railways. Theoretically, a curvature diagram must only identify horizontal alignments (with constant curvature) and oblique alignments, as shown in Fig. 11.7. However, in practical applications, slight offsets or errors can cause sharp changes in the curvature at measured points, leading to poor identification results of the curvature diagram, especially for circular curves and transition curve segments that are more difficult to identify.

Therefore, the azimuth change rate diagram (as shown in Fig. 11.8), which can also reflect curvature conditions, can be used for the rough segmentation of straight and curved elements: First, compute the azimuth of each measured point in sequence, then compute the azimuth

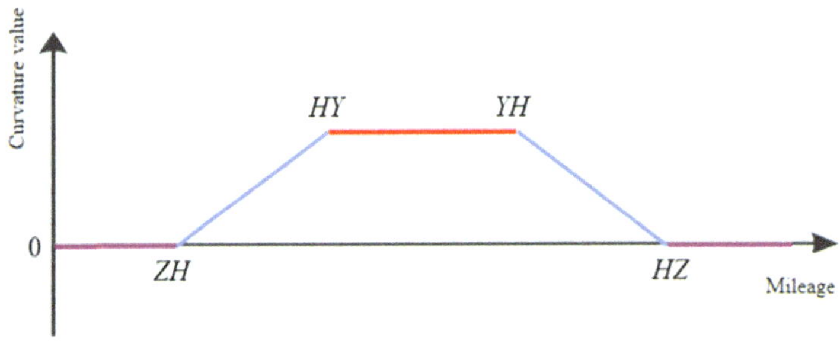

Figure 11.7 Diagram of curvature for a single intersection point.

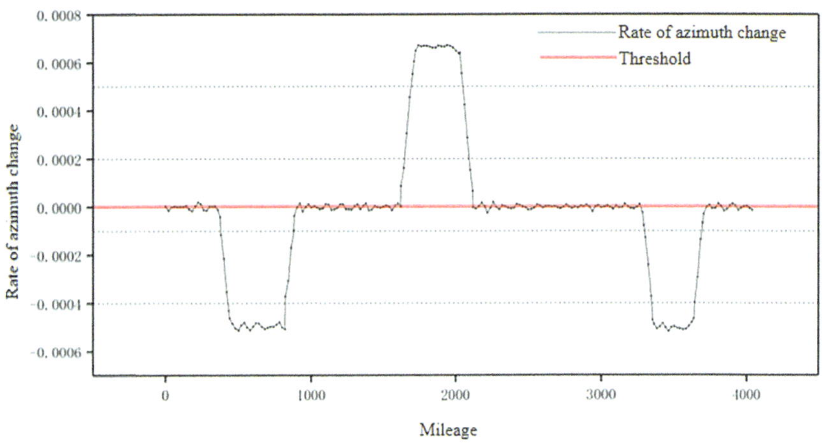

Figure 11.8 Diagram of azimuth change rate.

change rate at each measured point, perform initial segmentation of alignment elements based on the change rate values, and establish sets of measured points for straight segments and curved segments. Compared with the curvature diagram, the azimuth rate change diagram has the following advantages:

a. In terms of computational method, the azimuth change rate is theoretically equal to the curvature;

b. In terms of stability, the azimuth change rate is less affected by offsets and measurement errors, and generally does not have many noise points with sudden curvature changes;

c. In terms of identification method, only straight alignments and curves are identified during rough segmentation, while circular curves and transition curve segments, which are more difficult to identify, are accurately identified in the subsequent iteration process.

Assuming that the tangent azimuth change rate of the initial and final measured points is zero, the computation of the tangent azimuth change rate is using Eq. (11.20).

$$\theta_k = \theta_{(k-1,k)} \pm \arccos\left(\frac{b^2 + c^2 - a^2}{2bc}\right) \tag{11.20}$$

where θ_k is tangent azimuth of the k-th measured point; $\theta_{(k-1,k)}$ is azimuth of the chord $M_{k-1}M_k$; a,b,c are length of the chord $M_{k-1}M_k$, length of the chord M_kM_{k+1}, and length of the chord $M_{k-1}M_{k+1}$ respectively. To determine the azimuth change rate φ_k at the k-th measured point, it can be calculated by the following formula, as Eq. (11.21).

$$\varphi_k = \frac{\theta_k - \theta_{k-1}}{l_k} \tag{11.21}$$

where φ_k is tangent azimuth change rate at the measured point M_k; l_k is arc length between adjacent measured points $M_{k-1}M_k$.

If the maximum allowable curve radius in alignment design is set as $R_{H\max}$ it can be considered that points with $\varphi_k < 1/R_{H\max}$ belong to straight alignment elements, while the remaining measured points belong to curved alignment elements. Through this, straight and curved segments can be separated.

(2) Vertical alignment

For the vertical alignment of existing railways, the rail surface mileage data can be computed using the method introduced in Section 11.2.2 based on the relation between field-measured points and alignment elements. Combined with the measured elevation data, it forms the set of vertical rail surface measured points, which is used for recreating the vertical alignment.

There is an obvious difference in the grade change rate between straight segments and curved segments of the existing alignment's vertical alignment. Therefore, this chapter uses the grade change rate to separate measured points into straight and curved segments, as shown in Fig. 11.9. However, due to the different expressions of straight grade segments and curved segments, their function types are also discontinuous discrete functions. Hence, the approximate value of the second derivative is computed here through the second-order difference

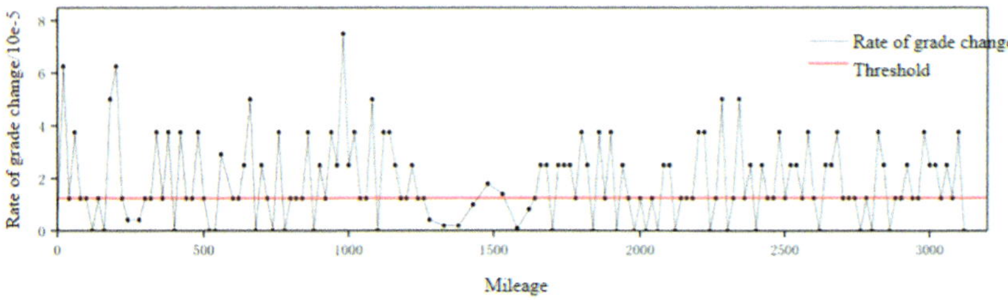

Figure 11.9 Diagram of grade change rate.

quotient of the measured point's elevation and mileage (i.e., the grade change rate) to classify the attributes of measured points. The grade change rate of measured points can be computed with Eq. (11.22).

$$\xi = h_i'' = \frac{\frac{h_{i+1} - h_i}{D_{i+1}} - \frac{h_i - h_{i-1}}{D_i}}{(D_{i+1} + D_i)} = \frac{D_i(h_i - h_{i-1}) - D_{i+1}(h_{i+1} - h_i)}{(D_i D_{i+1}(D_{i+1} + D_i))} \tag{11.22}$$

where h_i is the elevation of the i-th measured point; D_i is the mileage difference between the i-th and $(i-1)$-th measured points, as Eq. (11.23).

$$D_i = s_i - s_{i-1} \tag{11.23}$$

where s_i is the mileage of the i-th measured point.

The grade change rate of the straight segment measured point set tends to zero, while that of the vertical curve measured point set tends to the reciprocal of the radius. When ξ is less than the set threshold, the measured points can be classified into the straight grade segment measured point set; otherwise, they are classified into the curved segment measured point set. Based on engineering examples and extensive test analyses, the threshold can be set as $1/(10R_{V\min})$, where $R_{V\min}$ is the minimum vertical curve radius specified in the alignment design code.

(3) Principles for correcting measured point segments

After completing the traversal computation of all measured points, the number of alignment intersection points or vertical points of intersection can be initially determined based on the straight-curved segmentation results. However, there may still be misidentification of point attributes in each straight or curved measured point set, leading to cases where alignment element fitting fails to satisfy requirements. Therefore, it is necessary to judge and adjust the attributes of measured points to finally identify straight and curved alignment element segments that satisfy the requirements. The adjustment methods are as Fig. 11.10.

① **Number of points in measured point sets**

If a straight segment has less than 2 number of measured points, it cannot satisfy the requirements for straight alignment fitting. In such cases, the measured points of the current straight segment must be merged with those of the adjacent curved segment, and the point attributes should be corrected. Similarly, if a curved segment has less than 3 measured points, it cannot satisfy the requirements for circular curve fitting. The measured points of the curved segment should be merged with those of the adjacent straight segment, and the point attributes should be corrected, as shown in Fig. 11.10.

② **Lengths of straight alignments and circular curves**

The cumulative sum of distances between each measured point in a straight segment must satisfy the minimum length requirement for straight segments between curves. Otherwise, the straight segment should be merged with the adjacent curved segment, and the point attributes should be corrected. The cumulative sum of distances between each measured point in a curved segment must satisfy the

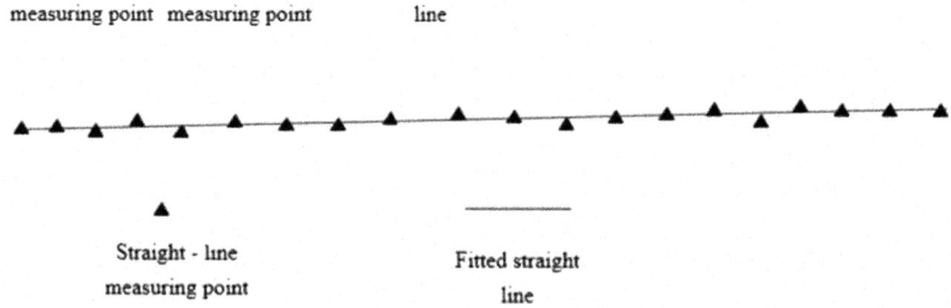

Figure 11.10 Diagram of adjusting the number of points in measured point sets.

minimum circular curve length requirement. Otherwise, the curved segment should be merged with the adjacent straight segment, and the point attributes should be corrected, as shown in Fig. 11.11.

③ **Intersection of fitted straight alignment elements**

The intersection point obtained by fitting the straight segment measured point set, which can be regarded as an alignment intersection point or vertical point of intersection, should lie within the curved segment. If it does not, the curved segment measured point set between two adjacent straight alignment elements should be merged into the straight segment, and the point attributes should be corrected, as shown in Fig. 11.11.

Through threshold identification and preliminary adjustment, the accurate number of intersection points or vertical points of intersection can be computed, laying a solid foundation for further achieving the convergence condition of point-alignment consistency in the iteration process.

11.3.2 Idea of point-alignment consistency

The idea of point-alignment consistency is applicable to both horizontal and vertical recreation. Taking the horizontal alignment as an example, the idea of point-alignment consistency is explained as follows:

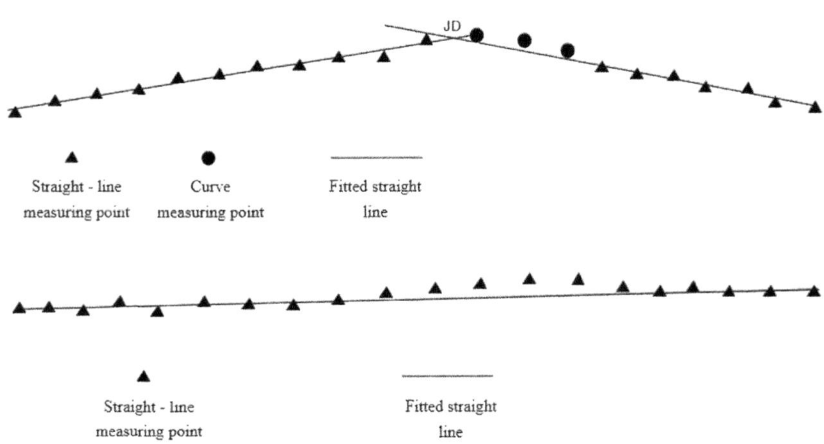

Figure 11.11 Diagram of intersection adjustment for fitted straight alignment elements.

The recreation methods for straight alignments, circular curves, and transition curves are completely different. During alignment recreation, it is first necessary to classify all measured points into three categories: Straight alignments, circular curves, and transition curves, and then adopt corresponding recreation methods for different groups of measured points. Typical measured point data of existing railways consists of a sequence of straight alignment measured point sets Q_T, preceding transition curve measured point sets Q_{SF}, circular curve measured point sets Q_C, followed by transition curve measured point sets Q_{SB}, and straight alignment measured point sets Q_T. However, during the recreation of existing railways, it is impossible to accurately determine to which type of alignment element (straight alignment, circular curve, preceding transition curve, or following transition curve) each measured point belongs. First, assume the alignment element attribution of each measured point, then use different methods to fit the alignment for the groups of measured points belonging to straight alignments, circular curves, and transition curves. After fitting, a measured point initially assumed to belong to category A may lie on the fitted category B alignment. In this case, the point's attribution is inconsistent with the alignment element type, indicating that the previous assumption is incorrect. "Point-alignment consistency" is achieved when the assumed type of all measured points exactly matches the type of the finally fitted alignment elements.

In the process of iterative fitting, when the initially assumed alignment element attribution of a measured point is exactly consistent with the alignment element attribution deduced during iteration, the measured point and the alignment element are consistent. However, after fitting alignment elements based on the rough segmentation results of measured points, there may be situations where some measured points used for fitting straight alignments lie within the curved segment (e.g., point ⑯) or some measured points used for fitting curves lie within the straight segment (e.g., points ⑦–⑩). This is known as inconsistency between point and alignment attributes, as shown in Fig. 11.12.

Based on the idea of point-alignment consistency, a swing iteration recreation optimization method is proposed. This method continuously adjusts the attributes of measured points through alignment element fitting results, achieving accurate division of the measured point set when point-alignment consistency is reached, thereby improving the recreation effect.

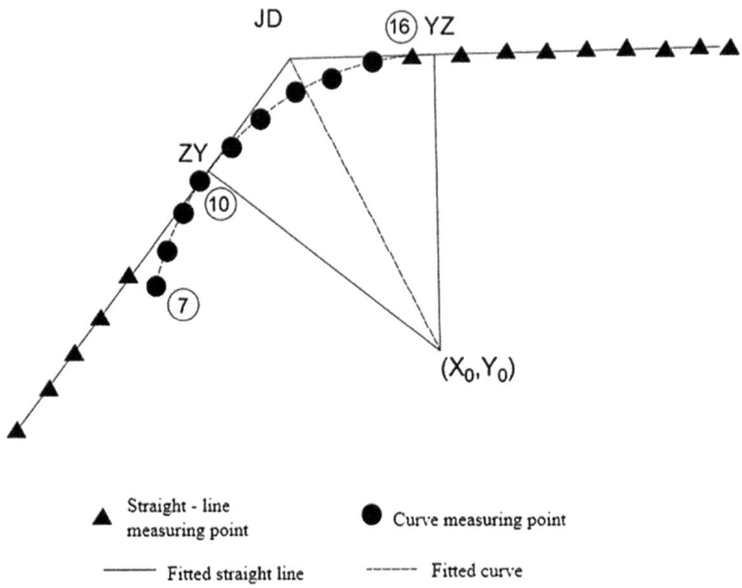

Figure 11.12 Inconsistency between measured point attributes and alignment element attributes.

11.3.3 Swing iteration recreation optimization method for horizontal alignments

(1) Swing iteration method for individual curves

A good horizontal alignment recreation effect is built on the accurate division of the measured point set. Therefore, it is necessary to consider adjusting the measured point set during the optimization process to improve the accuracy of its division and avoid poor recreation results due to inaccurate division.

To achieve accurate division of the measured point set, the idea of the swing iterations method for individual curves is to evaluate the accuracy of the measured point set division based on recreation results during horizontal alignment recreation, use this to correct the measured point set, and iteratively optimize alignment elements within the range of each intersection point. Based on the idea of point-alignment consistency, this method first reconstructs the horizontal alignment using the current division of the measured point set, then computes the types of alignment elements on which all measured points are projected in the recreated horizontal alignment to divide a new measured point set. The above process of horizontal alignment fitting and re-division of the measured point set is iterated continuously until the measured point set no longer change in consecutive iterations. At this point, the iteration is considered to have reached a steady state, and the measured point set has been accurately divided based on the horizontal alignment recreation results.

The overall computation process of swing iteration for individual curves is shown in Fig. 11.13, with specific steps as follows.

① **Initialize the measured point set**

Based on the preliminary straight-curved segmentation, for each intersection point, the measured point sets of the preceding and following straight segments between curves are taken as the straight alignment measured point sets Q_{TF} and Q_{TB}. Initially, the measured point set of the curve segment is treated as the circular curve measured point set Q_C, as shown in Fig. 11.14(a).

② **Fit each alignment element of the intersection point**

First, for the preceding and following straight segment measured point sets Q_{TF} and Q_{TB}, the least squares method is used to obtain the straight alignment element parameters. Then, for the circular curve measured point set Q_C, the circular curve least squares fitting is performed to obtain the circular curve radius R. Finally, the

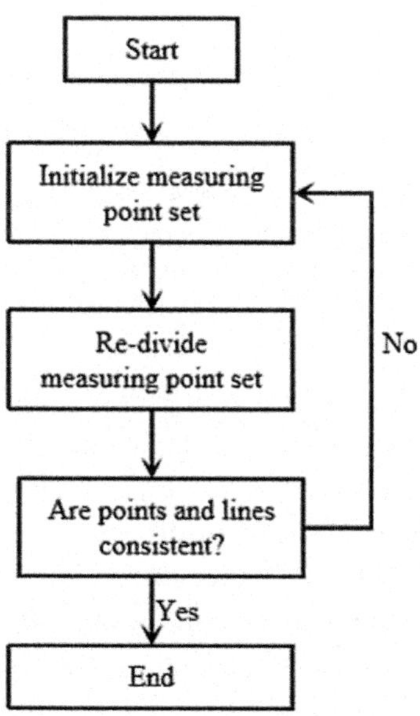

Figure 11.13 Computation flow of swing iteration for individual curves.

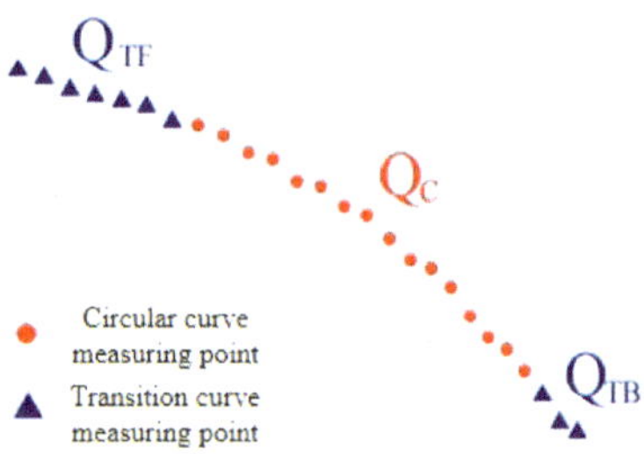

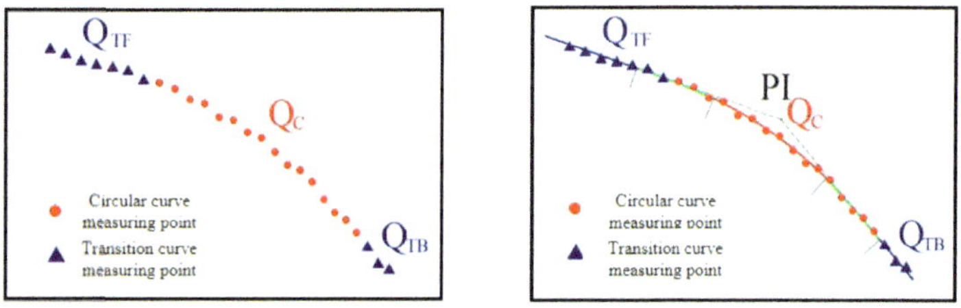

(a) Initialization of Measured Point Set (b) Fitting of Intersection Alignment elements

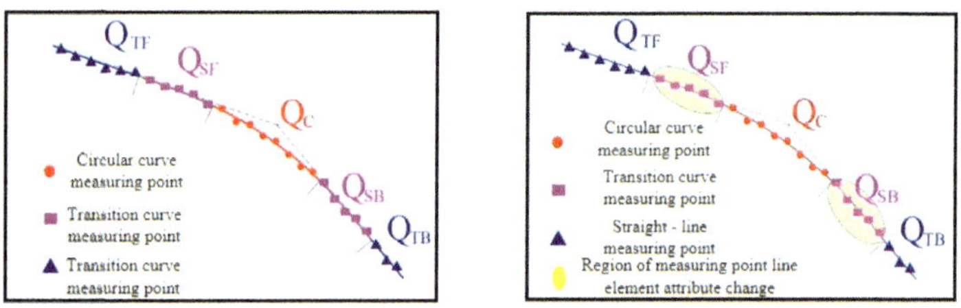

(c) Re-division of Measured Point Set (d) Detection of Point-Alignment Consistency

Figure 11.14 Example of swing iteration computation for a single curve. (a) Initialization of measured point set (b) fitting of intersection alignment elements (c) re-division of measured point set (d) detection of point-alignment consistency.

transition curve lengths l_1 and l_2 are inversely calculated using the offset distance, resulting in the transition curve measured point sets Q_{SF} and Q_{SB}, as shown in Fig. 11.14(b).

③ **Re-divide the measured point set**

After completing step (2), the horizontal alignment of the intersection point has been recreated in the current iteration. Based on the idea of point-alignment consistency, the alignment element range where each measured point lies in the recreated horizontal alignment can be computed, which serves as the new alignment element attribution of the measured point. After computing the new alignment element attribution for all measured points belonging to the alignment element at the intersection point, the measured point set can be re-divided according to the different alignment element attributions of the measured points, as shown in Fig. 11.14(c).

④ **Detect point-alignment consistency**

Check whether the new alignment element attribution of all measured points computed after horizontal alignment recreation is the same as the previous old alignment element attribution. If they are the same, the measured point satisfies the point-alignment consistency; otherwise, it does not, as shown in Fig. 11.14(d). When all measured points satisfy the point-alignment consistency, the iteration stops, and the optimal division of the measured point set is achieved; otherwise, return to step (2).

Compared with existing recreation methods that fix straight segments and only optimize curve segment parameters, the single-curve swing iteration method, based on the idea of point-alignment consistency, continuously recomputes the alignment element attribution of each measured point within the intersection range during iteration using horizontal alignment recreation results, realizing the optimization of straight segments during recreation. Compared with the former, the single-curve swing iteration method has stronger robustness and can achieve better recreation precision.

However, the swing iteration method for individual curves adjusts the straight alignments preceding and following a single curve during its recreation. When recreating the horizontal alignment for each intersection point sequentially, it fails to consider the mutual influence between adjacent intersection points, leaving room for further optimization of the recreation results. As shown in Fig. 11.15, steps (1) and (2) sequentially identify the alignment element attribution of measured points within the ranges of intersection points JD_1 and JD_2. In step (2), the range of measured points in the common straight segment between the two intersection points changes, but such a change in the measured points of the straight segment will not be fed back to intersection point JD_1 under the single-curve swing iteration method. Step (3) attempts to

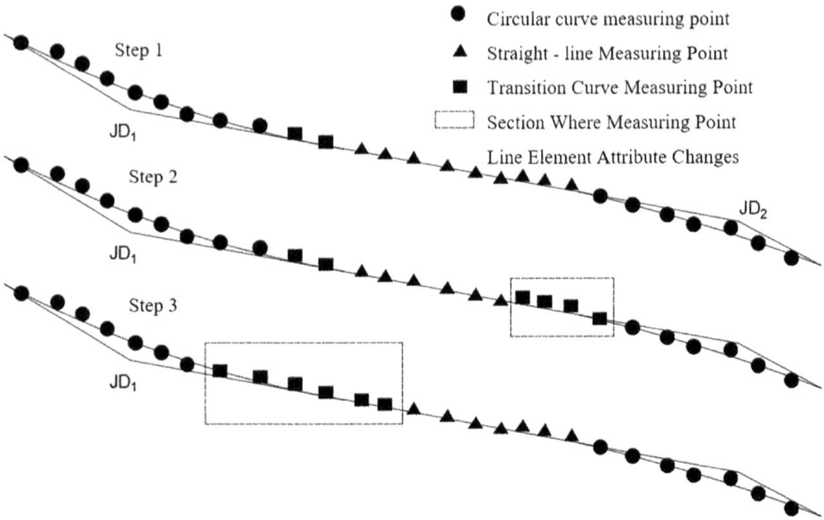

Figure 11.15 Influence between adjacent intersection points in single-curve swing iteration.

use the changing measured points of the straight segment as initial values and re-identify the alignment element attribution of measured points within the range of intersection point JD_1 using the swing iteration method. It can be observed that the following transition curve of intersection point JD_1 and the common straight segment between the two intersection points have undergone significant changing, which exert a substantial impact on the horizontally recreated alignment. Nevertheless, step (3) is not executed in the single-curve swing iteration method.

To solve the above problem, the entire recreation process can be adjusted on the basis of single-curve swing iteration and modified into an overall swing iteration method to achieve the optimal overall horizontal alignment recreation for the entire alignment.

(2) Overall swing iteration method for the entire alignment

Based on the idea of point-alignment consistency for the entire alignment, this section introduces an overall swing iteration method that fully considers the mutual influence between adjacent intersection points to reconstruct the overall alignment. In each iteration, the measured point set is first divided according to the alignment element attribution of measured points, each alignment element is recreated and optimized based on the measured point set, and then the alignment element range to which each measured point belongs is calculated to re-judge the alignment element attribution of the measured point. Through the overall optimization method, each measured point set of the straight segment between curves is jointly determined by the optimization results of the preceding and following curves. During continuous iterations, the alignment element attribution of measured points for the entire alignment will tend to be accurate. When all measured points satisfy the point-alignment consistency, the iteration stops, and the optimal overall horizontal alignment can be recreated based on each measured point set.

To ensure that the determination of the common straight segment between curves considers both the preceding and following curves, the overall swing iteration method adopts an overall iteration approach for the entire alignment to reconstruct and optimize each alignment element. In each iteration, the alignment parameters are first recreated and optimized based on all straight alignment measured point sets. Subsequently, under the condition that the alignment is fixed, the curve elements are recreated and optimized for each intersection point sequentially. Since the alignment element attribution of measured points near the straight-curve boundary has not been accurately identified, the measured point data of the straight segments preceding and following the intersection point will affect the optimization results. Therefore, when optimizing the curve elements of intersection point JD, the curve measured point sets Q_{SFi}, Q_{Ci}, Q_{SBi} within the range of JD_i, together with the straight alignment measured point sets Q_{Ti} and Q_{Ti+1} preceding and following JD_i, jointly participate in the curve element optimization process. After adjacent intersection points have been used to recreate and optimize their curve elements, the start and end points of the common straight segment element are determined by the curve elements of the preceding and following intersection points, respectively. Thus, the determination of the common straight segment at this point considers both the preceding and following curves.

In the process of recreating and optimizing each alignment element based on the measured point set, initial values of alignment element parameters are first obtained through the Least Squares Method and geometric element calculation, but these initial values are not optimal. Therefore, further optimization is required. For straight alignment element parameters, there are inequality constraints, e.g., that the deviation of controlling measured points in bridge and tunnel structure sections must not exceed the limit. The Augmented Lagrange Multiplier Method (ALMM) can be applied for optimization, which is relatively suitable. For curve element parameters, the Mesh Adaptive Direct Search algorithm (MADS) can be used for processing during optimization.

In summary, as shown in Fig. 11.16, the specific process of the overall swing iteration method for the entire alignment is as follows.

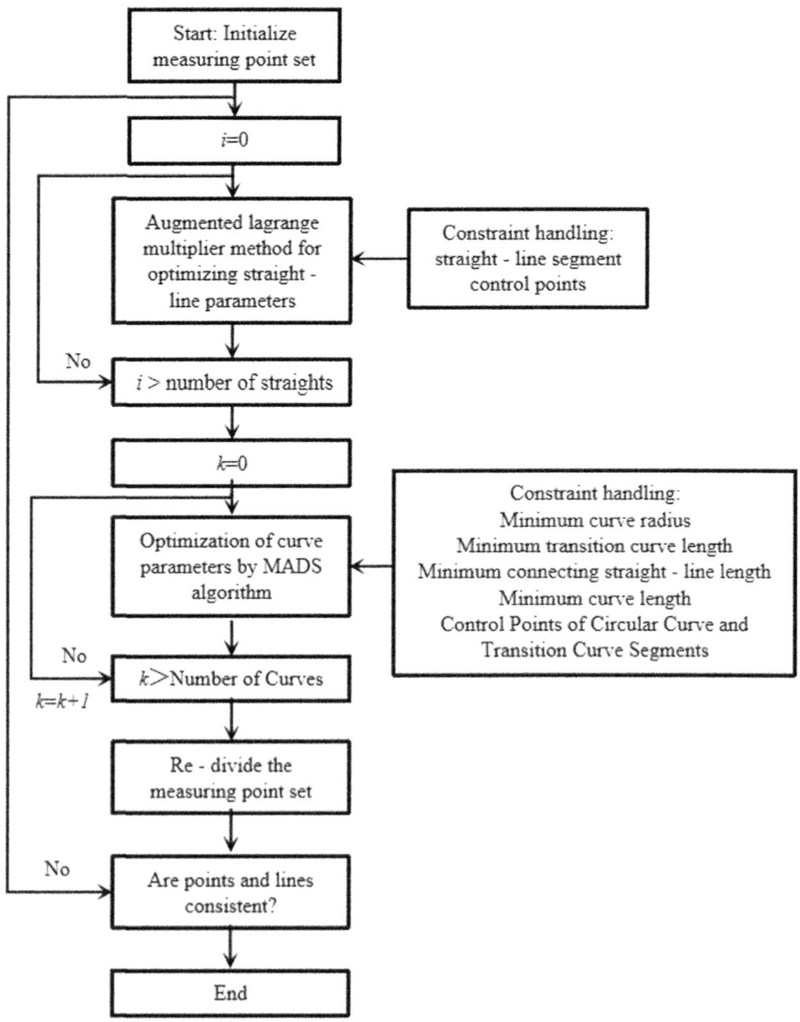

Figure 11.16 Computation process of the overall swing iteration method for the entire alignment.

① **Initialize the measured point set**
Based on the initial results of straight-curved segmentation for the measured point set, attribute classification is performed on the entire horizontal alignment's measured point set. This yields the preceding straight segment measured point set Q_{Ti} and the following straight segment measured point set Q_{Ti+1} for the *i*-th intersection point. At this stage, the curved segment measured point set is assigned to the circular curve measured point set Q_{Ci}, while the transition curve segment measured point set is preset as an empty set. Taking JD_1 and JD_2 as examples, this is illustrated in Fig. 11.17.

② **Reconstruct all straight alignment elements of the entire alignment**
The Orthogonal Least Squares Method (OLSM) and the Augmented Lagrange Multiplier Method (ALMM) are used to initially fit the straight alignment measured point sets of the entire alignment [11], obtaining the parameters of all straight alignment elements in the horizontal alignment. Based on the parametric equations of each adjacent straight segment, the coordinates of each intersection point of the alignment are further solved, as shown in Fig. 11.18.

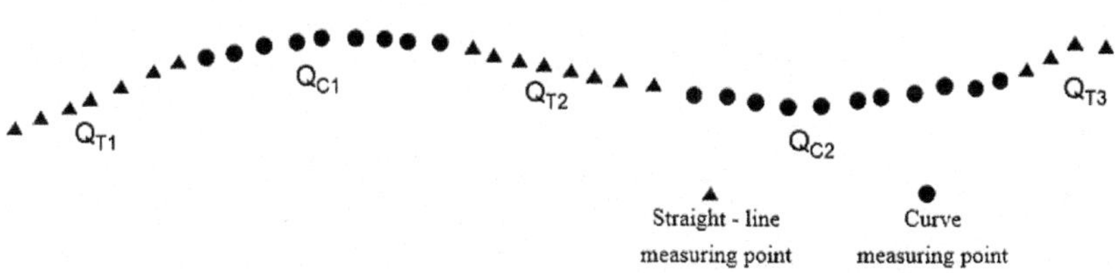

Figure 11.17 Diagram of rough straight-curved segmentation.

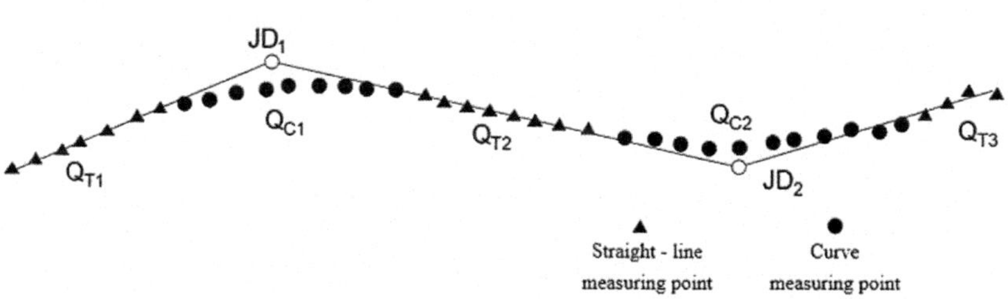

Figure 11.18 Straight alignment fitting for intersection computation.

③ **Reconstruct all curve elements of the entire alignment**

The Orthogonal Least Squares Method is applied to fit the circular curve measured point set, obtaining the center coordinates and circular curve radius. The preceding and following offset distances for each intersection point are solved using the fitting equations of adjacent straight segments and the circular curve measured points. The lengths of the preceding and following transition curves are inversely calculated from the preceding and following offset distances, and the Mesh Adaptive Direct Search algorithm (MADS) is used for constraint-aware optimization of curve parameters, as shown in Fig. 11.19. Let the center coordinates and curve radius of the i-th intersection point be (x_{Hi}, y_{Hi}) and R_{Hi} respectively; the preceding and following offset distances be p_{Fi} and p_{Bi} respectively; the lengths of the preceding and following transition curves be l_{Fi} and l_{Bi} respectively; and the equations of the preceding and following straight segments be $y_{i-1} = k_{i-1}x_{i-1} + b_{i-1}$ and $y_i = k_i x_i + b_i$ respectively. The initial solutions for the offset distances and the lengths of the preceding and following transition curves can be computed using the following formulas:

$$p_{Fi} = \left| \frac{k_{i-1}x_{Hi} + b_{i-1} - y_{Hi}}{\sqrt{k_{i-1}^{2} + 1}} \right| - R_{Hi} \tag{11.24}$$

$$p_{Bi} = \left| \frac{k_i x_{Hi} + b_i - y_{Hi}}{\sqrt{k_i^{2} + 1}} \right| - R_{Hi} \tag{11.25}$$

$$l_{Fi} = \sqrt{\left(8\sqrt{49 + 42p_{Fi}} - 56\right) R_{Hi}} \tag{11.26}$$

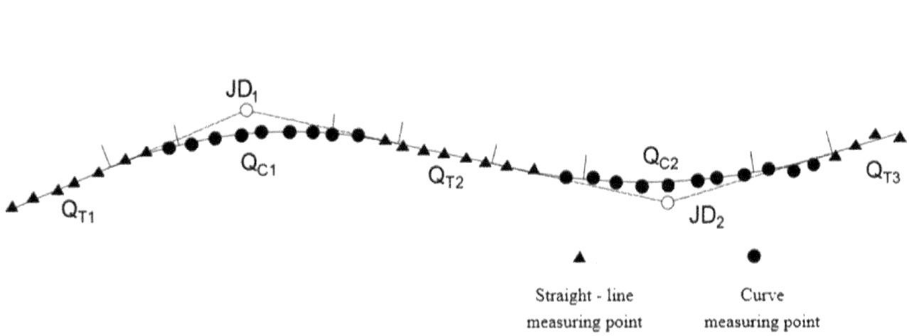

Figure 11.19 Diagram of curve parameter fitting.

$$l_{Bi} = \sqrt{\left(8\sqrt{49 + 42p_{Bi}} - 56\right)R_{Hi}} \tag{11.27}$$

④ **Re-divide the measured point set**

The parameter data of each intersection point are obtained from the above steps. The straight segment measured points and circular curve segment measured points are re-divided according to the range of transition curves. The original straight-curve-straight measured point set within a single intersection point is re-divided into a straight-preceding transition curve-circular curve-following transition curve-straight measured point set. Meanwhile, due to the re-division of each measured point set, the attributes of the measured points must be corrected, as shown in Fig. 11.20.

⑤ **Detect point-alignment consistency**

As shown in Fig. 11.21, the new alignment element attribution of all measured points is compared with their attribution before re-division. If there are measured points that do not satisfy the point-alignment consistency, jump back to step (2); otherwise, the iteration ends. As shown in Fig. 11.22, the obtained intersection point parameters at this stage are the best ones for the entire alignment.

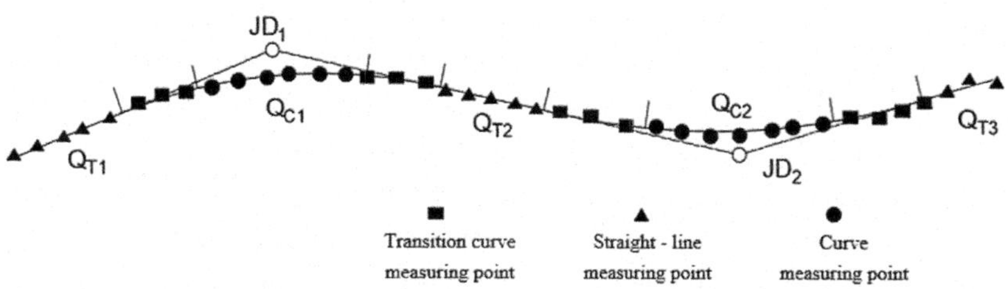

Figure 11.20 Re-division of measured point attributes.

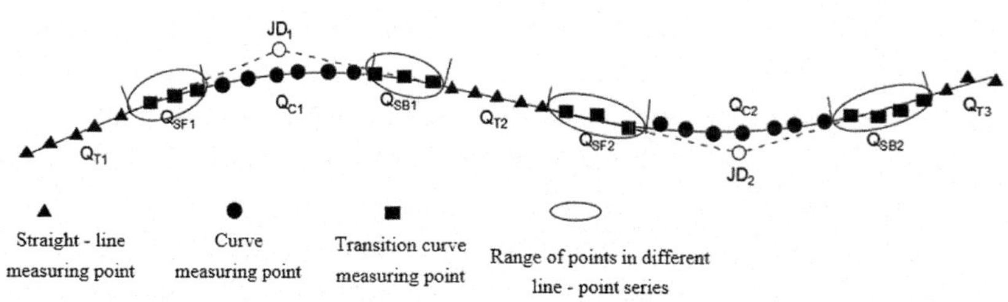

Figure 11.21 Detection of point-alignment consistency.

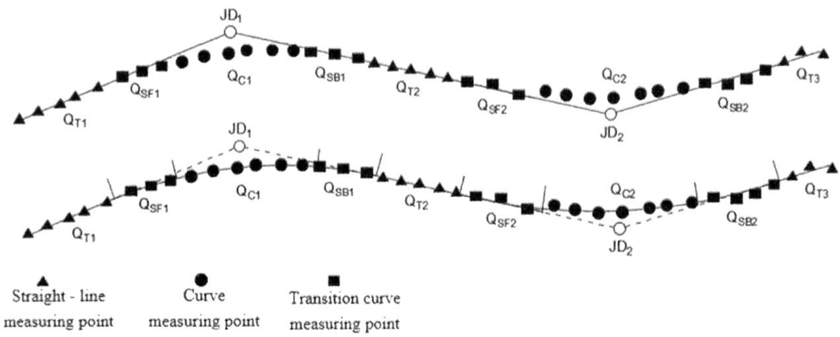

Figure 11.22 Secondary iteration satisfies point-alignment consistency for the entire alignment.

The alignment element types computed through steps (1) to (4) enable the first division of each measured point set. During subsequent iterations, the attributes of measured points near feature points will oscillate from left to right or right to left until accurate segmentation of measured points under the point-alignment consistency condition is achieved, thereby fitting the initial overall horizontal alignment. Compared with the swing iteration method for individual curves, the overall swing iteration method for the entire alignment adopts a completely different recreation strategy. In each iteration, the parameters of all straight alignment elements of the entire alignment are fitted first, followed by the parameters of all curve elements. This ensures that the common straight segment between curves is determined under the condition of simultaneous optimality of the preceding and following curves, achieving more accurate and reasonable division of the measured point set through continuous iteration. Additionally, corresponding optimization algorithms are applied in the optimization of straight alignment and curve elements, realizing constraint-aware optimization of alignment element parameters and ultimately obtaining the overall optimal horizontal alignment recreation results for the entire alignment.

(3) Curve parameter optimization based on MADS algorithm

The Mesh Adaptive Direct Search (MADS) algorithm is an extension of the Generalized Pattern Search (GPS) algorithm. It obtains approximate optimal solutions by iteratively searching for descent directions and step sizes, featuring global convergence and independence from the derivatives of the objective function. It can handle various complex constraints simultaneously but is not suitable for cases with excessive independent variables, making it more appropriate for accelerating local optimization of alignments. In this chapter, all intersection points along the alignment are traversed, and MADS is applied to each intersection point for constraint-aware curve parameter optimization [11].

The iterative search process of the MADS algorithm mainly consists of two steps: Search and poll. During the search process, the objective function of search points can be evaluated at any position within the set grid.

The process of using MADS for curve parameter optimization of the j-th horizontal intersection point is as follows:

① Objective function design: Let all measured points within the range of the j-th horizontal intersection point be P_i ($i = 1, \ldots, n$). The objective function can be expressed as Eq. (11.28).

$$F = \min \sum_{i=1}^{n} d_i^2 = \min \sum_{i=1}^{n} f_d^2 \left(x_j^k, P_i \right) \tag{11.28}$$

where $x^k_j = (l^k_{Fj}, R^k_{Fj}, l^k_{Bj})$ is the optimal grid point after the k-th iteration within the range of the j-th intersection point; l^k_{Fj}, R^k_{Fj}, l^k_{Bj} are the preceding transition curve length, curve radius, and following transition curve length of the j-th intersection point after the k-th iteration; P_i is the i-th measured point; and n is the number of measured points within the range of the j-th intersection point.

② Algorithm initialization: Initialize the current optimal grid point as $x^0_j = (l^0_{Fj}, R^0_{Fj}, l^0_{Bj})$, where l^0_{Fj}, R^0_{Fj}, l^0_{Bj} are set as the preceding transition curve length, curve radius, and following transition curve length obtained from swing iteration, respectively. To avoid local optima in the search, the initial grid size for R^0_{Fj} is set to 100 m, and the initial grid sizes for $l^0_{Fj} l^0_{Bj}$ are set to 10 m, considering the different variation ranges of circular curve radius and transition curve length.

③ Search process: Select test points on the grid in sequence and compute their function values. If a new grid point with $F(x^k_j) < F(x^k_j)$ is found during the search, the search step is successful, and the optimal solution x^k_j is updated. If no new grid points are found, proceed to the polling step.

④ Polling process: Polling primarily estimates the objective function of grid points within the adjacent range of the current grid point. If a new grid point with a better objective function value than the current point is found, the optimized solution x^k_j is updated for the next iteration. The set of grid points generated during polling is called the frame P^k. If the search process fails, a set of grid points within a certain neighborhood of the current grid point is generated on the frame P^k for optimization search.

⑤ If no better solution is found in both the search and polling processes, set $x_j^{k+1} = x_j^k$ and reduce the grid size parameter to search for a better locally optimal solution.

⑥ If a better solution is generated in both the search and polling processes, update the current optimal solution and expand or keep the grid size parameter unchanged to accelerate the search for the global optimal solution. The search terminates when the grid size reaches the set value.

For handling constraints, the MADS algorithm provides two basic methods: The extreme barrier method and the progressive barrier method. The extreme barrier method is used for handling non-negotiable constraints, which are applicable to mandatory constraints such as those in specifications. During the search process of the MADS algorithm, grid points are checked for constraint satisfaction; if a grid point violates any constraint, it is discarded as infeasible. The progressive barrier method is used for handling negotiable constraints, which are applicable to constraints that should be satisfied as much as possible in alignment recreation. During the search, if a grid point does not satisfy the constraints, it is adaptively reduced during iteration until a solution satisfying all control point constraints is found.

11.3.4 Swing iterative recreation optimization method for vertical alignments

The recreation method for vertical alignments is similar to that for horizontal alignments, which is also based on the idea of point-alignment consistency. The vertical alignment recreation is performed after the completion of horizontal alignment recreation. The specific process of vertical alignment recreation is as follows.

(1) Form a sequence of stake number measured points for the existing rail surface alignment

Based on the recreated horizontal alignment generated in Section 11.3.3, organize the set of measured points M on the existing rail surface alignment into a sequence of stake number measured points (each measured point data includes mileage and existing rail surface elevation information), providing a data basis for subsequent recreation processes.

(2) Initially identify the geometric attribution of measured points

Since the grade change rates of straight segments and vertical curve segments in the vertical alignment are significantly different, this chapter uses the grade change rate of rail surface measured points as the basis for dividing their geometric attribution [12]. Each measured point's grade change rate is compared with the set grade change rate threshold f_v. If the grade change rate of a measured point is less than f_v, it is classified as a straight alignment measured point; otherwise, it is classified as a vertical curve measured point.

(3) Adaptive fitting of straight and curved alignment elements

With the goal of minimizing the sum of squares of lifting/lowering amounts and considering various constraints mentioned in Section 11.2.2, first, the Least Squares Method is used to obtain the straight alignment element parameters for the preceding and following straight segment measured point sets. Then, the Least Squares Method for circular curves is applied to the vertical curves to obtain the vertical curve radius R. All alignment elements are then combined to restore an initial recreated vertical alignment.

(4) Iterative optimization process

After completing steps (1)–(3), further optimization of the initially recreated vertical alignment is required using the overall alignment swing iteration method based on the idea of point-alignment consistency. During this process, the straight alignments and vertical curves must be tangent at junction points during fitting to ensure alignment smoothness. Inconsistencies between measured points and alignment element attributes may also exist at the boundaries between vertical curves and straight alignments. The specific iterative process is as follows:

Step 1: Divide the measured point set according to the rough segmentation results to obtain the vertical straight segment measured point sets Q^t_{i-1}, Q^t_{i+1} and vertical curve segment measured point set Q^c_i for the entire alignment, as shown in Fig. 11.23.

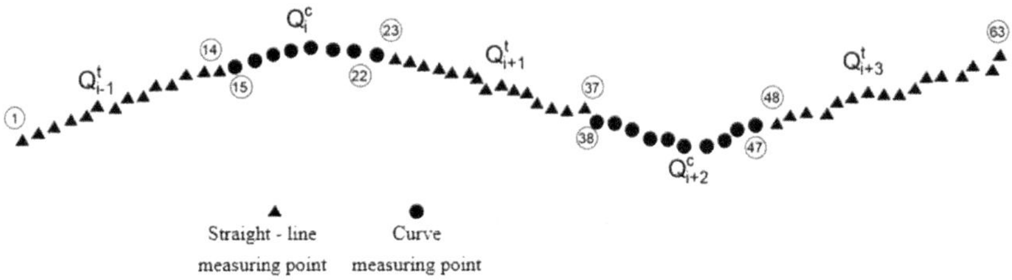

Figure 11.23 Rough straight-curved segmentation.

Step 2: Use the Orthogonal Least Squares Method and the Augmented Lagrange Multiplier Method (ALMM) to fit the straight alignment measured point sets of the entire alignment, obtain the straight grade segment parameters, and compute the intersections based on the fitting equations of adjacent straight grade segments to determine the position of each vertical point of intersection, as shown in Fig. 11.24.

Step 3: Perform vertical curve computation and fitting to determine the vertical curve radius that satisfies the constraint conditions. Simultaneously solve the parametric equations of straight alignments and vertical curves to obtain the mileages of the start and end points of vertical curves, and then determine the geometric ranges of straight alignment and curve elements, as shown in Fig. 11.25.

Step 4: Re-divide the attributes of measured points within the range of each vertical point of intersection and compare them with the initial measured point attributes to determine whether the measured point attributes have changed between two consecutive iterations, as shown in Fig. 11.26.

Step 5: If the measured point attributes are inconsistent, this indicates that the attributes are not yet stable; return to Step 2 until the point-alignment consistency requirement is satisfied. If the measured point

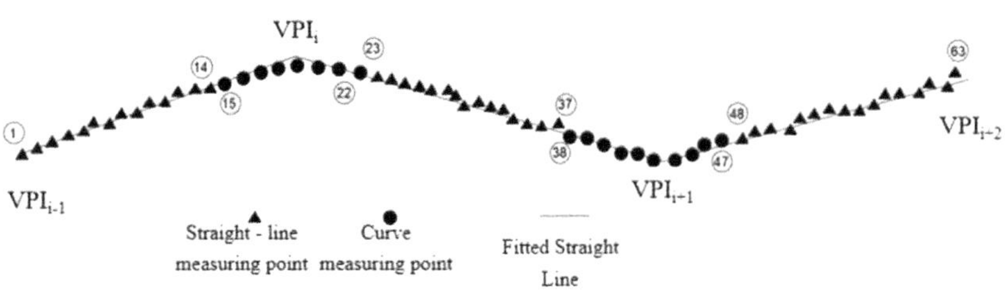

Figure 11.24 Straight alignment fitting for intersection calculation.

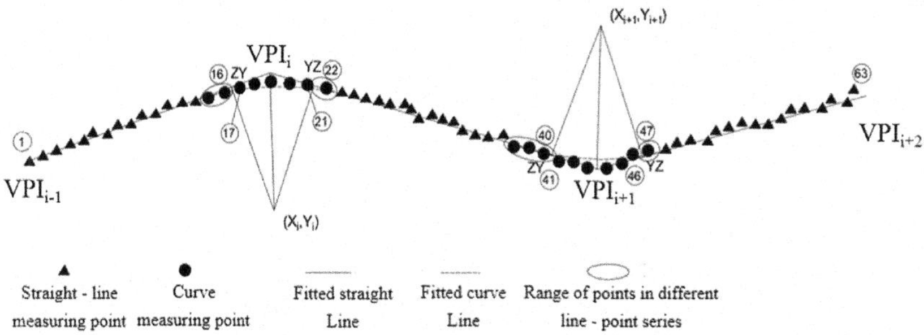

Figure 11.25 Dividing boundary point ranges.

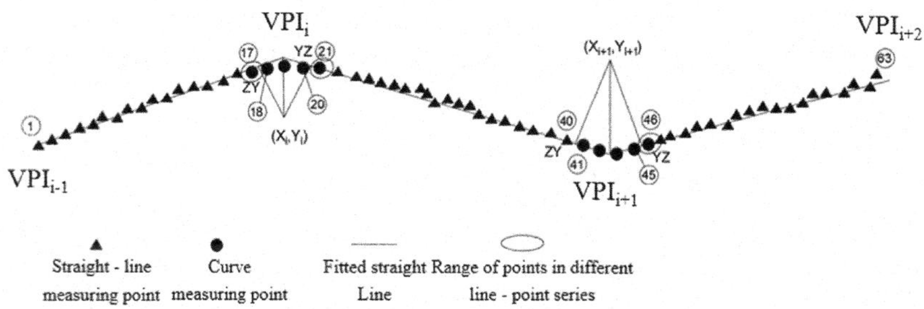

Figure 11.26 Re-division of measured point attributes.

attributes satisfy the point-alignment consistency requirement after adjustment, the iteration terminates, and the initial alternative of the vertical alignment reaches optimality, as shown in Figs 11.27 and 11.28.

Through the above overall alignment swing iteration process, accurate segmentation of measured points under the condition of point-alignment consistency can be achieved, thereby fitting the initial vertical alignment of the entire alignment. The changes in the number of measured points in each alignment element range are shown in Table 11.1.

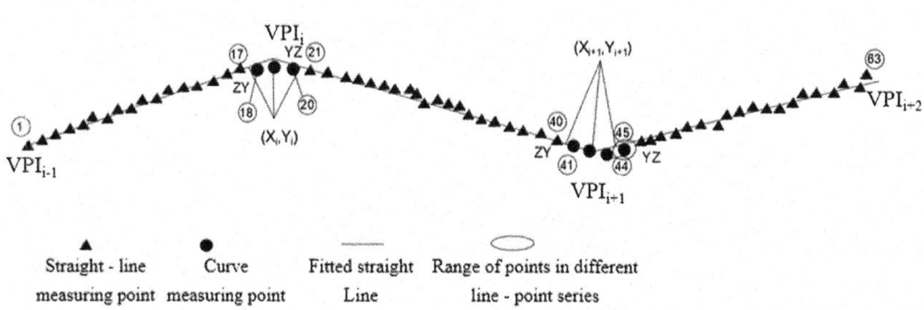

Figure 11.27 The *i*-th iteration.

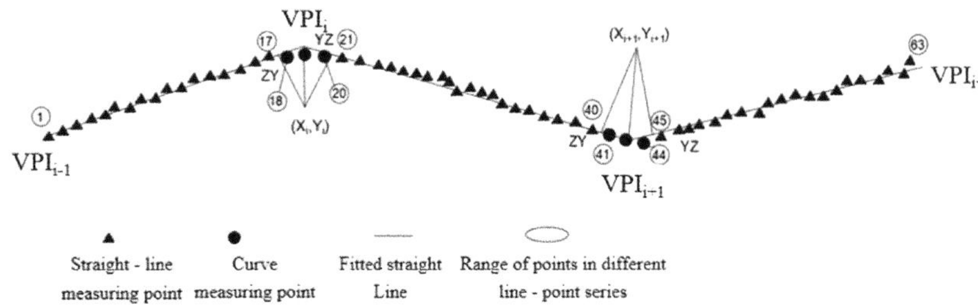

Figure 11.28 Convergence at the *n*-th iteration.

11.4 Adaptive recreation and optimization method for horizontal and vertical alignments of existing railways

11.4.1 Global adaptive horizontal recreation and optimization method for alignments

Most recreation methods, including the swing iteration discussed in the previous section, fit alignment elements in the order of straight alignments first and then curves after completing the initial identification of the geometric attribution of measured points. Then they combine them to form a complete recreated alignment. However, in complex existing railways, this method cannot be directly applied to sections with a high proportion of curves, showing certain limitations.

Table 11.1 Adjustment of measurement point attributes during the vertical section oscillation iterative recreation process.

Iteration	Q_{i-1}^t	Q_{i-1}^c	Q_{i+1}^t	Q_{i+2}^c	Q_{i+3}^t	Serial number of the measurement points with inconsistent points and alignments
1	1–14	15–22	23–27	38–47	48–63	-
2	1–16	17–21	22–40	41–46	47–63	15–16, 22, 38–40, 47
3	1–17	18–20	21–40	41–15	16–63	17, 21, 46
…	…	…	…	…	…	…
i	1–17	18–20	21–40	41–44	45–63	45
n	1–17	18–20	24–40	41–44	45–63	-

To address the limitations of the aforementioned methods and improve recreation accuracy, this section divides complex existing railways into straight-segment-dominated sections and curve-dominated sections, and proposes a global adaptive recreation strategy. Based on two alignment recreation methods—straight-first-then-curve method and curve-first-then-straight method—an initial horizontal alignment is recreated.

After completing the identification of alignment element attribution of measured points, the initial number of straight segments between curves and circular curves can be determined. The final identification of the geometric attribution of measured points is as follows:

$$\mathbf{M} = \left\{\mathbf{M}^{\mathbf{t}}, \mathbf{M}^{\mathbf{c}}, \mathbf{M}^{\mathbf{lp}}, \mathbf{M}^{\mathbf{ls}}\right\}$$

$$\begin{cases} \mathbf{M}^{\mathbf{t}} = \left\{\mathbf{M}_1^{\mathbf{t}}, \mathbf{M}_1^{\mathbf{t}}, \ldots, \mathbf{M}_{m-1}^{\mathbf{t}}\right\} \\ \mathbf{M}^{\mathbf{c}} = \left\{\mathbf{M}_1^{\mathbf{c}}, \mathbf{M}_1^{\mathbf{c}}, \ldots, \mathbf{M}_{m-2}^{\mathbf{c}}\right\} \\ \mathbf{M}^{\mathbf{lp}} = \left\{\mathbf{M}_1^{\mathbf{lp}}, \mathbf{M}_2^{\mathbf{lp}}, \ldots, \mathbf{M}_{m-2}^{\mathbf{lp}}\right\} \\ \mathbf{M}^{\mathbf{ls}} = \left\{\mathbf{M}_1^{\mathbf{ls}}, \mathbf{M}_2^{\mathbf{ls}}, \ldots, \mathbf{M}_{m-2}^{\mathbf{ls}}\right\} \end{cases} \tag{11.29}$$

where $\mathbf{M}$ is defined as in Section 11.2.1; $\mathbf{M}^{\mathbf{t}}$ is the subset of measured points on straight segments between curves; $\mathbf{M}^{\mathbf{c}}$ is the subset of measured points on circular curves; $\mathbf{M}^{\mathbf{lp}}$ and $\mathbf{M}^{\mathbf{ls}}$ are the subsets of measured points corresponding to the preceding transition curve and following transition curve, respectively; m is the total number of alignment intersection points including the start and end points, so the number of straight segments in the entire alignment is $m-1$, and the number of curve segments is $m-2$.

In general, the alignment features rapid changes in local sections, with many small-radius curves to bypass obstacles or overcome difficult terrain (defined as Case A, sections with a high proportion of curves). However, in a relatively simple surrounding environment, the alignment may be very straight to reduce its costs and ensure operational quality (defined as Case B, sections with a high proportion of straight alignments). For complex existing railways, both cases may occur (as shown in Fig. 11.29). Therefore, to reconstruct the horizontal alignment of a complex existing alignment, both Case A and Case B must be analyzed.

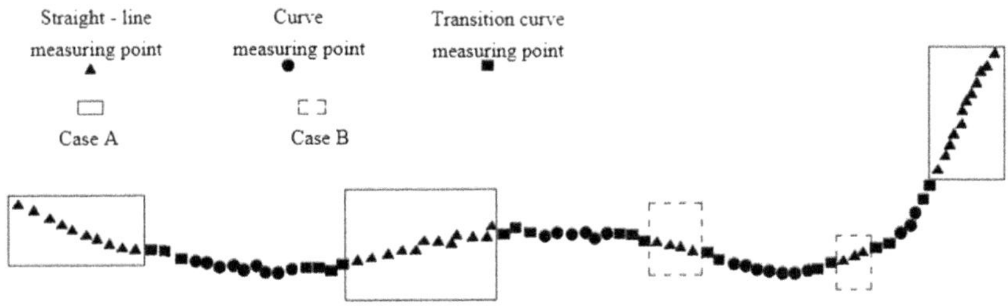

Figure 11.29 Actual distribution of measured points on existing railways.

In this research field, previous methods usually reconstruct a horizontal alignment by first fitting straight segments between curves and then circular curves. Therefore, the determination of geometric parameters of circular curves and transition curves is mainly affected by the parameters of straight segments between curves. This method is suitable for Case B but not for Case A. When this method is used in Case A, the recreation effect of straight segments between curves is relatively ideal, but curve elements may produce unacceptable offsets between measured points and the fitted alignment. In fact, the recreation order of geometric alignment elements of the alignment should be dynamically determined according to the geometric attribution and distribution of measured points. Therefore, this chapter introduces two alignment recreation methods, namely the straight-first-then-curve method and the curve-first-then-straight method.

(1) Straight-first-then-curve method

The straight-first-then-curve method was often used in previous alignment recreation studies. Based on the parameters provided in Fig. 11.30, the implementation process of the straight-first-then-curve method is as follows:

① Use the Least Squares Method to fit $\mathbf{M^t}_1$ into a straight segment between curves ($\boldsymbol{L_T}$) to obtain its slope (K) and intercept (B).

② Use the Least Squares Method to fit $\mathbf{M^c}_1$ and $\mathbf{M^c}_2$ into two adjacent circular curves (Cir_1 and Cir_2) to determine their radii (R_1 and R_2) and centers (C_1 and C_2).

③ Compute the offset $p^s{}_1$ between Cir_1 and $\boldsymbol{L_T}$ to determine the transition curve $L^s{}_1$ (Eq. 11.30). Similarly, compute the offset $p^p{}_2$ between Cir_2 and $\boldsymbol{L_T}$ to determine the transition curve $p^p{}_2$ (Eq. 11.31).

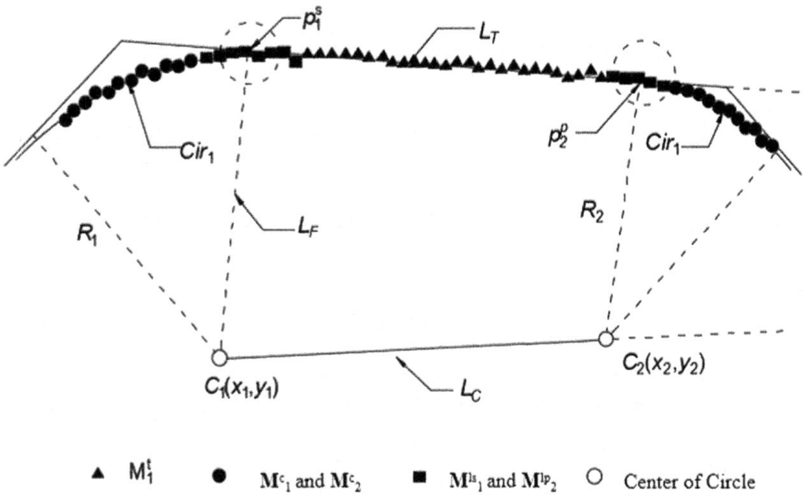

Figure 11.30 Straight-first-then-curve method.

$$p_1^s = \left(|Kx_1 - y_1 + B| \,/\, \sqrt{1+K^2}\right) - R_1, L_1^s = \sqrt{\left(8\sqrt{49+42p_1^s} - 56\right)R_1} \tag{11.30}$$

$$p_2^p = \left(|Kx_2 - y_2 + B| \,/\, \sqrt{1+K^2}\right) - R_2, L_2^p = \sqrt{\left(8\sqrt{49+42p_2^p} - 56\right)R_2} \tag{11.31}$$

where (x_1, y_1) are the coordinates of center C_1; (x_2, y_2) are the coordinates of center C_2; it is assumed that the equation of straight alignment $\boldsymbol{L_T}$ is $y = Kx + B$.

(2) Curve-first-then-straight method

Compared with the straight-first-then-curve method, the process of the curve-first-then-straight method is more complex. Based on Fig. 11.31, the implementation process of the curve-first-then-straight method is as follows.:

① Use the Least Squares Method to fit $\mathbf{M^c}_1$ and $\mathbf{M^c}_2$ into two adjacent circular curves (Cir_1 and Cir_2) to determine their radii (R_1 and R_2) and centers (C_1 and C_2).

② Based on $\mathbf{M^{ls}}_1$ and $\mathbf{M^{ls}}_2$, obtain p^s_1 and p^p_2 by using the Levenberg–Marquardt (LM) algorithm [13]. Taking $\mathbf{M^{ls}}_1$ as an example, the fitting process of the LM algorithm is shown in Fig. 11.32. Then compute p^s_1 and p^p_2 through Eq. (11.32).

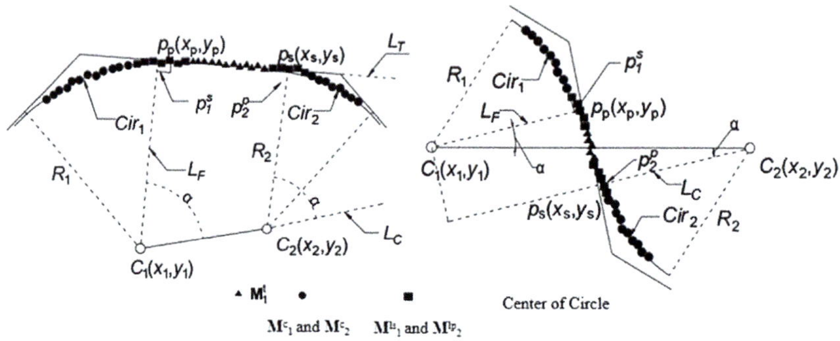

(a) Same - direction curve model (b) Reverse - direction Curve Model

Figure 11.31 Curve-first-then-straight method.

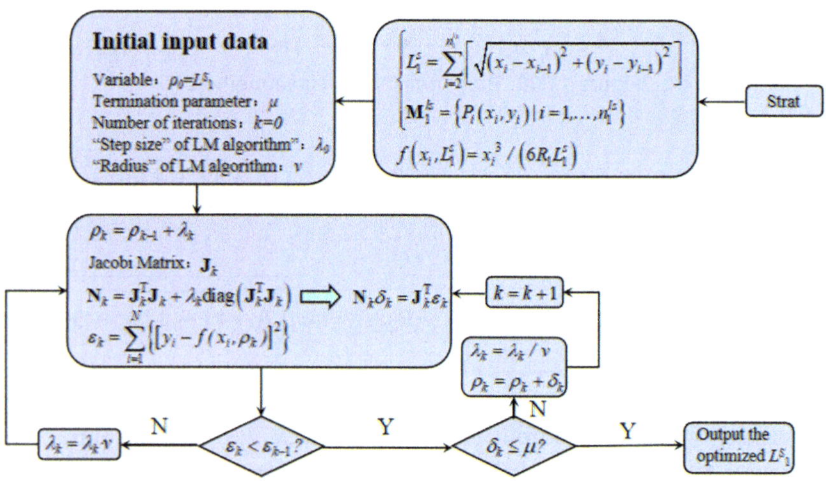

Figure 11.32 Fitting process of the Levenberg-Marquardt algorithm.

$$
\begin{aligned}
p_1^s &= \left(L_1^s\right)^2/(24\cdot R_1)+\left(L_1^s\right)^4/\left(2688\cdot R_1^2\right),\\
p_2^p &= \left(L_2^p\right)^2/(24\cdot R_2)+\left(L_2^p\right)^4/\left(2688\cdot R_2^2\right)
\end{aligned}
\tag{11.32}
$$

where: $p^s{}_1$ is the offset of Cir_1 due to the transition curve; $p^p{}_2$ is the offset of Cir_2 due to the transition curve.

③ Based on the same-direction and reverse curve models constructed in Fig. 11.31, determine the two feet of perpendicular P_p and P_s on the fitted straight segment $\boldsymbol{L_T}$ between curves to determine the position of $\boldsymbol{L_T}$, as Eqs. (11.33) and (11.37).

$$
x_p=x_1+\left(R_1+p_1^s\right)\cdot\sin\beta,\ y_p=y_1+\left(R_1+p_1^s\right)\cdot\cos\beta \tag{11.33}
$$

$$\begin{cases} x_s = x_2 + \left(R_2 + p_2^p\right) \cdot \sin\beta, y_s = y_2 + \left(R_2 + p_2^p\right) \cdot \cos\beta \rightarrow \text{Co-directional curve} \\ x_s = x_2 - \left(R_2 + p_2^p\right) \cdot \sin\beta, y_s = y_2 - \left(R_2 + p_2^p\right) \cdot \cos\beta \rightarrow \text{Reverse curve} \end{cases} \tag{11.34}$$

$$\beta = (x_2 - x_1)/(y_2 - y_1) - \alpha \tag{11.35}$$

$$\alpha = \begin{cases} arc\cos\left[\left(R_1 + p_1^s - R_2 - p_2^p\right)/D\right] \rightarrow \text{Co-directional curve } (R_1 \geq R_2) \\ arc\cos\left[\left(R_2 + p_2^p - R_1 - p_1^s\right)/D\right] \rightarrow \text{Co-directional curve } (R_1 < R_2) \\ arc\cos\left[\left(R_1 + p_1^s + R_2 + p_2^p\right)/D\right] \rightarrow \text{Reverse curve} \end{cases} \tag{11.36}$$

$$D = \sqrt{(x_1 - x_2)^2 + (y_1 - y_2)^2} \tag{11.37}$$

where (x_p, y_p) are the coordinates of the foot of perpendicular P_p; (x_s, y_s) are the coordinates of the foot of perpendicular P_s; β is the azimuth of $\boldsymbol{L_F}$; α is the turning angle between $\boldsymbol{L_C}$ and $\boldsymbol{L_F}$; D is the distance between C_1 and C_2, as shown in Fig. 11.31.

(3) Straight-curve adaptive recreation strategy

For a complex existing alignment, Case A and Case B may exist simultaneously. Therefore, to improve the recreation quality, the straight-first-then-curve method and the curve-first-then-straight method should be flexibly applied in the recreation process of complex alignments. In this section, the application of the straight-first-then-curve method and the curve-first-then-straight method is adaptively executed based on a new threshold (i.e., the straight segment threshold f_t). In addition, to improve recreation accuracy, the coupling relation between the geometric attribution of measured points in the set **M** and the recreated alignment is considered, and the results of attribution identification of measured points and the recreated horizontal alignment can be iteratively updated. The specific implementation process of the straight-curve adaptive recreation strategy is as follows (as shown in Fig. 11.33).

Step 1: For each iteration shown in Fig. 11.33, based on the current identification results, use the Least Squares Method to fit all subsets of the curve segment measured point set $\mathbf{M}^{c}$ to obtain the geometric parameters of all circular curves.

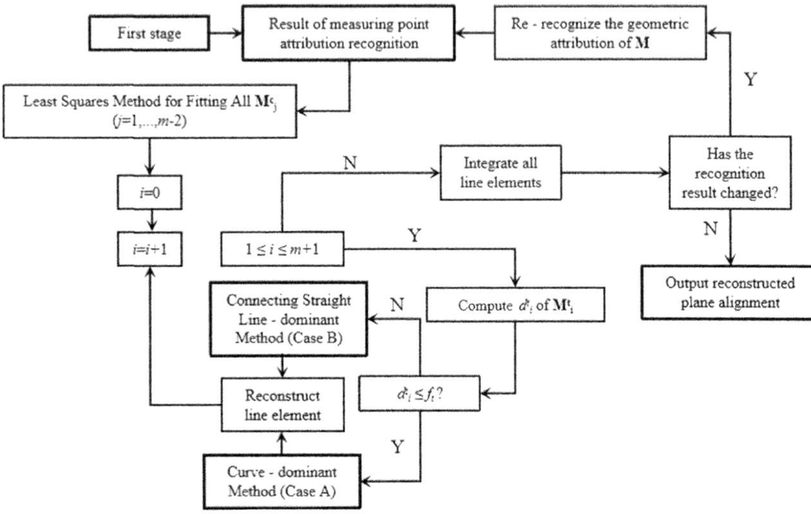

Figure 11.33 Adaptive recreation strategy.

Step 2: Further subdivide to obtain the geometric parameters of straight segments between curves and transition curves by dividing all subsets of the straight segment measured point set $\mathbf{M^t}$ into different cases. For the i-th straight segment measured point subset $\mathbf{M^t}_i$ ($i = 1, \ldots, m-1$):

a. Compute the cumulative sum of distances *dt i* between all adjacent measured points in $\mathbf{M^t}_i$, as Eq. (11.38).

$$d_i^t = \sum_{j=2}^{n_i^t} \left[\sqrt{\left(x_j - x_{j-1}\right)^2 + \left(y_j - y_{j-1}\right)^2} \right] \tag{11.38}$$

where *dt i* is the cumulative sum of distances between all adjacent measured points in the i-th straight segment measured point subset $\mathbf{M^t}_i$; n^t_i is the number of all measured points in $\mathbf{M^t}_i$; (x_j, y_j) are the coordinates of the j-th measured point in $\mathbf{M^t}_i$.

b. Compare *dt i* with the straight segment threshold f_t. If *dt i* is less than f_t, the geometric parameters of the local straight segment between curves and transition curves (corresponding to $\mathbf{M^t}_i$, $\mathbf{M^{ls}}_i$, and $\mathbf{M^{lp}}_i$ respectively) are determined by the curve-first-then-straight method (Case A); otherwise, they are determined by the straight-first-then-curve method (Case B).

Step 3: Combine all local alignment elements generated in the above steps and handle all constraints mentioned in Section 11.2.1 to reconstruct a complete horizontal alignment.

Step 4: Project **M** onto the recreated horizontal alignment and record the current geometric attribution of **M**. Then check whether the current attributes of **M** are consistent with the identification results obtained in the previous iteration. If consistent, terminate the program; otherwise, update the identification results based on the current attributes of **M**, return to Step ①, and continue the iteration.

Through the Above process, a Global Horizontal Alignment can be Recreated.

11.4.2 Recreation and optimization method for vertical alignments

The recreation method for vertical alignments is similar to that for horizontal alignments, which is still based on the idea of straight-curve adaptive recovery. The vertical alignment recreation is performed after the completion of horizontal alignment recreation. As shown in Fig. 11.34, the specific process of vertical alignment recreation is as follows.

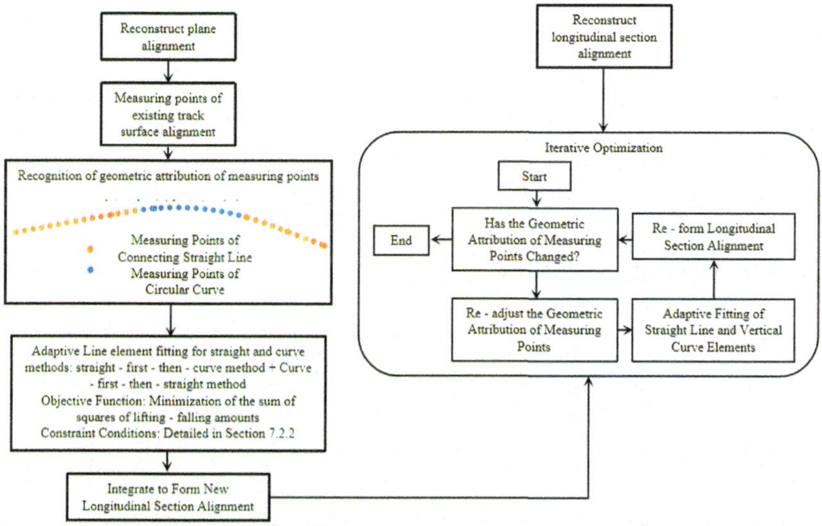

Figure 11.34 Recreation process of vertical alignment.

(1) Form a sequence of stake numbers for measured points on the existing rail surface alignment

Based on the recreated horizontal alignment generated in Section 11.4.1, organize the set of measured points **M** on the existing rail surface alignment into a sequence of stake numbers for measured points (each stake number includes mileage and existing rail surface elevation information), providing a data basis for subsequent recreation processes.

(2) Initially identify the geometric attribution of measured points

Since the grade change rates of straight segments and vertical curve segments in the vertical alignment are significantly different, the grade change rate of measured points is used as the basis for dividing their geometric attribution. Each measured point's grade change rate is compared with the set grade change rate threshold. If the grade change rate of a measured point is below the threshold, it is classified as a straight alignment measured point; otherwise, it is classified as a vertical curve measured point.

(3) Adaptive fitting of straight and curved alignment elements

After identifying the geometric attribution of measured points, the set of measured points **M** is divided into several ranges of straight segments and vertical curve segments. Similarly to horizontal recreation, sections belonging to Case A (using the curve-first-then-straight method) and Case B (using the straight-first-then-curve method) are first divided by the straight segment threshold f_t. Then, with the goal of minimizing the sum of squares of lifting/lowering amounts and considering various constraints mentioned in Section 11.2.2, the straight-first-then-curve method and the curve-first-then-straight method are adaptively used to fit alignment elements of different sections (ignoring the transition curve fitting sub-process). All alignment elements are then combined to restore a temporary vertical alignment.

(4) Iterative optimization process

After completing steps (1–3), further optimization of the recreated vertical alignment is required. Based on the idea that the geometric attribution of measured points should be consistent with the range of fitted alignment elements, in each iteration, check whether the recreated vertical alignment is consistent with the geometric attribution of measured points. If it is not, re-adjust the geometric attribution of measured points according to the latest temporarily recreated vertical alignment, then re-fit alignment elements using the adaptive recreation process based on the new range of measured point attribution, generate a new temporary vertical alignment, and enter the next iteration until "point-alignment consistency" is achieved. Exit the iteration process to obtain the final recreated vertical alignment.

11.5 Three-dimensional subway alignment recreation considering tunnel construction gauges

Subways are of vital importance in modern urban transportation. Their operational safety is closely related to alignment conditions. Ideally, for a completely constructed subway tunnel, its geometry and location should coincide with those of the originally-designed alignment. Nevertheless, due to the errors generated by, e.g., measurement, construction and permanent structural settlement, there are inevitable deviations between the subway tracks center line and the originally-designed alignment (Fig. 11.35(a)). When these Deviation Values (DVs) exceed allowable thresholds, the underground structures (e.g., tunnel linings, fans and power supply equipment) along the subway infringe upon the construction gauges. In these conditions, trains may impact the underground infrastructures during operation and, hence, greatly threaten safety.

In the real-world, to eliminate such potential threats, the subway alignment sections whose DVs do not satisfy design standards must be recreated based on the actually-measured point data of constructed tunnels (Fig. 11.35 (b)) before subsequent processes, e.g., laying tracks and installing communication equipment. During this process, the horizon and vertical subway alignments geometry is first fitted with the same design standards used for

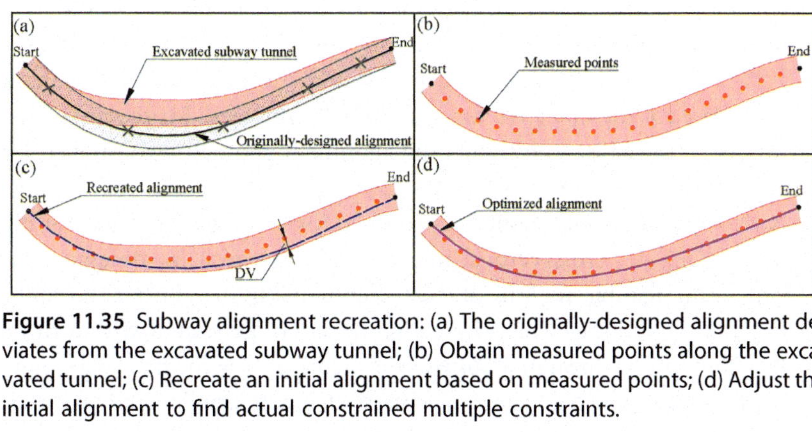

Figure 11.35 Subway alignment recreation: (a) The originally-designed alignment deviates from the excavated subway tunnel; (b) Obtain measured points along the excavated tunnel; (c) Recreate an initial alignment based on measured points; (d) Adjust the initial alignment to find actual constrained multiple constraints.

the original alignment (Fig. 11.35(c)). Then, using the recreated 3-D alignments information, detailed DVs between the recreated alignment and measured points can be computed and the locations where construction gauges are infringed can also be modified (Fig. 11.35(d)). After this secondary readjustment process, also called "tuning of line and slope" [14], the final subway alignment can be used as a reliable reference for further construction purposes. The above process is very important due to the high accuracy requirements of subway construction and operation.

However, the 3-D alignment recreation process is fairly complex. With higher design standards, the tunnel construction gauges are strict and, hence, impose difficult-to-satisfy constraints for subway alignment recreations. Furthermore, the horizontal and vertical alignments are closely-coupled during the recreation process. Specifically, the vertical alignment generation must be based on the horizontal alignment, whereas the horizontal solution evolution is also significantly affected by the vertical one. In general, after determining the horizontal alignment during the recreation process, the horizontal DVs can be obtained. Meanwhile, due to construction gauges, the vertical DVs will also be measured to a specific range that is difficult to adjust. Conversely, if the vertical DVs violate the construction gauge constraint severely, the horizontal alignment also becomes infeasible. It is actually conflicting and challenging when minimizing horizontal and vertical DVs simultaneously to find 3-D DV trade-offs within highly-constrained search spaces.

To address this problem, an optimization model and a solution method for recreating as well as optimizing 3-D subway alignments are proposed by our team. The main contributions are as follows:

1. An optimization model for 3-D subway alignment recreation is developed. After defining the alignment design variables, the sum of DVs between the actually-measured points and the recreated alignment is formulated as the objective function to be minimized. Moreover, spatial tunnel construction constraints and alignment geometric constraints are also specified.
2. A two-stage solution method is proposed for solving the model. For the first stage, a "cross-vertical-integration" approach is designed for generating feasible regions of 3-D alignments by handling constraints. For the second stage, recreated alignments are generated within those feasible regions through a customized heuristic search, i.e., swing iteration, within which a "horizontal-vertical-integration" search process is incorporated.

3. Finally, a complex real-world case is employed for demonstrating the proposed method. Through the detailed output data analyses regarding the comparison among the computer-generated solution, a GA-optimized solution and the best manually-designed solution, the effectiveness of proposed method is verified.

11.5.1 Optimization model

(1) Design variables

According to previous studies [7,15], a 3-D subway alignment can be described using the information about its Points of Intersection (*PIs*). More specifically, as shown in Fig. 11.36, the Horizontal *PIs*, (*HPIs*) information includes coordinates (*Xi, Yi*), the circle curve radius (*RHi*) and transition curve length (*L01i, L02i*), while the vertical *PIs*, (*VPIs*) information includes station point (*Kj*), design elevation (*Hj*) and vertical curve radius (*RVj*).

Therefore, a 3-D subway alignment can be described as Eq. (11.39).

$$[\mathrm{HPI}, \mathrm{VPI}] = [\mathrm{X}, \mathrm{Y}, \mathrm{L}_{01}, \mathrm{R}_{\mathrm{H}}, \mathrm{L}_{02}, \mathrm{K}, \mathrm{H}, \mathrm{R}_{\mathrm{V}}] \tag{11.39}$$

(2) Objective function

In this study, the total DVs (i.e., sum of deviation values) between the actually-measured points and recreated alignments is used as the objective function, which is computed based on the geometric characteristics of tangents, circular curves and transition curves of the recreated alignment, as shown in Fig. 11.37.

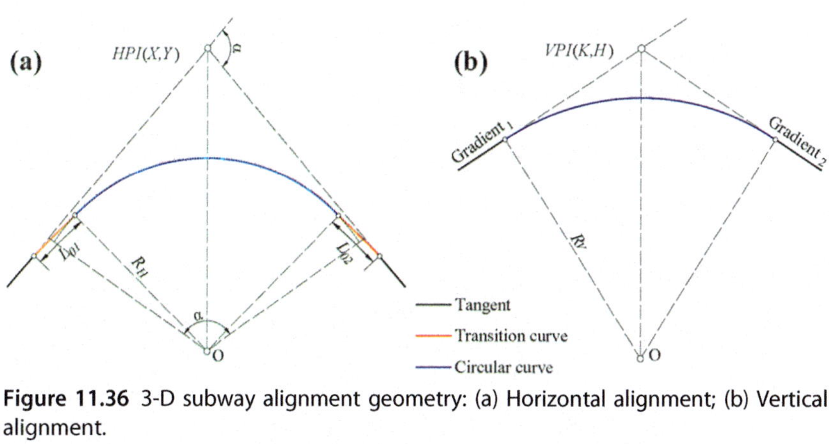

Figure 11.36 3-D subway alignment geometry: (a) Horizontal alignment; (b) Vertical alignment.

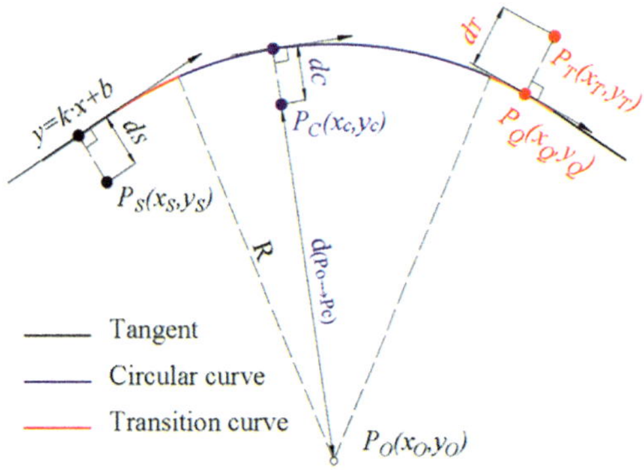

Figure 11.37 DV computation.

① **Tangents**

If a measured point *PS* (x_s, y_s) is projected on a straight alignment (i.e., tangents) of the subway alignment, which is represented by $y = kx + b$, the corresponding DV can be expressed as Eq. (11.40).

$$d_S = \left|kx_s - y_s + b\right|\sqrt{1 + k^2} \qquad (11.40)$$

② **Circular curves**

If a measured point P_C (x_c, y_c) is projected on a circular curve whose radius is R and center point is P_O (x_o, y_o), its DV can be computed as Eq. (11.41).

$$d_C = |d_{(P_O - P_C)} - R| = \left|\sqrt{(x_o - x_c)^2 - (y_o - y_c)^2} - R\right| \qquad (11.41)$$

③ **Transition curves**

The transition curve is usually a parabola in subway alignment designs according to the *Code for Design of Metro* [16]. If a measured point P_T (x_T, y_T) is projected on a transition curve at point P_Q (x_Q, y_Q), which is determined by using the "golden section method" as previously applied in Refs. [7,17], the DV is computed as Eq. (11.42).

$$d_T = \left|\sqrt{(x_T - x_Q)^2 - (y_T - y_Q)^2}\right| \qquad (11.42)$$

Thus, the objective function is formulated as Eq. (11.43).

$$f(\mathbf{HPI}, \mathbf{VPI}) = Sum_H + Sum_V$$
$$= \sum_{i=1}^{I} \left(d_{SH}^{i}\right)^2 + \sum_{j=1}^{J} \left(d_{TH}^{j}\right)^2$$
$$+ \sum_{k=1}^{K} \left(d_{CH}^{k}\right)^2 + \sum_{n=1}^{N} \left(d_{SV}^{n}\right)^2 + \sum_{m=1}^{M} \left(d_{CV}^{m}\right)^2 \quad (11.43)$$

where Sum_H and Sum_V represent the quadratic sum of DVs along horizontal and vertical alignments, respectively; I, J and K are the numbers of measured points that are projected, respectively, on tangents, transition curves and circular curves of a horizontal alignment; N and M denote the number of measured points that are projected, respectively, on tangents and circular curves of a vertical alignment.

(3) Constraints

① Tunnel design gauge

According to the Standard for Metro Gauges [18] and Code for Urban Rail Transit Engineering Survey [16], the cross section of subway tunnels can be described by using information on eight key points, as shown in Fig. 11.38. It is important to note that the design gauge considered here is the total gauge, including the vehicle gauge and equipment gauge. During the alignment recreation process, for simplicity, the design boundary is assumed to be a

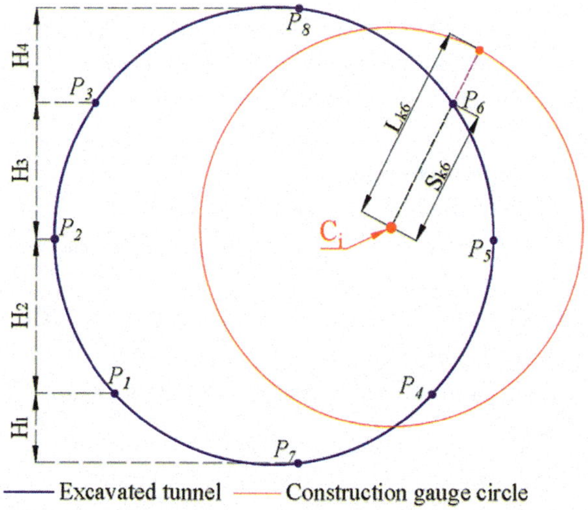

Figure 11.38 The subway design gauge constraint of the *i*th cross section.

circle [18]. Besides, since currently most deep subways are excavated by using the shield tunneling method, in this study, the shape of tunnel cross sections is also assumed to be circular.

Based on the above discussions, in the proposed model, the tunnel design gauges are analyzed and formulated using the following steps: In Fig. 11.38, the elevations of P_1 and P_4 are consistent with the elevation of originally-designed alignment. The remaining six key points are determined according to the elevation differences of H_1, H_2, H_3 and H_4. Detailed values of H_1, H_2, H_3 and H_4 can be found in the above two authoritative specifications. Specific values of these parameters in the case study are shown in Section 11.5.3.

In general, the constraints of tunnel design gauge are complex and difficult-to-satisfy. Thus, directly eliminating solutions that violate the gauge constraint would overlook many promising solutions. Therefore, in this chapter, the design gauge is treated as a soft constraint, which is expressed as a penalty function (i.e., Sum_G) by modifying the objective function, as Eqs. (11.44) and (11.45).

$$f(x) = Sum_H + Sum_V + Sum_G \tag{11.44}$$

$$Sum_G = \sum_{i=1}^{I} \sum_{k=1}^{8} (l_{ki}) \text{ where } \begin{cases} l_{ki} = 0 & L_{ki} \leq S_{ki} \\ l_{ki} = (S_{ki} - L_{ki})^2 & L_{ki} > S_{ki} \end{cases} \tag{11.45}$$

where I denotes the number of subway cross sections; L_{ki} is the radius of design gauge circle that is determined based on the recreated alignment; S_{ki} represents the distance between the *kth* key point and the design gauge circle's center C_i.

② **Geometric constraints**

According to the Code for Design of Metro [16] and based on Section 11.5.1, subway alignments should satisfy the following geometric constraints:

- Horizontal geometric constraints $C_H(\boldsymbol{HPI})$, as Eqs. (11.46–11.49).
  (a) The minimum circular curve radius ($R_{H\min}$):

$$R_{Hi} \geq R_{H\min} \tag{11.46}$$

(b) The minimum transition curve length ($L_{\min}$):

$$L_{01i} \geq L_{\min}, L_{02i} \geq L_{\min} \tag{11.47}$$

(c) The minimum circular curve length ($L_{C\min}$):

$$\alpha_i \cdot R_{Hi} \geq L_{c\min} \tag{11.48}$$

(d) The minimum tangent length between two adjacent horizontal curves ($T_{H\min}$):

$$T_H \geq T_{H\min} \tag{11.49}$$

where T_H is the tangent length between the ith and $(i+1)^{\text{th}}$ horizontal curves.

- Vertical geometric constraints $\boldsymbol{C}_V(\boldsymbol{VPI})$, as Eqs. (11.50) and (11.53).
  (a) The minimum sloped section length between adjacent *VPIs* ($T_{V\min}$):

$$\left|K_{j+1} - K_j\right| \geq T_{V\min} \tag{11.50}$$

  (b) The maximum gradient ($G_{\max}$) of sloped sections:

$$\left|H_{j+1} - H_j\right| / \left|K_{j+1} - K_j\right| \leq G_{\max} \tag{11.51}$$

  (c) The maximum vertical curve radius ($R_{V\max}$) and minimum vertical curve radius ($R_{V\min}$):

$$R_{V\min} \leq R_{Vj} \leq R_{V\max} \tag{11.52}$$

  In sum, the optimization model can be expressed simply as:

$$\begin{cases} \min(Sum_H + Sum_V + Sum_G) \\ \text{s.t.}\mathbf{C_H}(\mathbf{HPI}), \mathbf{C_V}(\mathbf{VPI}) \end{cases} \tag{11.53}$$

11.5.2 Optimization method

(1) Main framework

In solving the above model, there are two major problems:

① The available input data are generally limited. In most cases, information on only eight key points of a tunnel cross section is available (see Fig. 11.38). Generally, it is costly and time-consuming to obtain more point data. However, the alignment geometry and engineering constraints are complex. Thus, it is difficult to determine the alignment solution and handle constraints based on these limited data.

② The horizontal and vertical alignments are highly-intertwined. Moreover, with the complexity of the alignment geometric characteristics, it is difficult to simultaneously recreate horizontal and vertical alignments by reducing 3-D DVs.

Consequently, a two-stage approach is proposed, as shown in Fig. 11.39.

A "cross-vertical-integration" method for generating 3-D feasible regions of recreated alignments is developed for handling complicated constraints. During this process, the feasible regions

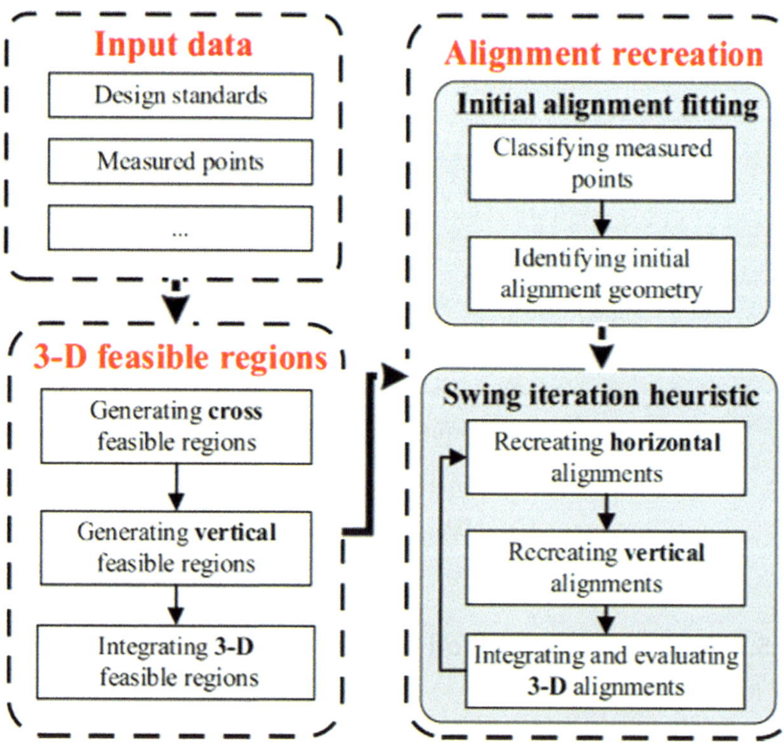

Figure 11.39 Flowchart of the solution method.

within tunnel cross sections are first determined and then the feasible regions for vertical alignments are generated. Finally, the 3-D feasible regions are integrated and used to identify original alignment segments that require recreations.

A "horizontal-vertical-integration" search process for recreating 3-D subway alignments is developed. During this process, continuous alignments are first initialized based on discretized measured points. Afterward, based on a swing iteration heuristic, horizontal as well as corresponding vertical alignments are optimized sequentially. Then, the horizontal alignment and corresponding vertical one are integrated as a 3-D solution for detailed evaluation and iteration.

(2) Generation of spatial feasible regions for recreating alignments

To handle the complex constraint of tunnel design gauges, a novel approach is proposed for determining the 3-D feasible regions of recreated alignments. Given the limited available data, the originally-designed alignment is used as a reference. The detailed implementation process is as follows.

① Determine the target alignment center

For subway alignment recreation, the theoretical location of the actual alignment, which is defined as the "target alignment center" in Fig. 11.40, must first be determined based on the eight groups of measured points. The target alignment center is then used for DVs computation by being projected on the recreated alignment (as shown in Fig. 11.37).

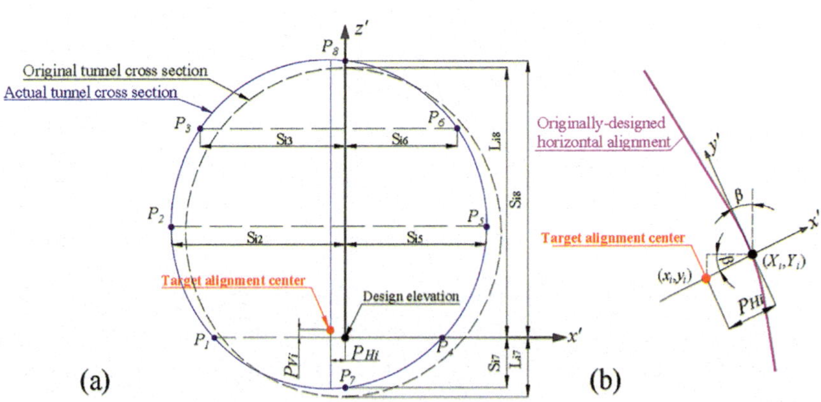

Figure 11.40 Diagram for determining the target alignment center: (a) Tunnel cross section; (b) Horizontal view of the originally-designed alignment.

As shown in the cross view of Fig. 11.40 (a), based on the 3-D Cartesian coordinate system (X, Y, Z), a local coordinate system (x', y', z') is established by using the originally-designed alignment's center point (i.e., design elevation point) as the origin. Specifically, the z' axis coincides with Z axis and y' points to the heading direction of the horizontal alignment (as shown in Fig. 11.40 (b)).

Then, the horizontal offset P_{Hi} in Fig. 11.40 (b)) between the target alignment center and the originally-designed alignment center is computed using Eqs. (11.54) and (11.55). Note that for cross sections on circular and transition curves, additional deviations (i.e., p_i in Eq. (11.54) due to superelevation must be considered. For measured points that correspond to tangent sections, p_i is 0.

$$p_i = h_0 h_{ac} / s \tag{11.54}$$

$$P_{Hi} = \left\{ \left[\frac{(S_{i1} - S_{i4})}{2} + \frac{(S_{i2} - S_{i5})}{2} + \frac{(S_{i3} - S_{i6})}{2} \right] / 3 \right\} - p_i \tag{11.55}$$

where i denotes the ith cross section; h_0 is the distance from the originally-designed gauge's center to the x' axis (mm); h_{ac} is the value of track superelevation; s is a constant of 1500 mm according to the Standard for Metro Gauges [18]; s_{i1}−s_{i6} are defined as in Fig. 11.40(a).

Afterward, the vertical offset (P_{Vi}) is computed with a similar method based on the data of P_7 and P_8, as Eq. (11.56).

$$P_{Vi} = [(S_{i8} - L_{i8}) + (L_{i7} - S_{i7})] / 2 \tag{11.56}$$

where S_{i7} and S_{i8} represent the distance from P_7 and P_8 to the x' axis, respectively; L_{i7} and L_{i8} are defined as in Fig. 11.39 (a).

Finally, the coordinates of the target alignment center (z_i, y_i, h_i) can be obtained as Eqs. (11.57–11.59).

$$x_i = X_i - P_{Hi} \cos(\beta_i) \tag{11.57}$$

$$y_i = Y_i - P_{Hi} \sin(\beta_i) \tag{11.58}$$

$$h_i = H_i + P_{Vi} \tag{11.59}$$

where (X_i, y_i) denotes the horizontal coordinates of the originally-designed alignment's center point; β_i indicates the heading direction of the ith cross section; H_i is the design elevation of the originally-designed alignment's center point.

② **Generate 3-D alignment feasible regions**

Using the target alignment center as a new origin for every cross section, the 3-D feasible regions can be generated. This is done by moving the origin clockwise toward eight directions which are 0 degrees, 45 degrees, 90 degrees, 135 degrees, 180 degrees, 225 degrees, 270 degrees, and 315 degrees based on the positive x' axis (as shown in Fig. 11.41). It is noteworthy that, with the considered directions increase (e.g., from eight directions to 16 directions), the area of generated feasible region (i.e., Fig. 11.41(b)) also increases and, hence, the search spaces can be enlarged. However, such increases in the search space have no significant effects. According to our experience, eight directions suffice to determine the feasible regions for this problem. Specifically, 0° is used in the following as an example.

As shown in Fig. 11.41(a), centered at C1, the design gauge circle (the red one) corresponding to the target alignment center is moved by Δ mm in sequence along the 0 degrees direction. After each movement, we check whether the design gauge circle intersects with the actual tunnel circle (the blue one in Fig. 11.40(a)). If not, the process continues until the intersection point is found after n movements, as indicated by the ellipse in Fig. 11.41(a). Note that Δ should be small enough (in this study $\Delta = 1$ mm) to ensure that an accurate intersection can be found for the feasible region.

After traversing the eight directions in the cross section, an octagon can be generated, as shown in Fig. 11.41(b). Thus,

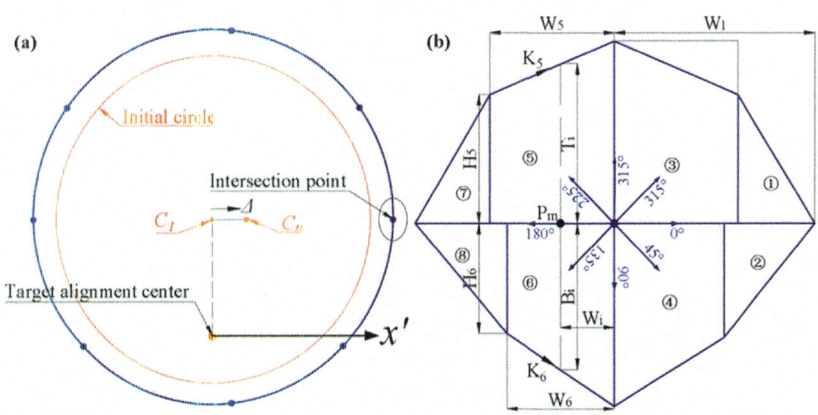

Figure 11.41 Generating feasible regions for cross sections.

recreated alignments should be located within this octagonal feasible region. Moreover, to generate vertical feasible regions, the octagon is then divided into eight subregions. As shown in Fig. 11.41(b), if the recreated alignment center (i.e., P_m) corresponds to subregions 5 and 6, its vertical feasible range can be computed as Eq. (11.60).

$$\begin{cases} T_i = H_5 - W_i K_5 \\ B_i = H_6 + W_i K_6 \end{cases} \tag{11.60}$$

where T and B are the upper and lower elevation thresholds of the recreated alignment at P_m; K and H denote the slope and intercept, respectively, of every subregion's boundary alignment, as shown in Fig. 11.41(b).

The vertical feasible region along subway is shown in Fig. 11.42. After the above "cross-vertical-integration" process, 3-D feasible regions can be generated, which will be used to screen infeasible sections of the originally-designed alignment that violate design gauge constraint. As shown in Fig. 11.43, this is done through the following steps:

1. Check every alignment geometric component (i.e., tangent, circular curve and transition curve) from the start point to end, as numbered in Fig. 11.43(a). Identify all the geometric components located partially or entirely outside the feasible region. If several successive alignment components are outside the feasible region (e.g., components No. 5 to 11 in Fig. 11.43(a)), they are combined as a "component integration". The geometric components at the integration's two ends are defined as "terminal components" (i.e., components No. 5 and 11).

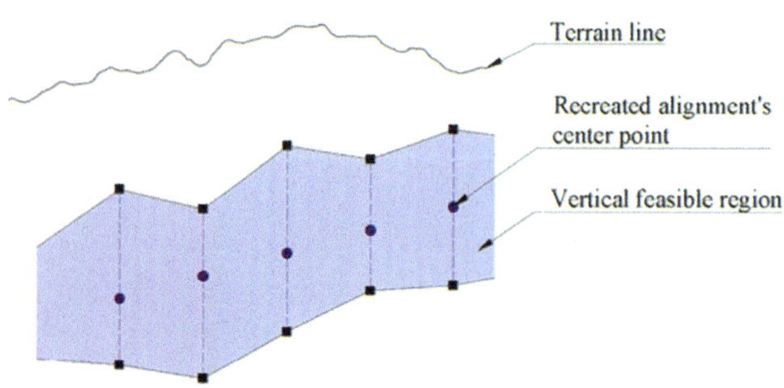

Figure 11.42 Vertical feasible region.

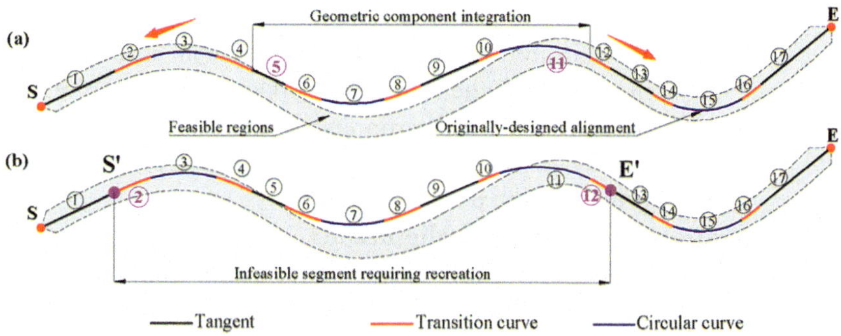

Figure 11.43 Screening infeasible sections of the original alignment for recreation: (a) Identify geometric components that violate design gauge constraint; (b) Expand the components to determine the infeasible segment that requires alignment recreation.

2. As shown in Fig. 11.43 (b), according to the two terminal components, the component integration will then be expanded to determine the alignment segment that requires recreation. More specifically:
   a) If the terminal component is a tangent (e.g., component No. 5), the component integration is expanded by including the entire feasible curve connecting the tangent to find the corresponding tangent-curve demarcation point (i.e., S' in Fig. 11.43(b)). The terminal component is also relabeled as component No. 2.
   b) If the terminal component is a (circular or transition) curve (e.g., component No. 11), the component integration is expanded to the corresponding tangent-curve demarcation point within the feasible region (i.e., E' in Fig. 11.43(b)). The terminal component is also relabeled as component No. 12.
   c) If component No. 2 or 12 is also infeasible, the expansion continues until the feasible tangent-curve demarcation point is found to guarantee the recreated alignment can be connected with the original tangent sections smoothly.

   The final segment between S' and E' is used for alignment recreation. It is noteworthy that if several such segments exist along the original alignment, they are recreated sequentially from S to E.

The aims of proposing the above process are to preserve the originally-esigned alignment as much as possible as well as to reduce the recreation workload and the cost of recreating tracks. After determining the infeasible sections of the originally-designed alignment, the recreation process will be explained in the next section.

(3) Recreation of 3-D subway alignments

Based on the spatial feasible regions, two steps are developed to obtain the recreated subway adjustment: First, a global exploration process is developed to fit discrete measured points into a continuous initial alignment. Then, a specifically-ustomized swing iteration method is adopted for local exploitation and optimization of recreated alignments.

① Global exploration

In this step, the tangent, circular curve and transition curve sections of an alignment are fitted. As shown in Fig. 11.44, the heading direction is used as the basis for distinguishing tangents and circular curves. By defining the change rate of heading direction at the start and end points as 0, the *i*th measured point's (P_i's) heading direction (δ_i) and change rate of heading direction (C_i) are computed as Eqs. (11.61 and 11.62) [16].

$$\delta_i = \theta_{(i-1,i)} \pm \arccos\left[\frac{\left(S^2_{(i,i+1)} S^2_{(i-1,i+1)} - S^2_{(i-1,i)}\right)}{\left(2S_{(i,i+1)} \cdot S_{(i-1,i+1)}\right)}\right] \tag{11.61}$$

$$C_i = (\delta_i - \delta_{i-1}) / l_{(i-1,i)} \tag{11.62}$$

where $\theta_{(i-1,i)}$ represents the azimuth of the string connecting P_{i-1} and P_i; $S_{(i,i+1)}$; $S_{(i-1,i+1)}$, and $S_{(i+1,i)}$ are the lengths of strings connecting P_i and P_{i+1}, P_{i-1} and P_{i+1}, as well as P_{i+1} and P_i, respectively; $l_{(i-1,i+1)}$ denotes the arc length between P_{i-1} and P_i.

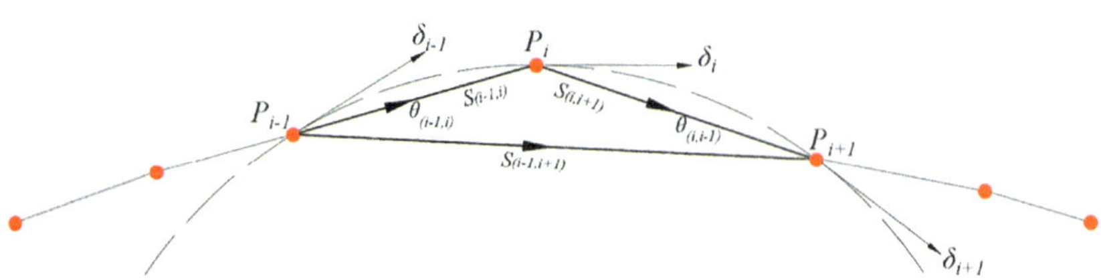

Figure 11.44 Heading direction computation of P_i.

Taking $1/R_{\min}$ as the threshold for identifying tangents and circular curves (where $R_{\min}$ denotes the minimum circular curve radius): When C_i is lower than $1/R_{\min}$, the measured point is assigned to a tangent; otherwise, the point is deemed to correspond to a circular curve. Thus, the overall tangent and circular curve segments ψ can be obtained as Eqs. (11.63) and (11.64).

$$\begin{cases} \psi_T = \{P_i | C_i < 1/R_{\min}\} \\ \psi_C = \{P_i | C_i \geq 1/R_{\min}\} \end{cases} \tag{11.63}$$

$$\psi = \{\psi_{T1}, \psi_{C1}, \psi_{T2}, \psi_{C2} \cdots \psi_{T_{m-1}}, \psi_{C_{m-1}}, \psi_{T_m}\} \tag{11.64}$$

where m is the number of tangents; Φ_T is the tangent point set; Φ_C is the curve point set.

Then, considering geometric constraints explained in Section 11.5.1.3, tangents and circular curves can be fitted with a least-squares method using the point data in Φ_T and Φ_C. According to Ref. [7], the sequence in this step fits tangents before circular curves. Moreover, for horizontal alignments, there exist unrecognized transition curve points at the junction of Φ_T and Φ_C. Additional steps are thus needed to compute the corresponding transition curve length (L_0), as shown in Fig. 11.45(a). In this chapter, this is implemented by computing as Eq. (11.65).

$$l_{0i} = \sqrt{24R_i p_i} = \sqrt{24R_i(d_i - R_i)} \tag{11.65}$$

where di is the distance from the circular curve center to the tangent alignment. Afterward, the measured points corresponding to the transition curves can be obtained, as shown in Fig. 11.45 (b). Eq. (11.64) is also modified to Eq. (11.66)

$$\psi = \{\phi_{T1}, \phi_{TB1}, \phi_{C1}, \phi_{TA1}, \phi_{T2}, \phi_{TB2}, \phi_{C2} \cdots \phi_{Tm1}, \phi_{TBm-1}, \phi_{Cm-1}, \phi_{TAm-1}, \phi_{Tm}\} \tag{11.66}$$

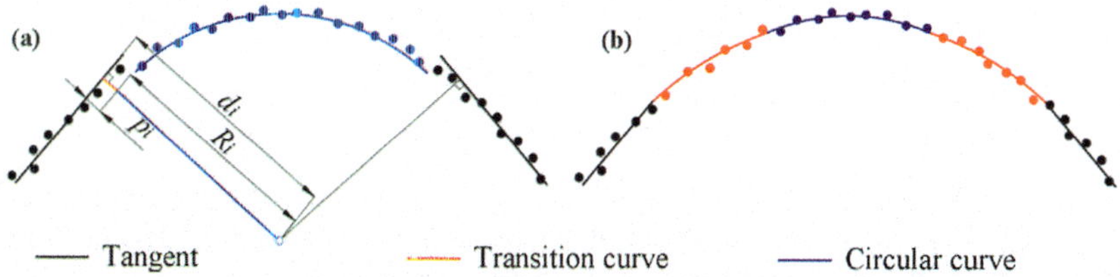

Figure 11.45 Determination of transition curves: (a) Compute the length of a transition curve; (b) Obtain a complete horizontal alignment.

where Φ_{TB} and Φ_{TA} are transition curve point sets.

Similarly, the vertical alignment consisting of tangents and circular curves can be fitted by using the above method. Thus, an initial 3-D alignment satisfying geometric constraints can be obtained based on discrete measured points. This initial alignment will be further optimized using the method proposed in the next section.

② **Local exploitation**

The generation of the initial alignment is a global recreation process but may still deviate from measured points significantly or violate tunnel design constraints in many local sections. To address this problem, a specifically-customized swing iterative method (Section 11.3) for local exploitation is used to further optimize the 3-D alignment. Taking point P_i in Fig. 11.46 as an instance, this method is explained below.

As shown in Fig. 11.46(a), after the segmentation process for tangents and curves (Eqs. (11.63), (11.64) and (11.66)), it is preliminarily determined that P_i corresponds to a horizontal transition curve. However, after the alignment fitting process, P_i actually corresponds to a fitted circular curve (Fig. 11.46(b)). Thus, the "measured point" attribute of P_i conflicts with its "alignment

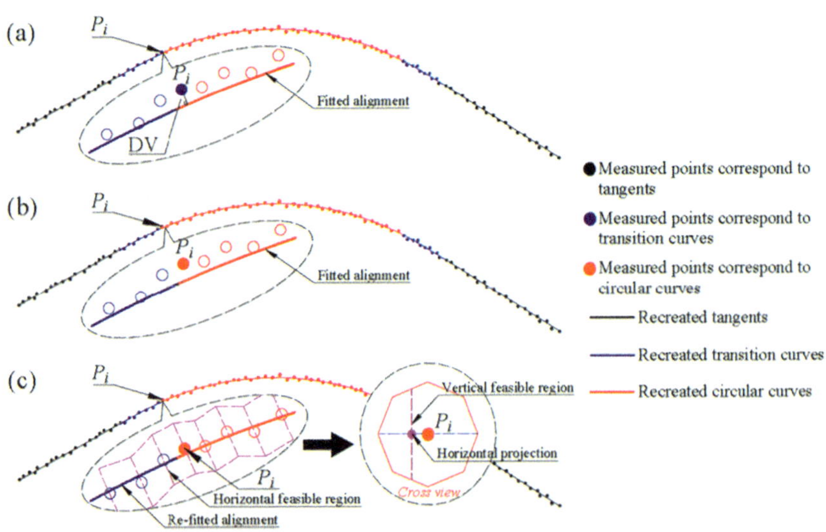

Figure 11.46 The swing iteration process: (a) The "measured point" attribute of P_i conflicts with its "alignment geometry" attribute; (b) Re-adjust P_i's "measured point" attribute; (c) Re-fit the alignment based on 3-D feasible regions.

geometry" attribute, which is defined as "point-alignment inconsistency" in the swing iteration method that results in large DVs or violations of design gauges.

In this condition, as shown in Fig. 11.46(b), P_i's "measured point" attribute is re-adjusted as corresponding to a circular curve to be consistent with its "alignment geometry" attribute. Thus, Φ_T, Φ_C, Φ_{TB} and Φ_{TA} in Eqs. (11.64) and (11.66), i.e., the measured point sets, are updated. This step is the heuristic operator of the swing iteration algorithm. That is, the optimization process is driven by the "point-alignment consistency" rationale we applied to every measured point to reduce DVs and degree of design gauge violation. Thus, the objective function can be minimized through successive iterations.

Afterward, as shown in Fig. 11.46(c), based on the updated measured point sets, the horizontal feasible region corresponding to P_i can be determined by referring to Section 11.5.2.2 and the horizontal alignment can be re-fitted by referring to Section 11.5.2.3 (1). Also, the vertical feasible region can then be interpolated and the vertical alignment can be re-fitted. Thus, a 3-D alignment is recreated and evaluated with respect to its DVs and degree of design gauge violation by using Eq. (11.44).

Next, if there are other points whose "measured point" attribute conflicts with their "alignment geometry" attribute, the above process is executed iteratively until the number of those points falls below 5% of the total measured points.

It is worth noting that, during every iteration, the horizontal alignment is first checked and re-fitted. Based on the horizontal alignment, the corresponding vertical alignment can then be recreated. Afterward, an integrated 3-D alignment can be evaluated and optimized. The proposed model and method have been explained above. They will be applied to a real-world example in the next chapter.

References

[1] D. Ben-Arieh, S. Chang, M. Rys, et al., Geometric modeling of highways using global positioning system data and B-Spline approximation, J. Transport. Eng. 130 (5) (2004) 632–636.

[2] M. Banihashemi, Using quasi-horizontal alignment in the absence of the actual alignment, Accid. Anal. Prev. 95 (2016) PA 33–PA 41.

[3] H.J. Gao, K. Miao, A method for recreating the planar alignment of existing railways based on wavelet denoising, J. Railw. Sci. Eng. 19 (12) (2022) 3600–3611.
[4] M. Imran, Y. Hassan, D. Patterson, GPS-GIS-Based procedure for tracking vehicle path on horizontal alignments, Comput. Aided Civ. Infrastruct. Eng. 21 (2006) 5 383–394.
[5] S. Othman, R. Thomson, G. Lannér, Using naturalistic field operational test data to identify horizontal curves, J. Transport. Eng. 138 (2012) 9 1151–1160.
[6] Z. Song, F. Yang, P. Schonfeld, et al., Integrating segmentation and parameter estimation for recreating vertical alignments, Comput. Aided Civ. Infrastruct. Eng. 36 (2021) 4 472–488.
[7] W. Li, H. Pu, P. Schonfeld, et al., A method for automatically recreating the horizontal alignment geometry of existing railways, Comput. Aided Civ. Infrastruct. Eng. 34 (2019) 1 71–94.
[8] J. Wang, J.X. Liang, W. Li, et al., An automatic recreation method for the overall iteration of the planar alignment of existing railway alignments based on point-alignment consistency, J. Railw. Sci. Eng. 20 (3) (2023) 890–899.
[9] W. Li, Z.Y. Zhang, L. Zhao, et al., Automatic recreation of the planar alignment of existing railways based on the GA-MADS hybrid algorithm, J. Railw. Eng. Soc. 38 (1) (2021) 19–24.
[10] National Railway Administration, Code for the Design of Railway Alignments (TB 10098-2017), China Railway Publishing House, Beijing, 2017.
[11] S. Cellmer, J. Rajdński, M. Skala, et al., New approach to arc fitting for railway track realignment, J. Survey Eng. 142 (2) (2015), 06015005.
[12] W. Li, Y. Zhou, J. Wang, et al., Automatic recreation method for the longitudinal section of existing railway alignments based on point-alignment consistency, J. Railw. Sci. Eng. 16 (2019) 2684–2691.
[13] A. Coelho, Optimum levenberg-marquardt constant determination for nonlinear least-squares, J. Appl. Crystallogr. 51 (2018) 428–435.
[14] S.Y. Chen, H.J. Wang, Q. Wang, et al., Fine-tuning of line and slope based on evolutionary mechanism, Int. J. Mach. Learn. Cybern. 11 (2020) 10 1631–1641.
[15] T.R. Song, H. Pu, P. Schonfeld, et al., Simultaneous optimization of 3D alignments and station locations for dedicated high-speed railways, Comput. Aided Civ. Infrastruct. Eng. 37 (2021) 4 405–426.
[16] Ministry of Housing and Urban-Rural Development of the People's Republic of China, Code for Design of Metro (GB 50157-2013), China Architecture and Building Press, 2013.
[17] W.H. Press, S.A. Teukolsky, W.T. Vetterling, et al., Golden section search in one-dimension. The Art of Scientific Computing, Cambridge University Press, 2007.
[18] Ministry of Housing and Urban-Rural Development of the People's Republic of China, Standard for Metro Gauges (CJJ/T 96-2018), China Architecture and Building Press, 2018.

CHAPTER TWELVE

Building information modeling (BIM) for railway alignments and stations

Abbreviations and corresponding full texts

AEC Architecture, Engineering and Construction
API Application Programming Interface
BIM Building Information Modeling
CAD Computer-Aided Design
CDT Constrained Delaunay Triangulation
CPU Central Processing Unit
DEM Digital Elevation Model
GIS Geographic Information System
GPU Graphics Processing Unit
IFC Industry Foundation Classes
MCSRS Multi-Component Subgrade in a Railway Station
OSG Open Scene Graph
RSS Railway Station Subgrade
VPI Vertical Point of Intersection
XML Extensible Markup Language

With the continuous advancement of surveying and design technologies, as well as the widespread application of BIM and 3D visualization techniques for railway alignments and stations, high-quality and high-precision design has become an industry demand. Traditional 2D CAD-based alignment design can no longer meet current requirements, and 3D alignment and station design have emerged as a research hotspot.

Since the concept of BIM was proposed by American scholar Chuck Eastman in the 1970s [1], it has undergone rapid development, evolving from an emerging idea to a mature technology, and expanding from the building sector to the field of infrastructure. BIM is not merely a three-dimensional modeling tool, but rather a comprehensive information management approach that integrates geometric, physical, regulatory, and semantic data throughout the entire project lifecycle. In railway engineering, the application of BIM has gradually extended from specific components such as stations, bridges, and tunnels to overall alignment design, becoming a key driver of digital transformation in the railway industry.

Intelligent Railway Alignment Design
ISBN: 978-0-443-45923-8
https://doi.org/10.1016/B978-0-443-45923-8.00011-X

Railway alignment serves as the spatial framework and design foundation of an entire railway project. The defined horizontal alignment, vertical profile, and cross-sections form the spatial constraints and reference for all subsequent design majors. Alignment parameters such as route direction, gradient, and curve radius directly determine the span of bridges, the depth of tunnels, the layout of stations, and the configuration of signaling systems. Therefore, the alignment BIM model is the first comprehensive model to be created in a project, acting as the "pioneer" and "fundamental coordinate system" for all other professional designs.

The railway station serves as the nerve center and functional anchor of railway engineering, playing a crucial role in linking and coordinating different systems. It embodies core functions such as transportation organization, passenger flow distribution, and train operations. The layout of the station (including arrival-departure tracks, throat areas, and station buildings) is not only constrained by the alignment but also, in turn, influences local alignment directions and traffic dispersion alternatives. Therefore, a railway station BIM model must be developed in parallel with the alignment model during the early planning stage, jointly shaping the overall framework of the project.

Design changes in alignments and stations often trigger a "chain reaction." A minor adjustment in station location may necessitate modifications to tens of kilometers of alignment, while an optimization of alignment gradient may impact station yard drainage design, signaling systems, and engineering quantities. This strong coupling highlights the critical need for BIM technology in management and coordination.

The parametric and associative update features of BIM play a vital role in this context. On an integrated BIM collaboration platform, modifications to the alignment model can be transmitted in real time to station yard, bridge, tunnel, and other related major models. This process also triggers clash detection and recomputations of engineering quantities, thereby avoiding frequent design changes and errors caused by information silos in traditional design workflows, and significantly improving both quality and efficiency.

This chapter will focus on the BIM methods for railway alignments and stations.

BIM is also closely linked with intelligent alignment selection. Intelligent alignment relies on GIS, DEM, geological databases, and other multi-source data to generate safe, economical, and environmentally friendly alternatives through optimization algorithms. BIM, in turn, provides an information integration platform and visualization tools. For instance, BIM-based alignment models can be embedded into intelligent selection systems for real-time 3D visualization and dynamic adjustment, while algorithm-generated alternatives can be imported into platforms such as Revit or Civil 3D for further refinement and clash detection. Moreover, alignment geometry, mileage, elevation, and curve elements stored in BIM can serve as structured input for intelligent algorithms, enhancing optimization accuracy and engineering quantity estimation.

BIM thus not only acts as a full-lifecycle information carrier for railway alignments, but also supports the evolution of intelligent alignment from a "2D drawing + algorithm" mode to a "3D information model + intelligent decision-making" approach. The integration of the two drives railway alignment design toward greater efficiency, integration, and intelligence.

12.1 BIM for railway alignment

Railway alignment BIM must integrate both geometric form and engineering semantics, making it a multi-step and detail-intensive process. Its essence lies in fusing traditional geometric alignments with rich semantic information, such as design parameters, technical standards, and functional requirements to create a comprehensive digital model that supports the entire lifecycle of design, construction, and operation. This section systematically introduces three key stages of railway alignment BIM, which are interdependent and together form a complete technical chain:

1. Geometric foundation. The first step is to construct the three-dimensional spatial framework of the alignment based on horizontal alignment and vertical alignment (curves and gradients). This provides the physical carrier for subsequent semantic enrichment, and its accuracy directly determines the usability of the model.
2. Extension of the IFC standard. The next step is to extend the Industry Foundation Classes (IFC) standard to incorporate railway-specific entities such as "track axis," "superelevation," and "sleeper." This ensures that semantic attributes are embedded into the geometric framework, enabling lossless data exchange and interoperability across platforms and lifecycle stages.
3. Dynamic updating and maintenance. Finally, mechanisms are required to maintain the model's consistency and maximize its long-term value. Railway projects often face changes in design alternatives or geological conditions. A dynamic updating method-capable of automatically or semi-automatically propagating modifications to associated

geometry, quantities, and semantic data-ensures real-time consistency between the BIM model and project reality, and is key to BIM-driven project management.

12.1.1 Spatial alignment modeling

The essence of railway alignment BIM lies in creating an information-rich model that integrates precise geometric configuration with comprehensive semantic information. This process goes far beyond simple 3D visualization; it is a digital construction workflow that strictly follows engineering standards, incorporates multi-source data, and represents multiple levels of detail. The core objective is to enable a seamless transition from linear alignment elements to parameterized 3D entities, establishing a spatial reference framework and an information foundation for downstream engineering applications.

12.1.1.1 Spatial alignment modeling: Parametric definition of 3D path

The spatial alignment serves as the "skeleton" of the entire railway project and provides the fundamental reference for positioning all structures. Its modeling process integrates mileage (1D), horizontal alignment (2D), and vertical profile (1D) into a precise 3D spatial curve. Horizontal alignment modeling involves decomposing the complex horizontal geometry into a series of connected basic alignment segments, including tangents, transition curves (clothoids), and circular curves. Each segment is defined by its start and end mileages, geometric parameters (such as tangent direction, curve radius, and transition length), and corresponding mathematical formulation. Vertical alignment modeling decomposes the longitudinal profile into gradients (tangent segments) and vertical curves. Vertical curves are typically represented by quadratic parabolas or circular arcs to ensure passenger comfort and safety. The mileage and elevation of each vertical point of intersection (VPI) are critical parameters controlling the vertical profile.

After completing the modeling of the horizontal alignment and vertical profile, the focus shifts to their integration and the synthesis of the 3D spatial alignment. The core logic is to combine the horizontal coordinates of any mileage point with its corresponding elevation to obtain the final 3D coordinates. In BIM platforms, this is typically achieved by creating "Alignment" and "Profile" objects and establishing their associations, with the platform kernel automatically performing the 3D synthesis.

Special or complex alignment combinations require refined treatment. For example: Horizontal tangent + vertical curve: Hermite spline interpolation can be applied to ensure vertical smoothness; Horizontal transition curve + parabolic vertical curve: while curvature varies along the clothoid, the elevation change must still follow a parabolic law. This can be solved through scripting or using specialized civil BIM plugins such as Infraworks or OpenRail to achieve high-precision results.

12.1.1.2 Modeling of railway structures: Parametric alignment-driven approach

Once the spatial alignment is established, all associated structures are created and positioned with reference to it, ensuring parametric associativity where "a change in alignment drives changes in all dependent models." Railway structures can be broadly categorized into linear structures and point structures, each requiring specific modeling approaches.

Linear structures include tracks, subgrade, bridges, tunnels, and guardrails. Their BIM models are generated using a sweep-fusion method [2], which extrudes cross-sectional templates along the 3D alignment. For example, the track cross-section includes rails, sleepers, and ballast, while the subgrade template incorporates the embankment body, drainage ditches, and slopes. These templates can vary with mileage (e.g., subgrade width transitions, slope gradient changes, ballast thickness variations). The BIM software interpolates between templates at key mileages to automatically produce smooth 3D entities.

Point structures include sleepers, signals, and other discrete components. They are modeled based on batch placement strategies. Using design spacing, insertion points are computed along the mileage sequence of the alignment. Each instance's orientation is determined by the local tangent (for Z-axis rotation, i.e., heading) and the normal vector (for X/Y-axis rotation, e.g., cant of catenary masts). This ensures accurate spatial positioning and orientation for large-scale repetitive components.

12.1.1.3 Information integration

The ultimate goal of BIM is not only to represent geometry but also to serve as a carrier of comprehensive information. In a BIM environment, this is achieved through two complementary aspects:

Parameter association: All dimensional parameters of structures (e.g., rail type, sleeper material, mast model) are embedded into the model as instance or shared parameters. This enables efficient statistics, filtering, and tabulation within BIM schedules.

Semantic enrichment: Each model element is supplemented with non-geometric attributes, such as material grade, manufacturer, construction date, and design standards. This transforms the geometric model into a complete digital asset, providing reliable data support for subsequent construction and operation–maintenance phases.

12.1.2 IFC extension for railway alignment

IFC (Industry Foundation Classes), as an international standard supporting BIM information exchange, is relatively mature in the architecture domain. However, its application in railway alignment engineering is still under development. Railway alignments involve complex reference relations, mileage systems, and lifecycle-specific characteristics, making the general IFC4×1 standard difficult to apply directly.

Although IFC4×1 introduces *IfcAlignment* and related entity classes (e.g., *IfcAlignment2DHorizontal, IfcAlignment2DVertical*) to represent spatial alignments, several limitations remain: (1) Incomplete information representation: Railway alignment design typically relies on intersection points and vertical curve control points rather than direct use of alignment segments, which are not fully supported by the existing standard. (2) Lack of reference relations: Railway projects often involve multiple reference designs (e.g., parallel tracks referencing a baseline). Current IFC lacks the ability to represent multi-track, multi-temporal reference relations. (3) Lack of support for mileage system: Key railway-specific information, such as chainage breaks and marked mileages, is not explicitly defined in IFC4×1.

To address these issues, domain-specific extensions of IFC4×1 for railway alignments are required. Such extensions do not create a separate standard but define new entities, properties, and relationships within the existing IFC framework to meet domain-specific needs while ensuring seamless interoperability with existing IFC entities. These extensions follow the modular architecture of IFC and are typically implemented using plugins such as IfcOpenShell. The specific IFC extensions for railway alignments mainly include the following entities:

IfcRailwaySite: A subclass of *IfcSite* used to represent the spatial scope and management unit of a railway as a specialized linear site. It aggregates all railway facilities within its scope (such as tracks, stations, bridges, and tunnels) via *IfcRelAggregates*.

IfcRailwayAlignment: A subclass of *IfcRailway* designed to integrate and represent all geometric and semantic information of the alignment specialty.

It can represent multiple tracks, including main lines, station tracks, and sidings.

IfcRailwayAlignmentPoint: An abstract parent class from which three types of key control points are derived. These point entities are connected to *IfcRailwayAlignment* via *IfcRelConnects* or other specific positional associations (e.g., *IfcRelContainedInSpatialStructure*) to provide precise geometric and parametric control. The relations among these points form the control network of the railway alignment design:

IfcRailwayIntersectingPoint: Horizontal intersection points (corresponding data model: coordinates of horizontal intersection points, radius, front and near transition curve length of horizontal curve);

IfcRailwaySlopePoint: VPI (mileage and elevation of VPI, radius of vertical curve);

IfcRailwayUnconnectedLink: Chainage break information.

The extended IFC standard can define the entity structures explicitly using the EXPRESS language. Data exchange is facilitated through neutral IFC files (.ifc), which contain two sections: HEADER and DATA. Each instance in the DATA section is uniquely identified by a #number, enabling lossless transfer of railway alignment BIM information across different platforms, such as Revit or the Open Infra Platform.

12.1.3 Dynamic updating method for BIM model of railway alignment

Railway alignment BIM models need to support dynamic adjustment of design parameters and linked model updates, which is particularly important in applications such as design alternatives comparison and renovation simulation.

12.1.3.1 Parameter-driven update mechanism

The BIM model should establish relations among parameters to form a dynamic response chain. Key parameters include intersection coordinates, curve radii, transition curve lengths, vertical curve control points (chainage and elevation), and chainage breaks. Associated constraints are defined to ensure dependent updates, e.g., changes in the spacing between double tracks adjust S-curve parameters, or modifications to longitudinal gradients adjust vertical curve lengths and elevations. Real-time model regeneration is achieved through APIs or visual programming tools that monitor parameter changes and trigger updates to geometry and attributes.

12.1.3.2 *IFC-based incremental updates and version management*

For large-scale railway projects, an IFC-based differential update strategy can be adopted. Change extraction identifies added, modified, or deleted entities by comparing IFC file versions. Local reload updates only the changed components, avoiding full-model regeneration and improving efficiency. Transactional submission ensures atomicity and consistency of updates on platforms such as Revit.

Railway alignment BIM technology is evolving toward full lifecycle and full information integration. By extending the IFC standard, it is possible to establish an information model framework that satisfies complex railway requirements. Parameterized and component-based modeling enables rapid generation and precise representation of alignments and associated structures. Dynamic update mechanisms ensure real-time synchronization between the BIM model and the actual design state. In the future, with further refinement of IFC standards, improved algorithm efficiency, and enhanced collaborative capabilities of BIM platforms, railway alignment BIM will play an increasingly central role in intelligent design, smart construction, and digital operation and maintenance.

12.2 BIM for railway stations

Railway stations in vital locations usually have a complex building structure and play critical roles in modern national transportation systems. In a railway station, there are various components and facilities, such as platforms, yards and subgrades. From a structural and BIM perspective, these facilities are typically categorized into two main parts based on their spatial relation to the track: the superstructure (or above-track facilities, such as platforms, canopies, and roofing) and the substructure (or subgrade, the earthworks and foundations below). The superstructure will be introduced in Section 12.2.4, this section focuses on the modeling of the substructure. Among them, Multi-Component Subgrade in a Railway Station (MCSRS) is one of the most important structures for supporting other facilities in a railway station. The traditional approach for designing the MCSRS is mainly based on the results of cross-section sampling. After the central location of the railway station is determined, cross-sections perpendicular to a baseline are inserted in the design according to distributions of railway facilities [3]. Based on these cross-sections, design work is conducted whose result is treated as the basis for further design and future construction of the railway station. A sample of an MCSRS cross-section is shown in Fig. 12.1.

As Fig. 12.1 illustrates, this cross-section includes multiple subgrade components and filler sections which are designed to support various facilities, such as tracks and platforms. With the rapid development of modern transportation, and especially the advances in high-speed railways, increasing numbers of large railway stations with complex structures are being designed and built [4]. Hence, MCSRS cross-sectional structures are getting increasingly complex in order to satisfy the support requirements in those railway stations. Due to the above features, MCSRS design is more complicated and needs more attention than ordinary simple subgrades, such as for roads.

In recent years, Building Information Modeling (BIM) has provided many powerful design tools in the Architecture, Engineering and Construction (AEC) industry [5,6]. The design of railway stations is also affected by this trend. As the basis for the whole station, obtaining a correct 3D model of the subgrade in a railway station is the first task in the entire procedure of BIM application. Consequently, MCSRS modeling based on the cross-sections in which the design data are recorded is a wise choice.

In performing BIM, the data interoperability is a widespread issue for BIM application in various domains [7]. To solve this problem, the Industry Foundation Classes (IFC) was proposed by buildingSMART International in 1994 [8], which is extendable to express the components in many civil engineering domains [9]. Hence, it is also appropriate to use the IFC entities as a foundation to describe the necessary structures of a railway station including the MCSRS. In fact, a series of IFC entities has been extended for describing the railway-related elements [10]. However, despite some existing subgrade-related entities [10,11], the current IFC is still incapable

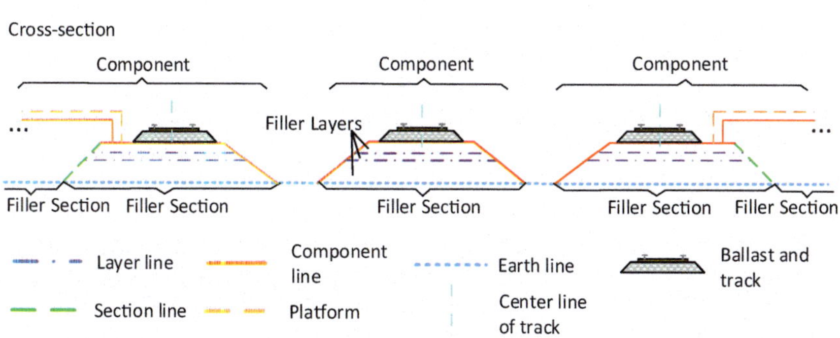

Figure 12.1 A sample of a cross-section of the MCSRS with multiple components, filler sections and filler layers.

of expressing the hierarchical and semantic information of an MCSRS due to the aforementioned MCSRS features. Considering the importance of the MCSRS in a railway station, there is a big gap in the BIM data standard for the essential structures in the railway station domain [12], which hinders the related BIM application. Furthermore, since the railway stations are key structures in a railway line, this problem also leads to an incomplete railway BIM standard.

12.2.1 BIM for railway station subgrade

12.2.1.1 Semantic model of MCSRS

As the foundation of the whole railway station, the MCSRS can be treated as a consolidated unit and required to accommodate itself to the variations of railway facilities supported by it, e.g., tracks and buildings in different locations, in order to offer dependable support. To achieve this purpose, designers usually divide the MCSRS into a group of continuous segments by defining cross-sections according to the distributions and variations of facilities, as shown in Fig. 12.2. Widths and lengths of these segments vary according to the locations and types of railway facilities.

These cross-sections serve as the basis for design work. In some large railway stations, the width of some cross-sections may reach hundreds of meters in order to contain various facilities. In addition, to reduce costs, the support requirements for different facilities are also different. Therefore,

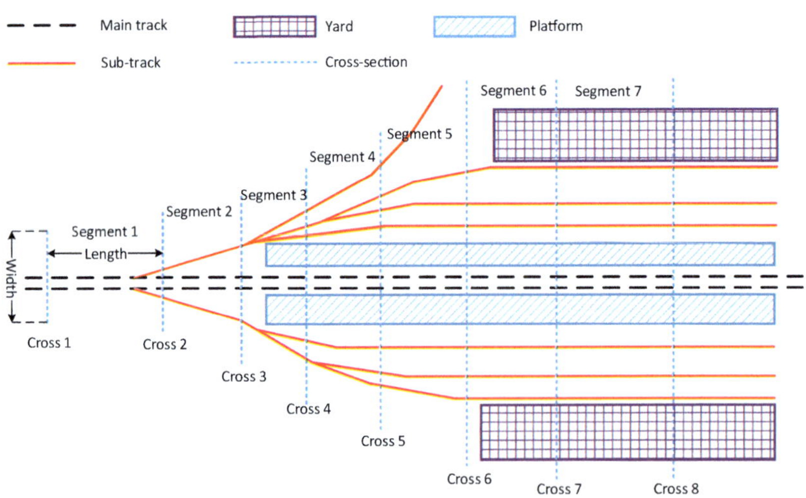

Figure 12.2 Horizontal design of a railway station with eight cross-sections and seven segments.

it is unnecessary and inadvisable to build the MCSRS uniformly with the same filler standard in every segment. Hence, the cross-sectional designs in these cross-sections comprise several subgrade components which support multiple facilities. In a component, there may be multiple filler sections with different filler standards for different facilities supported by this component.

As mentioned above, the MCSRS is constructed to satisfy different load bearing requirements for various facilities in a railway station. Several features are demonstrated in the following sections. The most common and widely used cross-sectional structure of subgrade for supporting a railway track is a kind of single-component structure shown in Fig. 12.3. For a single-component subgrade, all subgrade cross-sections are designed as single-component structures. The single-component subgrade is usually used for supporting a track and two adjacent cross-sections have very similar cross-sectional structures.

As shown in Fig. 12.3, contours and filler layers of subgrades in two cross-sections are almost identical. Therefore, the modeling result can be obtained by connecting each pair of filler layers. Unlike the subgrade with the simple structure mentioned above, most segments of the MCSRS are characterized by including multiple components.

In Fig. 12.4, a sample of MCSRS cross-sectional structure is illustrated, in which an obvious hierarchy exists. Part (a) in Fig. 12.4 shows that the cross-section has more than one component, and each component is

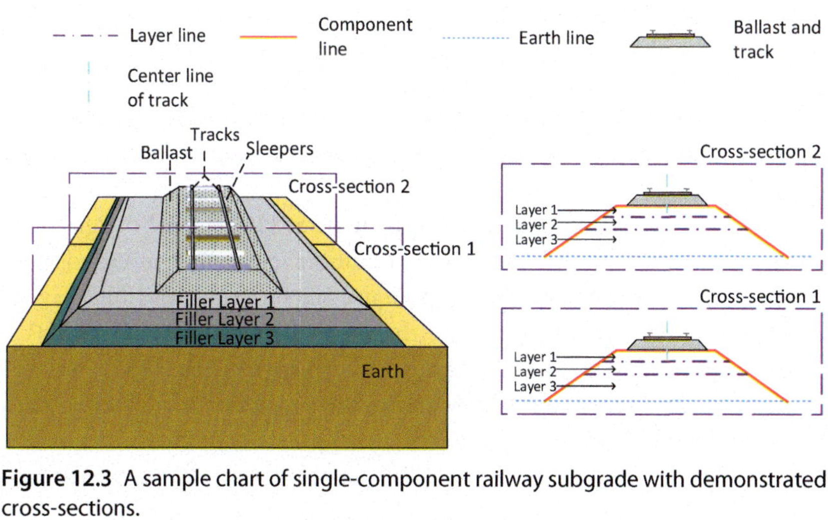

Figure 12.3 A sample chart of single-component railway subgrade with demonstrated cross-sections.

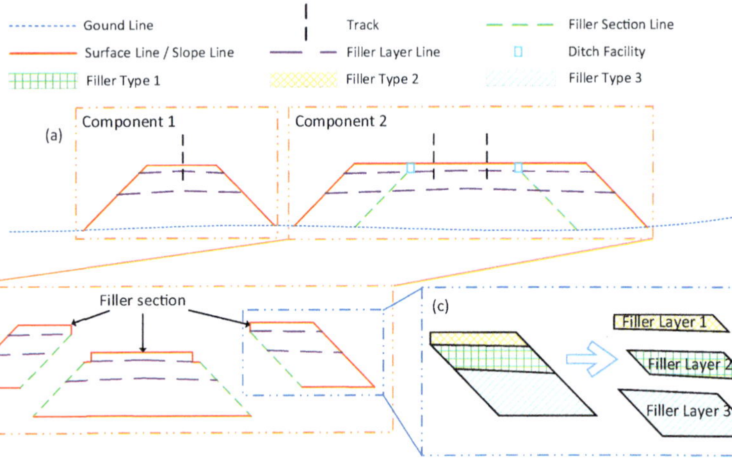

Figure 12.4 Cross-section of a typical multi-component segment of the MCSRS: (a) two subgrade components; (b) a component with three filler sections; (c) a filler section with three filler layers.

mapping to one or multiple railway facilities. Different railway facilities are characterized by their priorities, which depend on their different requirements for bearing capacity. To satisfy these requirements, many countries have developed a series of filler standards [13] to provide explicit filler definitions to support various railway facilities. Moreover, the construction cost for a subgrade is highly related to the used filler standard [14].

Some segments of an MCSRS are responsible for supporting multiple tracks and other buildings. An example of this kind of segment is shown in Fig. 12.5, along with the design of two cross-sections. Actually, the cross-sections are the foundation and data source for BIM. In Fig. 12.5, each cross-section has three components, which requires that matching relations among components of these two cross-sections be confirmed before modeling.

The greatest challenge of the conversion from MCSRS 2D design to 3D BIM lies in properly processing the transition segments among adjacent cross-sections for three levels: components, filler sections and filler layers. These transition segments compose the main BIM body for an MCSRS. The core issue in handling transitions is the mapping of relations for elements in these three levels. Furthermore, the procedure for mapping relations should satisfy a hierarchical order: components → filler sections → filler layers.

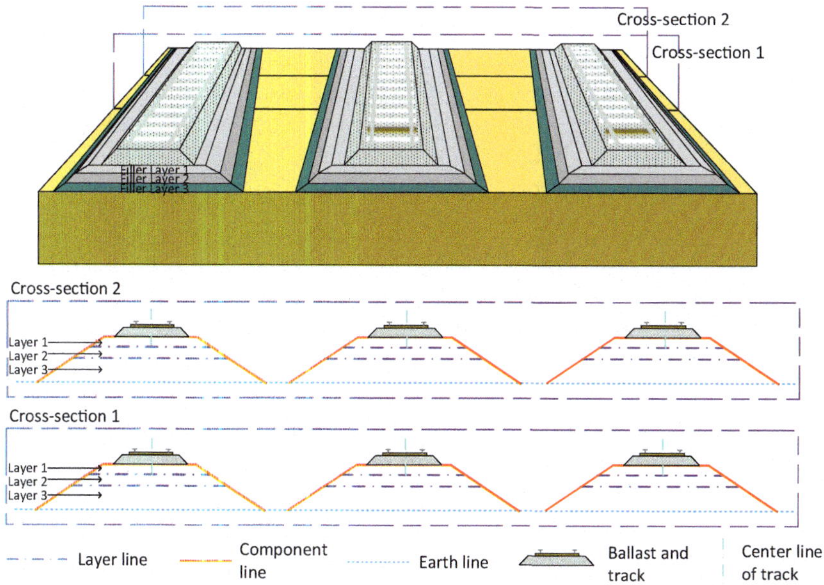

Figure 12.5 A sample of the multi-component segment and the corresponding cross-sectional designs.

The situations for matching adjacent components can be classified into three categories: single-to-single, single-to-multiple and multiple-to-multiple. In fact, the multiple-to-multiple case is a combination of the single-to-single and single-to-multiple cases.

In the filler section level, there are two situations for matching filler sections in a pair of matched components, which are shown in Fig. 12.6b. For filler layers, the situations are attributed to two types. The first type indicates whether the relation between the number of filler layers in two filler sections is or is not equal, as shown in Fig. 12.6c. In Fig. 12.6d, the situations in second type show specific cases of matching a pair of filler layers, which are influenced by filler level and filler type.

According to the basic hierarchical structure and features of MCSRS described, there should be several steps in conducting BIM: 1) confirming the related cross-sections to be modeled, 2) finding matching relations for the hierarchical elements in the adjacent cross-sections, and 3) generating 3D solids based on the matching relations.

By conducting the above steps, the following series of semantic issues should be solved to obtain the final BIM model:

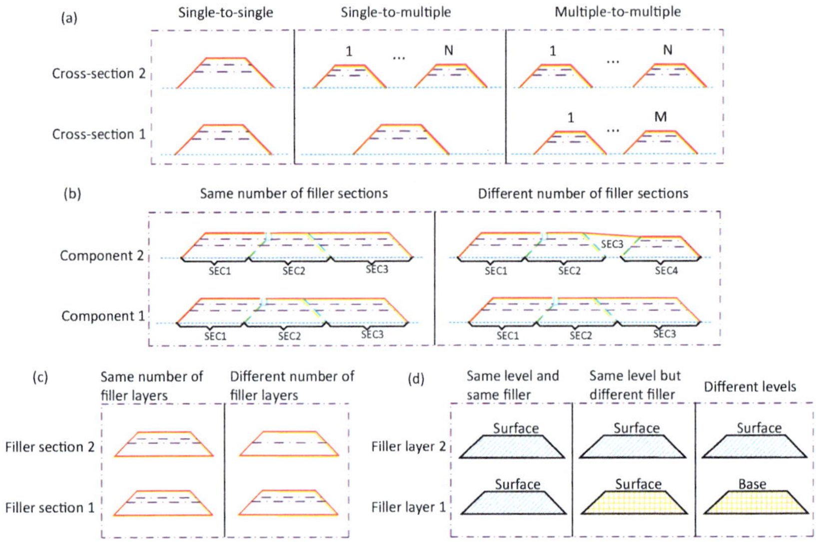

Figure 12.6 Different situations of element formations in adjacent cross-sections: (a) components; (b) filler sections; (c) and (d) filler layers.

- ➢ Hierarchical representation of the related cross-sections, which is the source data for performing 3D modeling. Moreover, the 3D BIM model of an MCSRS should also have a correct 3D representation for the hierarchy. The hierarchical information in the cross-sections of an MCSRS has been discussed in Section 12.2.1.
- ➢ Correct matching relations among the hierarchical elements. In other words, a component, filler section or filler layer in a cross-section should be matched with another corresponding element in an adjacent cross-section, based on which the 3D solids can be obtained by connecting the matched elements. Hence, a series of matching rules should be defined in order to correctly locate the matching relations for the hierarchical elements.
- ➢ Proper way of generating the 3D solids between two cross-sections.

In this study, the concept "control facility" refers to a facility that determines the design of filler standard for a component or filler section. As shown in Fig. 12.7a, a component has two facilities on it, namely a main track and a secondary track. According to the filler standard for supporting a main track, a three-layer structure S3 should be used for the filler standard, as demonstrated in Fig. 12.7b. Meanwhile, a two-layer structure S2 can be used for supporting a secondary track, as shown in Fig. 12.7c. As the bearing

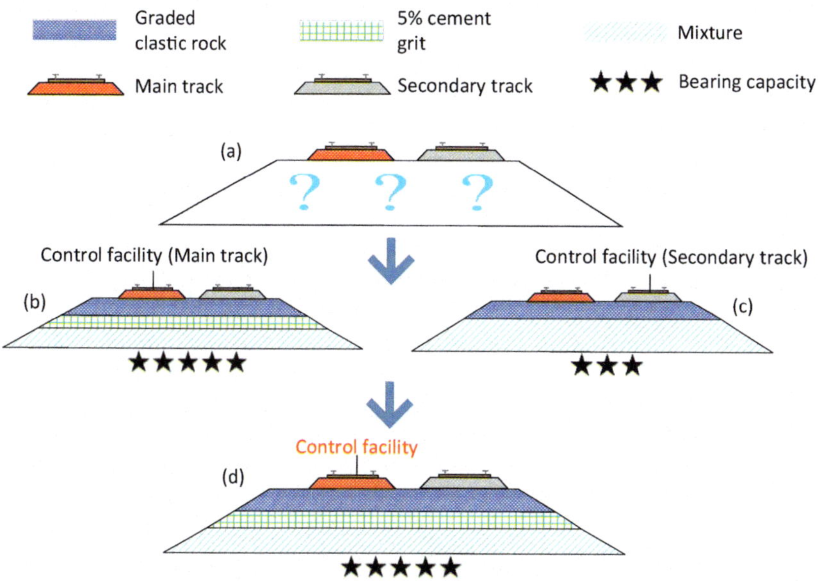

Figure 12.7 A demonstration of the control facility of a component.

capacity of S3 exceeds the capacity of S2, the S3 is selected for the filler standard of this component and this main track is the control facility in this component, as presented in Fig. 12.7d.

At the component level, the matching rules consist of a connecting rule and a separating rule. The principle of the connecting rule is that a pair of components in two adjacent cross sections should be matched only when they are crossed or contained by the same control facility, and these two components only match each other. For instance, in Fig. 12.8a these two components are crossed by a track and a ditch. The track is the control facility that matches these two components since it has a higher priority than the ditch. Two pairs of filler sections are matched by their control facility separately. In Fig. 12.8b, two components and their filler sections get matched because they are contained by a yard, which is treated as the control facility.

The separating rule decomposes a component that is matched with multiple components in another cross-section. This rule is applied in the single-to-multiple case. It separates a single component into a number of sub-components to turn a single-to-multiple problem into several single-to-single cases.

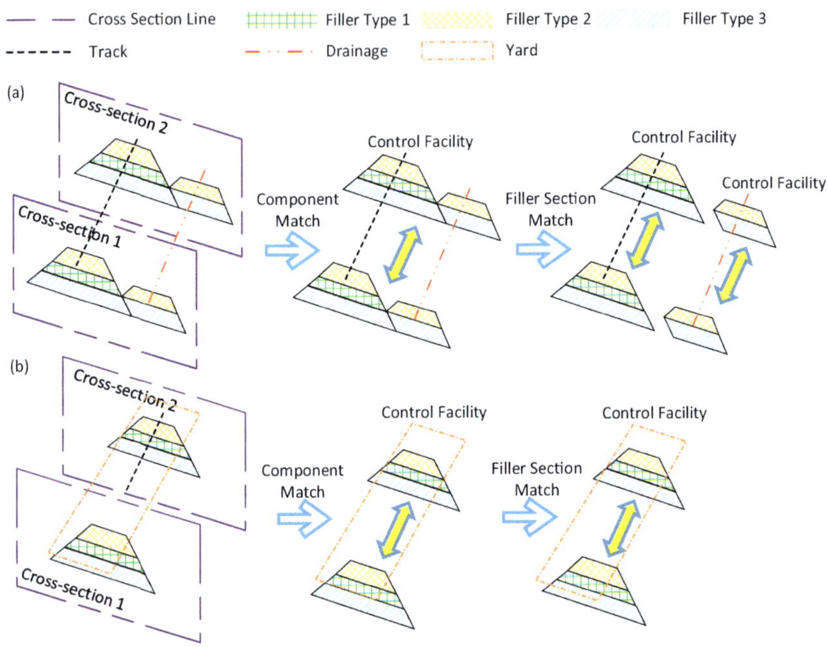

Figure 12.8 Two situations of matching components and filler sections with the control facility: (a) two components and their filler sections get matched with a track; (b) two components and their filler sections get matched with a yard.

For example, in Fig. 12.9 a component C1 in the cross-section 1 and two components C2, C3 in the cross-section 2 are crossed by two tracks T1 and T2. According to the matching rule, both C2 and C3 can match with C1. However, filler layers resulting from the model will intersect if the whole C1 is used for matching. To avoid such collision, the separating rule is utilized to divide the C1 into two sub-components according to two tracks connecting related filler sections. The sub-components consist of the filler sections in the original component, and are treated as new components for matching. In Fig. 12.9 two filler sections in the C1 are treated as two new components SC1 and SC2 for matching.

Similarly to the procedure for matching components described above, matching relations among filler sections are also obtained through the connecting rule. Two filler sections crossed by one control facility are matched. Unlike matching for components, a filler section can only match one filler section, which means the separating rule is not required. In some cases, the number of filler sections in two matching components is different.

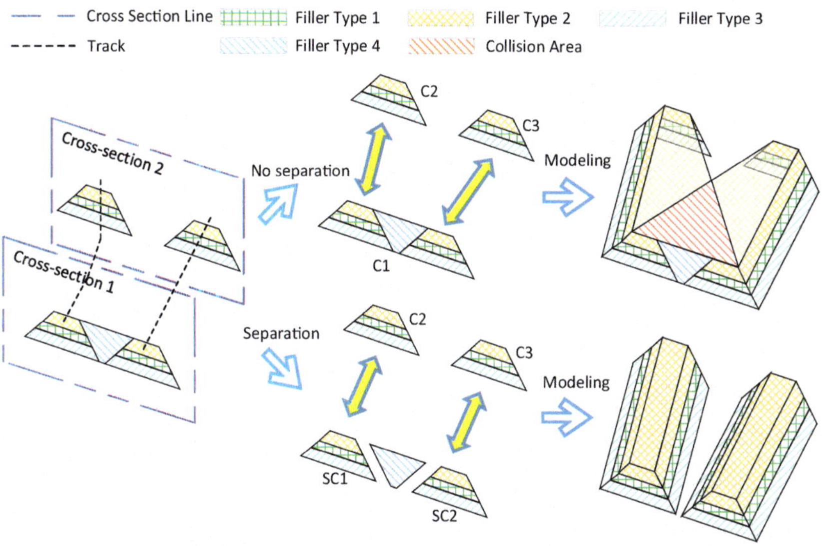

Figure 12.9 The separating rule for the components divides the component in the cross-section 1 into two sub-components to avoid a modeling collision.

Consequently, there will be some unmatched filler sections in this situation, and they will not be processed for matching filler layers.

Ordinarily, a component of the MCSRS has different filler standards which are represented by several filler sections in this component. Each standard offers the definition for the number of filler layers and specific filler type in each filler layer. There are two key attributes for a filler layer: level and filler.

The level of a filler layer indicates a relative position in the cross-sectional structure of the subgrade while the filler specifies a type of filler, as shown in Fig. 12.10a. The definition of levels includes surface, base and subbase layer. Here, these two attributes are the base for matching two filler layers in a pair of matching filler sections. Based on the matching result, the transition between two layers is generated, and presented as follows:

1) a pair of filler layers with the same level and filler is fully matched as shown in Fig. 12.10b;
2) a pair of filler layers with the same level but different filler is matched but connected with cross merging, which is illustrated in Fig. 12.10c;
3) a pair of filler layers with different levels is not matched. After the matching process for all filler layers, there may be some unmatched filler

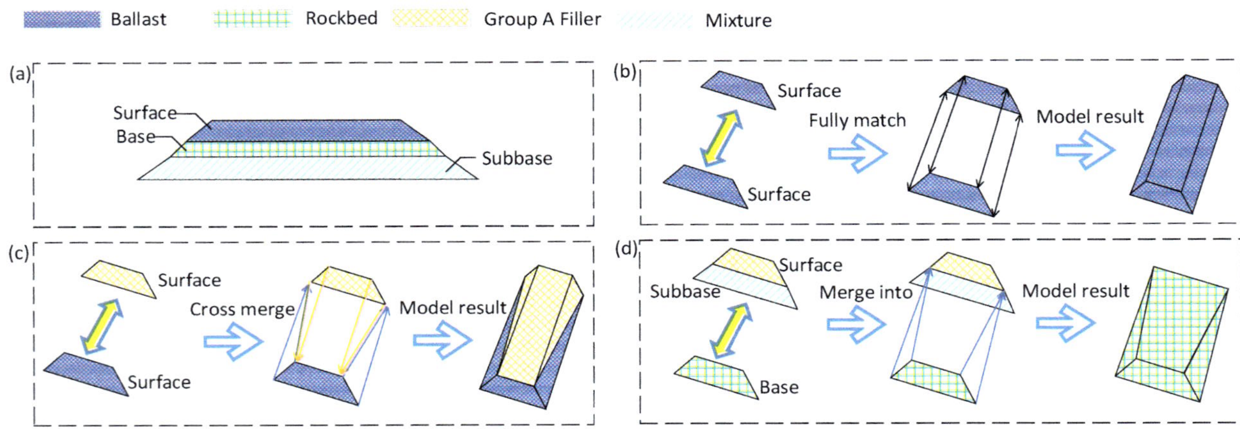

Figure 12.10 Matching rules for filler layers: (a) a typical standard for filler in a railway station; (b) matching two groups of filler layers with the same level and filler; (c) matching two groups of filler layers with the same level but different filler; (d) an unmatched base layer is merged into another layer.

layers. The unmatched layers are merged into the neighbor layer of the matching filler section. Fig 12.10d shows an e.g.

After obtaining matching relations for filler layers, the next step is to correctly "connect" them, i.e., convert these pairs of cross-sectional filler layers into 3D transition solids. The method introduced here for this conversion is called generalized hexahedron [2]. The central idea of this method is based on B-Rep, a famous boundary representation achieved by constructing every mesh face of a 3D object with lines. In this study, every solid component of the subgrade model is constructed from eight basic polylines, as shown in Fig. 12.11b. Geometric data of these eight polylines are obtained from the original cross-sectional design of MCSRS, which provides a cross-sectional geometry description of the filler layer being modeled. The geometric method uses the eight polylines of a filler layer to generate six sets of faces (front, back, left, right, top, bottom) by matching and connecting vertices on each pair of polylines, as shown in Fig. 12.11c. The matching strategy for the eight polylines is simple: a pair of lines in front and back

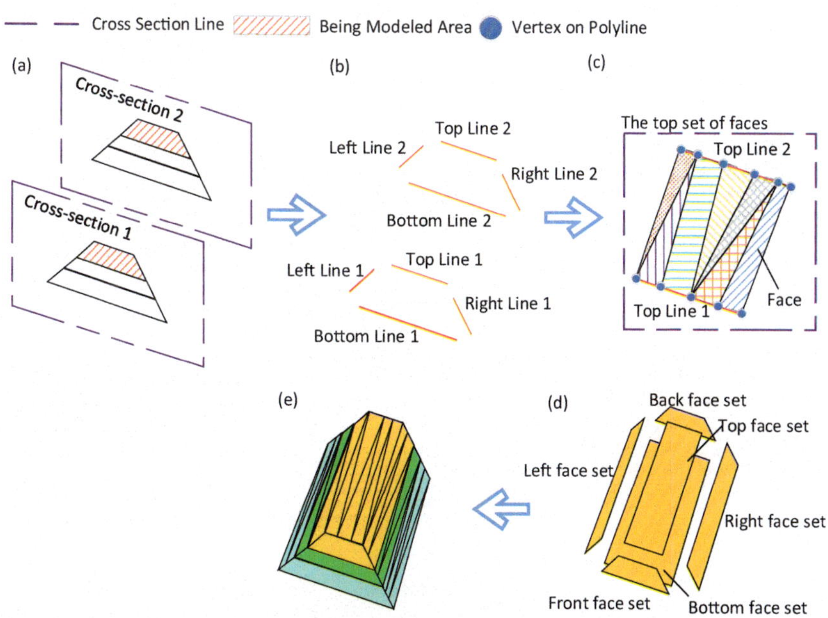

Figure 12.11 Modeling using the generalized hexahedron method: (a) cross-sectional geometric data; (b) eight polylines of filler layers on the first level; (c) connecting vertices on two polylines to generate a set of faces; (d) generated six sets of faces; (e) final modeling result.

cross-section of the same type match each other to get a set of faces, e.g., front top line matches the back top line.

Thus, the final modeling result can be obtained, through which the transition part of subgrade between two cross-sections is obtained and expressed. Fig 12.11 shows the procedures.

12.2.1.2 IFC extension of railway station subgrade

This section demonstrates the details of the proposed extension, which is shown in the UML diagram in Fig. 12.12. In the proposed extension, an entity *IfcRailwaySubgradeElement* is defined as a root to offer public attributes to the entities inheriting it. As the IFC standard is characterized by its internal inheritance relations among entities just like a hierarchy of classes in object-oriented programming. The first step in performing an extension is determining a parent entity from which the root entity is inherited. The entity *IfcCivilElement* is an existing definition in the current IFC

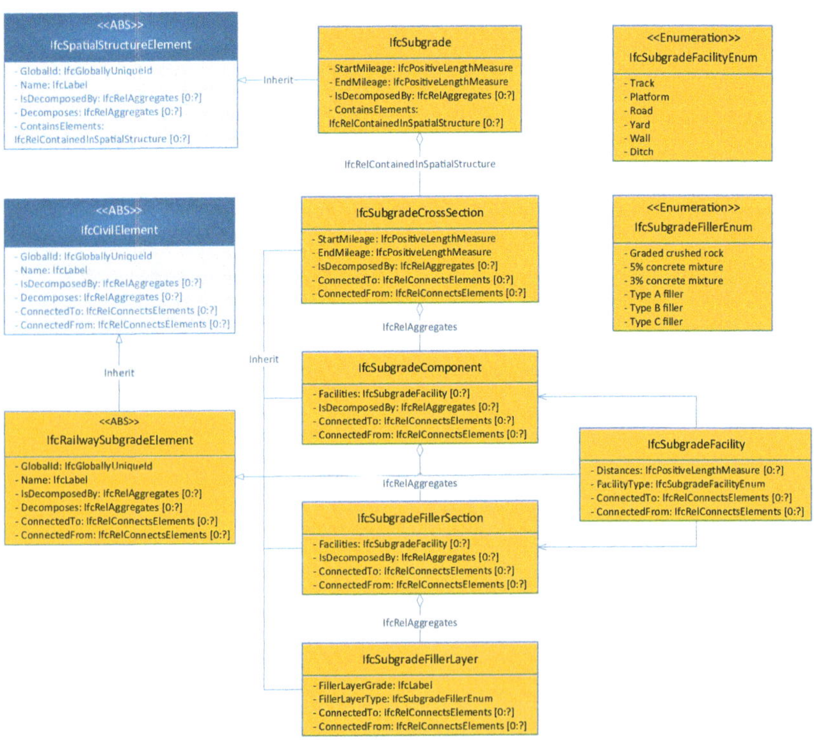

Figure 12.12 The UML diagram of the proposed IFC extension. Boxes in yellow color represent newly added entities.

standard, which can be utilized for the expression of common civil engineering objects, such as the segments of road, bridge and tunnel. Therefore, the entity *IfcRailwaySubgradeElement* is added by inheriting the *IfcCivilElement*, as shown in Fig. 12.12.

For modeling MCSRS, there are three basic requirements in the use of IFC: 1) an appropriate way of expressing hierarchy; 2) full records of matching relations among components, filler sections and filler layers; 3) a complete storage of the geometric data. Approaches to fulfill the above three requirements are presented in the following three sections.

a) In Fig. 12.12, the entities *IfcSubgradeCrossSection*, *IfcSubgradeComponent*, *IfcSubgradeFillerSection* and *IfcSubgradeFillerLayer* are extended by inheriting *IfcRailwaySubgradeElement* to describe the hierarchical relations of cross-section → component → filler section → filler layer inside the MCSRS. These relations are recorded and expressed by the attribute *IsDecomposedBy* which is inherited from the upper entity *IfcObjectDefinition* and uses the entity *IfcRelAggregates* to describe a composition relation.

The hierarchical data composing the cross-sections is expressed by an entity *IfcSubgrade* which is extended from *IfcSpatialStructureElement.* Because the entity *IfcSpatialStructureElement* describes the spatial elements and relations in a project, we extend the *IfcSubgrade* from it.

b) As mentioned above, the matching relations among components, filler sections and filler layers are vital for creating a BIM model of the MCSRS. These important matching relations can be recorded by two attributes, *ConnectedTo* and *ConnectedFrom*, both of which are inherited from the entity *IfcElement*. These two attributes are described by the entity *IfcRelConnectsElements* to express connecting relations among objects.

c) The whole MCSRS model can be obtained by correctly modeling every filler layer. The model of a filler layer consists of a series of mesh faces, all of which include three or four vertices. These faces and vertices are recorded by the property Representation inherited from the entity *IfcProduct*, which is the parent entity of the extended entity *IfcRailwaySubgradeElement.* As a result, the entity *IfcSubgradeFillerLayer* can use the property Representation to store the geometric data for modeling.

d) Another important aspect of information conveyed by a BIM model is the semantic properties of a domain. In this study, the final result is generated by modeling pairs of matching filler layers, which are characterized by a series of properties such as filler level and filler type. Therefore, these important properties are included in the definition of entity *IfcSubgradeFillerLayer*, as shown in Fig. 12.12.

12.2.2 Microprism-based layered BIM for railway station subgrade

This section contributes to the existing knowledge by providing a microprism-based layered representation of internal and spatial information for the hierarchical structure of a Railway Station Subgrade (RSS). For an RSS project, the volumetric data mainly consists of two aspects—various fillers used in fill operation and the soils excavated in cut operation. Moreover, the volume of filler on the bottom layer is influenced by the surface of terrain. Hence, there are totally three kinds of layer structures which should be considered in the volumetric modeling procedure for an RSS:

➢ Filler layers which are used for supporting other railway facilities and structures.
➢ Geology layers which are excavated in the cut sections of RSS.
➢ Terrain layer which lies between the filler layers and geology layers.

An example of the above layers is shown in Fig. 12.13.

To acquire volumetric data for the filler layers used in the project, it is essential to have accurate measurements of the quantity and type of fill materials being used. Various standard and parametric components, such as pre-mixed concrete, asphalt, or construction aggregates, may have well-

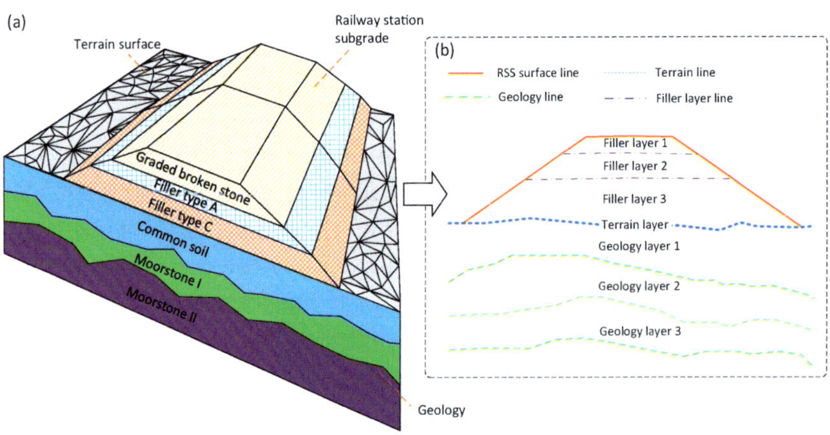

Figure 12.13 Related RSS layers in this chapter: (a) 3D view of a segment of RSS, and (b) cross-sectional view of this segment in which filler layers, a terrain layer and geology layers are included.

defined geometric shapes and volumes, making their acquisition relatively easy. On the other hand, the acquisition of volumetric data for the excavated soils is more challenging, since it is related to the geologic layers. The irregular geometric shape of the excavated soil makes it difficult to accurately estimate its volume. Traditional methods, such as manual soil sampling and laboratory testing, can be time-consuming and may not accurately represent the overall volume. Thus, the volumetric expression for an RSS comprises the volumetric spaces among a multiple-layer structure composed of several filler layers and geologic layers.

Therefore, this chapter utilizes a number of microprisms to develop an internal spatial expression for the RSS, which is based on a multi-layer model for depicting the related layers of filler and geology. In this model, a unified modeling method is proposed in order to express various related layers based on the original data of RSS design, terrain, and geologic condition. The generated microprisms act as a voxel representation of the used fillers and excavated soil, allowing a more accurate 3D volumetric expression. The high-density microprisms are designed to capture the irregularity and complexity of the surfaces representing the filler layers and geologic layers.

The main procedures of the proposed method are shown in Fig. 12.14. The multi-layer RSS model provides a unified expressing method for various layers, through which the triangular meshes of these layers can be generated in a convenient way. Next, the original terrain model is enriched by integrating the vertexes and edges of other layers with the triangles of terrain, by which the base triangles of the microprisms are generated. According to the base triangles, the microprisms are finally established by computing the triangles projected on other layers. To realize an efficient management and organization of the microprisms, a grid-based spatial retrieval method is proposed, with which a quick search of the target spatial information can be performed.

12.2.2.1 Multi-layer model for RSS

In an RSS project, multiple physical layers are involved in preparing a volumetric expression, which includes the terrain surface layer, filler layers and geologic layers, as shown in Fig. 12.15. Moreover, the data sources for these layers are heterogeneous in data structure and format. For instance, the terrain layer can be recorded as the point clouds from LIDAR or the terrain lines from field measurement, while the geologic layers can be described by the data from borehole or satellite remote sensing. Furthermore, previous

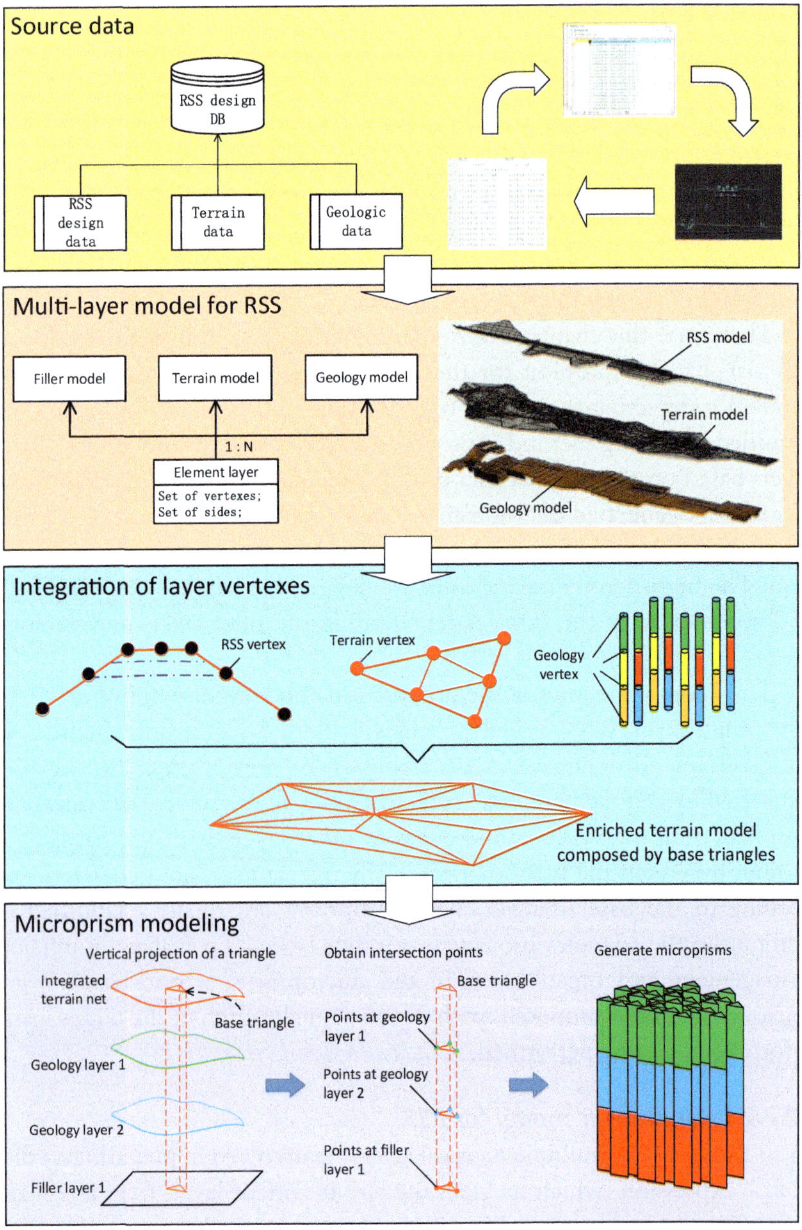

Figure 12.14 Procedures of the proposed method, through which the microprism-based volumetric model is built.

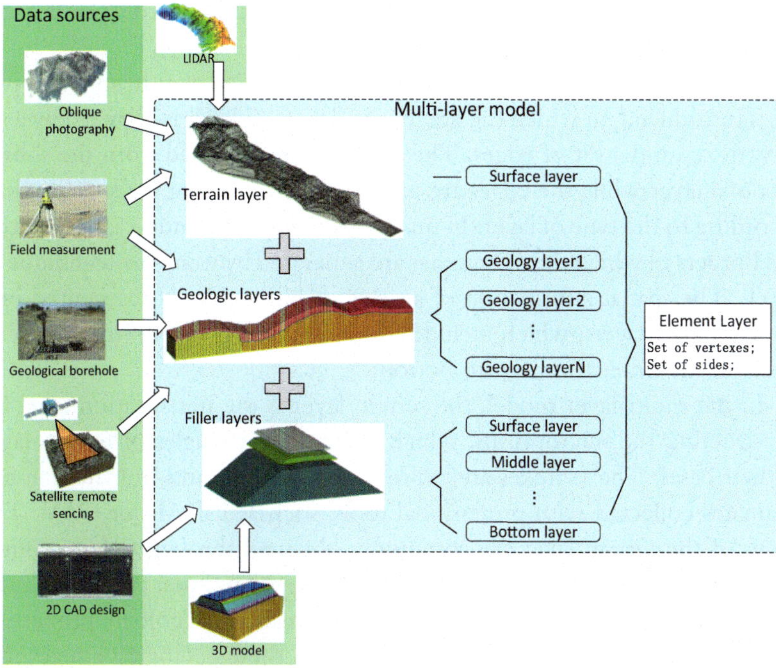

Figure 12.15 Proposed multi-layer model for depicting various layers in RSS volumetric modeling.

studies usually developed the specific methods for tackling the 3D modeling issues for these layers [2,15]. Thus, the heterogeneous data sources and modeling methods complicate the expression of various layers in the domain of subgrade BIM.

To provide a unified modeling solution based on different data sources, a multiple-layer data model for RSS is proposed in this chapter, in which the geometric data for every physical layer are abstracted as a set of vertexes and a set of edges, as shown in Fig. 12.15. The edges define a series of control borders in which the 3D facets can be generated based on the vertexes in the borders by using the method of Constrained Delaunay Triangulation (CDT) [16,17]. Thus, a standard process for conducting 3D modeling for the involved layered structures is provided.

In this section, the layered structures must be explicitly expressed before building microprisms. The data sources of these layers, e.g., the borehole and field management data for geology, as well as the LIDAR and UAV data for terrain surface, are heterogeneous. Furthermore, the past studies commonly used or developed the specific methods for modeling these layers. Applying a different method for expressing the related layered structures of an RSS is a tedious and error-prone process.

Hence, a unified geometric expression for various layers in an RSS project is introduced, in which the geometric data for a layer is depicted by a set of vertexes and a set of edges. The vertexes are obtained from the source data of a layer while the edges are produced by following the specific rules according to the type of layer. In practice, the edges provide a series of control borders in which the 3D meshes are generated by using the technique of CDT. The spatial information of an RSS is hierarchically segmented into three types of layers, which include a terrain layer, geology layers and filler layers in the order from top to bottom.

In the multi-layer model, the terrain layer is the initialization level for constructing the microprisms, which utilizes the triangles of terrain layer as their bases. The vertexes are expressed as spatial points in Cartesian coordinates collected with professional tools, such as LIDAR or UAV. The edges of the terrain layer are commonly obtained through field measurement, and can be categorized into ridge lines, valley lines, and scarp lines.

For geology layers, the vertexes store the necessary geologic properties, such as depth, soil type and density. Based on these geologic vertexes, the surface meshes of a geologic layer can be generated by using the method of Delaunay Triangulation or Inverse Distance Weighting [15]. In this chapter, the edges of each geology layer provide a series of control borders for generating the triangular meshes of this layer, which avoid incorrect mappings to the vertexes of other geologic layers. To obtain these edges, a filtering process of vertexes is performed based on the related semantic properties aiming to secure the vertexes on the same layer. Then, a geometric process is implemented to locate the convex hull of this layer [18], through which the edges are confirmed. An example is shown in Fig. 12.16.

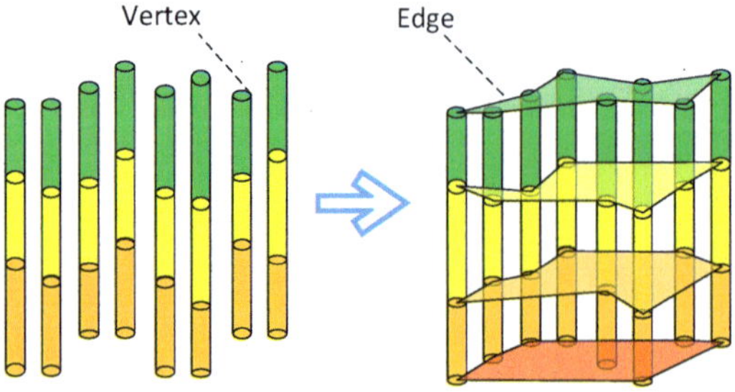

Figure 12.16 Procedures for locating vertexes and edges of geologic layers.

A filler layer refers to a structural stratum that is filled with a specific material and serves to provide support and stability to the structures above and below it [13]. To support these facilities, several filler layers are set by following the related design criteria and rules to generate the geometric shape and semantic properties of these layers. To express the geometric shapes in a cross-section, a series of geometric vertexes which can be mainly categorized into surface vertexes, changing slope vertexes, and slope vertexes, are defined, as shown in Fig. 12.17. Therefore, a set of vertexes for the filler layers can be obtained after the design of layers in all cross-sections is finished.

The edges of filler layers are composed by the cross-sectional edges and the longitudinal edges. The former ones can be secured based on the vertexes in a cross-section while the latter ones can hardly be confirmed unless the 3D modeling of filler layers is performed. Unlike the 3D modeling for an ordinary single road, the element matching among multiple components, filler sections and filler layers should be handled as in our previous study [19]. With this method, the 3D RSS model is obtained in the form of 3D filler layers, and the set of edges is successfully extracted, as demonstrated in Fig. 12.18.

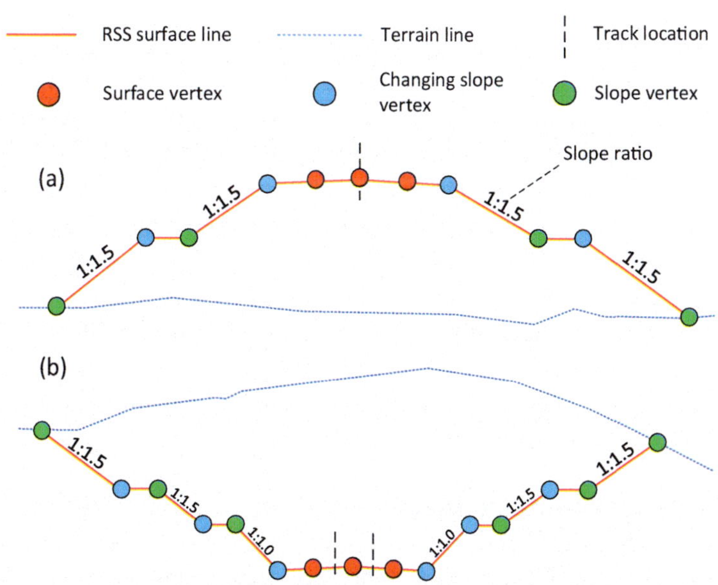

Figure 12.17 An example of various geometric vertexes constituting the cross-sectional design of RSS, in which (a) is a fill case and (b) is a cut case.

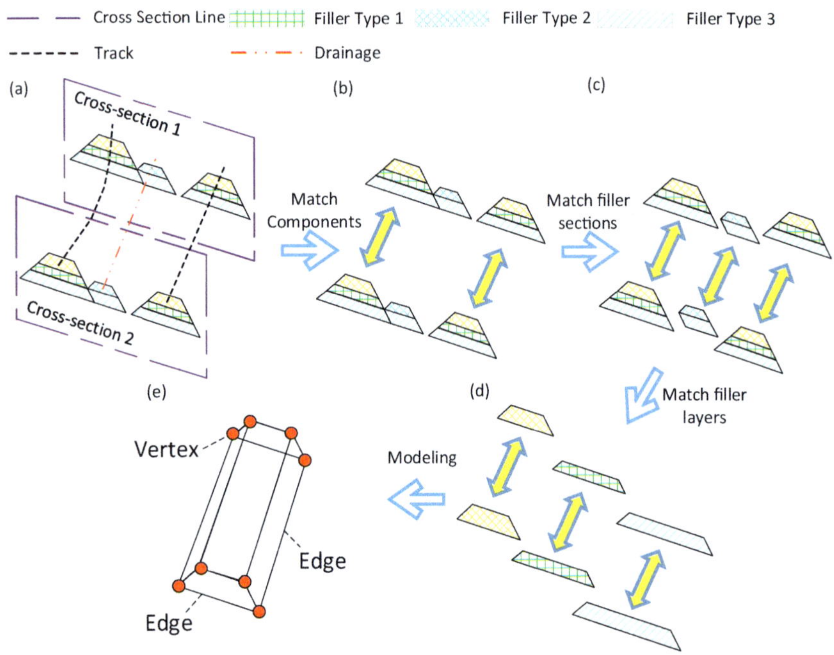

Figure 12.18 Procedures for obtaining edges and vertexes of filler layers.

In this step, the triangular meshes are generated for each layer in the aforementioned multi-layer model by using CDT. The CDT is highly efficient in converting a batch of scattered vertexes to triangular meshes by considering a series of constraints, which can be useful when dealing with complex or specific geometric requirements in a triangulation [17,20]. The advantages of using CDT include the consistency and uniqueness of the obtained result based on the given geometric conditions and constraints.

Hence, this chapter makes full use of the CDT to generate the meshes of the multi-layer model, through which a unified way of 3D transformation for the layers of filler, terrain and geology is provided. As shown in Fig. 12.19a and b, the original data of a layer consists of a set of vertexes and a set of edges. The triangular meshes of this layer are generated by using the CDT, as shown in Fig. 12.19c.

12.2.2.2 Microprisms modeling

In this section, a number of microprisms are generated based on the obtained meshes of the layers in the multi-layer model. The generation of microprisms starts from a triangle in the enriched terrain layer, based on which

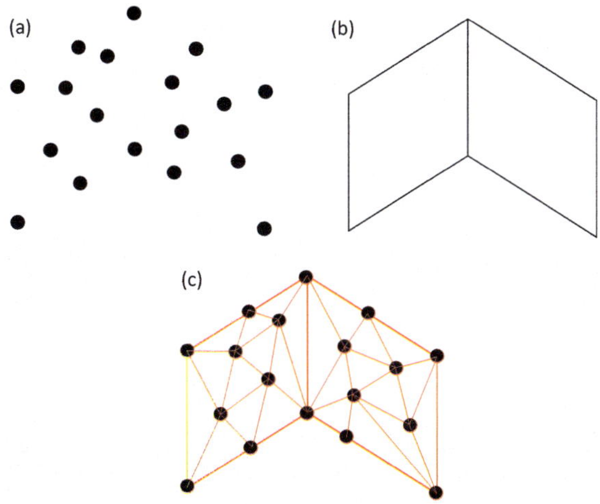

Figure 12.19 Result of mesh generation with CDT for related layers: (a) set of vertexes of layer, (b) set of constrained edges, (c) final triangular meshes generated with CDT.

the vertical projection lines from the three vertexes are created to obtain the intersection points on other layers. These intersection points provide necessary partitioned information for building a microprism of multiple segments with different types of soil or filler. To efficiently manage the generated microprisms, a specific spatial index from a grid to the related microprisms is created, based on which a grid-based spatial retrieval method is proposed to realize a quick query of volumetric information.

All the microprisms are generated by using a triangle of the terrain layer as the base triangle. Three vertical lines are extended from the vertexes of a base triangle, by which the intersection points are located on the related layers of filler and geology. Finally, the 3D microprism is constructed according to the base triangle and the intersection points, as shown in Fig. 12.20. Nevertheless, the original equally-spaced vertexes of the terrain layer are collected from the real terrain by using a given spacing value, due to which some geometric features of the RSS surface may be ignored if the microprisms were built based on the original vertexes. Furthermore, the geologic layers are also considered in the process of the microprism modeling, and the space between the terrain model and RSS model is divided into several sub-spaces by them.

In this situation, the original triangles of the terrain layer are enriched by integrating the vertexes and constrained edges from other layers in order to provide more base triangles for building the microprism. The main process

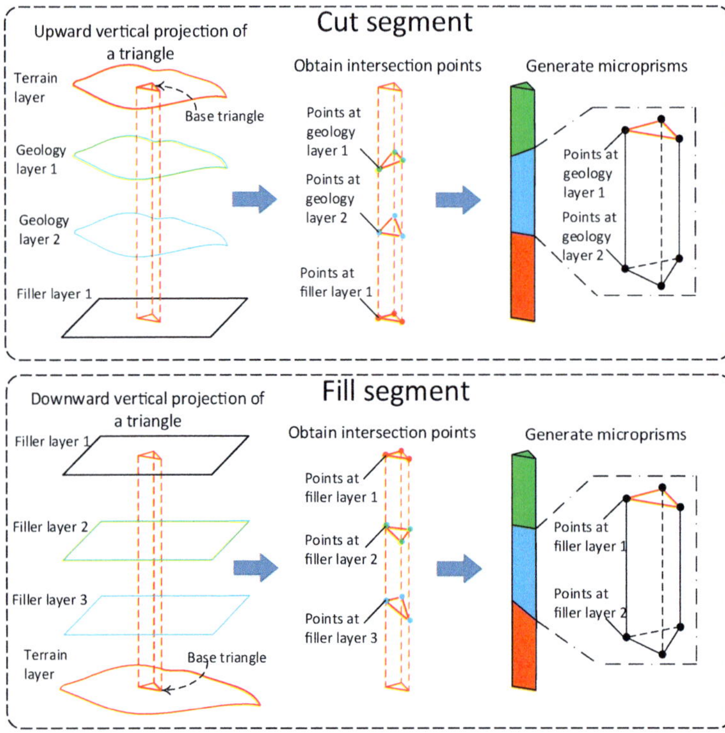

Figure 12.20 Obtaining intersection points for generating microprisms in cut and fill segments.

of this integration is shown in Fig. 12.21. First, the vertexes of a layer to be integrated are vertically projected to the triangles of terrain layer. Then, the triangles are rebuilt in the terrain layer by considering the Delaunay criterion which ensures the consistency and uniqueness of the generated net of triangles [21]. Next, the constrained edges are vertically projected to the newly generated triangles. For integrating a constrained edge, an influenced domain composed of the triangles intersecting this edge is first located, in which the original triangles are removed. Then, the new triangles are rebuilt in the influenced domain by treating the constrained edge as a triangle edge. Thus, the number of triangles before and after the integration stays unchanged, while the constrained edge is embedded into the net of triangles.

To construct microprisms, every triangle of the terrain layer is first processed by generating three vertical projection lines based on its vertexes, with which the other vertexes of a microprism are located on the filler

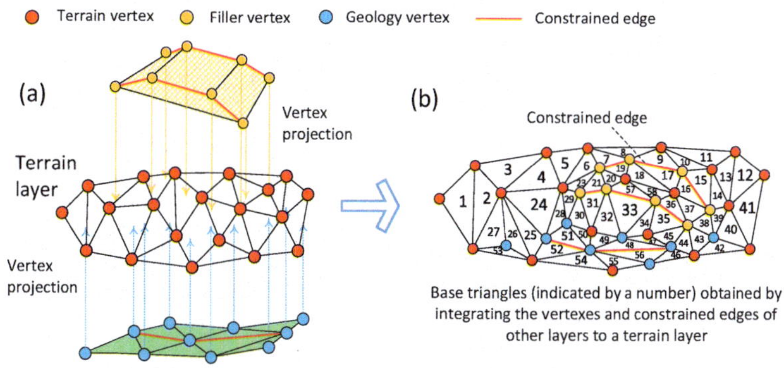

Figure 12.21 Procedures for generating base triangles by projecting the vertexes and edges of other layers on the original triangles of a terrain layer: (a) vertical projections of vertexes and edges on the terrain layer, and (b) obtained net of base triangles after the integration process.

or geology layers, as shown in Fig. 12.20. The direction of the projection line (up or down) depends on the type of earthwork, i.e., cut or fill. For both cut and fill, the targeted surface refers to the uppermost filler layer of RSS. During this process of finding a projection point, a projection line may have multiple intersections with different filler layers or geologic layers, according to which a whole prism is divided into several levels associated with a specific filler or soil. Besides, the sharing of a vertex among multiple microprisms is so common that considerable storage space may be wasted if the related vertexes are completely recorded in every microprism. Therefore, a specific table for indexing all the vertexes is created in the integrated set of layer point, which can be simply described as (*pID*, *vertex*), where *pID* is an integer and the coordinates of this point are assigned to *vertex*.

To record and organize the volumetric information expressed by microprisms, a specific data structure composed of several structs is shown in Fig. 12.22, in which a multi-level index from the grid to a microprism is presented. Furthermore, this data structure lays the foundation for realizing a quick spatial search in an RSS BIM project. For creating the spatial index, the planar area of a railway station is first divided into a number of 2D grids by using a given width, which turns the whole area into M * N sub-areas.

According to Fig. 12.22, there is a defined structure called *rssGrid* that represents the grids. Each grid has a *triangleID* field that only records a base triangle within the grid, and this relation is initialized in the construction of grids. The triangles within the grid are represented by the structure

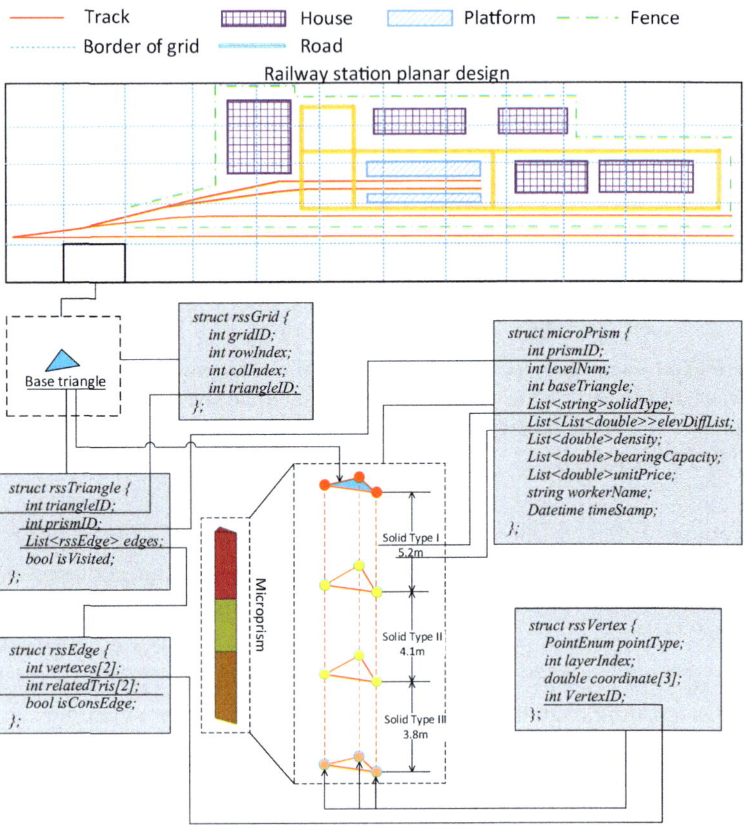

Figure 12.22 Proposed data structure for recording and organizing microprisms, in which the chain index of grid-triangle-microprism is created in order to provide a quick retrieval shortcut.

rssTriangle, which contains a *prismID* field to establish the relation with the microprism created based on that triangle. Moreover, the related edges of the triangle are described by the structure *rssEdge* and stored in the edges field. According to field *relatedTris*, the neighbor triangles of this triangle can be located, through which the adjacent microprisms are also confirmed. In the structure *microPrism*, the *levelNum* field indicates the number of segments in this prism while the specific types of filler or soil are specified by the field *solidType*. An attribute *elevDiffList* is set to save the elevation differences among the basic and projected vertexes, based on which the exact segment in this microprism can be easily located according to a given z value.

For modern civil engineering projects, the major purpose of spatial modeling is to provide precise spatial information for the subsequent project activities, such as cost estimation and bearing capacity analysis. Hence, several semantic attributes are also attached to the generated microprisms. The attribute *unitPrice* can be used for estimating the costs of fill or cut operation while the attributes *density* and *bearingCapacity* provide related information for conducting stress analyses. For recording the progress-related details, the attributes workname and timestamp are attached to mark the associated personnel and time information. Moreover, this chapter utilizes eXtensible Markup Language (XML) to store and format the semantic properties of the provided volumetric information, with which the attributes can be extended based on the actual need.

In the proposed data structure, a chain spatial index of grid-triangle-microprism is created, through which a target microprism can be located based on a given point at any place. The target grid can be quickly found to greatly narrow the search area. Nevertheless, the search process is inefficient if it is only conducted by sequentially checking the neighbor triangles recorded on triangle edges. For a further improvement, a grid-based spatial retrieval method is introduced in the next section.

In this section, the spatial expression of RSS consists of massive microprisms containing a large number of vertexes and facets, which leads to a tedious retrieval process for the microprism near the target location if the search process is conducted with a sequential algorithm. To solve this problem, a Grid-Based Spatial Retrieval Method (GBSRM) is proposed, which includes two major steps: the planar search for locating the base triangle by using the (x,y) coordinates of a given point, and the vertical search for locating the exact segment of microprism by using the z value. The detailed workflow is exhibited in Fig. 12.23.

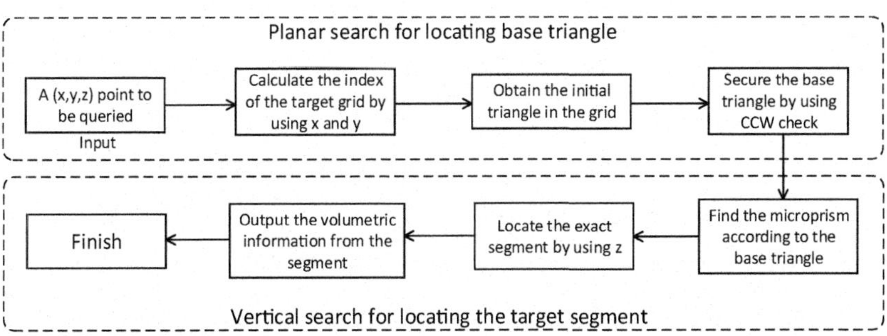

Figure 12.23 Procedures of the proposed GBSRM retrieval method.

To conduct the first step, we employ a directional search technique to locate the target triangle starting from the initial triangle of the grid. The structure *rssGrid* is introduced to describe the grids. By setting a standard width of grid, the whole RSS area can be gridded. For every grid, a base triangle is recorded as the starting point for searching the target triangle. In the initial triangle, a Counter-Clockwise Check (CCW) is performed on its three edges to determine a direction for searching the target triangle. The CCW uses the relative positions from the triangle's centroid and the target point to an edge to obtain the search direction: if the centroid and the target point are on the different sides of an edge, the CCW passes and the search direction is confirmed-the next object for conducting a CCW check is the adjacent triangle sharing this edge. If all three edges of a triangle fail to pass the CCW (the centroid and the target point are on the same edge for all edges), the search process ends and this triangle is the final target.

Next, by checking the *prismID* of the target base triangle, the microprism created from this triangle can be determined. In it a vertical search is performed based on the z value of the given point. After this step, the final result of spatial retrieval is provided in the target microprism. A diagrammatic demonstration of the procedures for finding the target microprism is shown in Fig. 12.24.

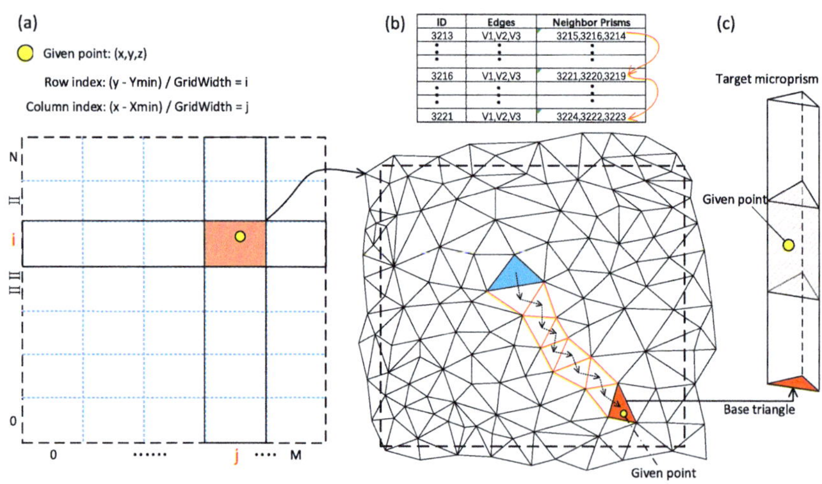

Figure 12.24 Procedures for finding a target microprism based on a given point: (a) locate a grid by computing indexes of row and column, (b) secure a target base triangle by using (x,y) and (c) obtain the exact segment of microprism according to the z of given point.

In addition to a spatial coordinate, the retrieval of spatial information can also be conducted based on mileage-offset and several semantic attributes in this chapter. The retrieval based on mileage and offset refers to the query according to a given mileage value and an offset value, as shown in Fig. 12.25. The mileage value specifies the longitudinal position of a cross-section while the offset value describes the cross-sectional distance to the center point in the cross-section.

By specifying one or multiple semantic attributes (a filler or geologic type), the associated 3D microprisms can also be obtained. Moreover, the above conditions can be combined to launch a further spatial query. For instance, the exact microprisms can be located by using a mileage range and a specific filler type, as shown in Fig. 12.26.

In summary, the proposed method allows multiple inputs for launching a spatial retrieval, namely spatial coordinates (x,y,z), mileage-offsets, and semantic conditions.

12.2.3 Dynamic BIM approach

In this section, a dynamic BIM approach is developed for achieving efficient BIM and updating for RSS, which fully utilizes the proposed IFC extension. The proposed framework consists of two major parts: a quick modeling approach and a partial updating algorithm.

12.2.3.1 Quick modeling approach

Modeling Principle. In this study, the source data of RSS mainly includes a set of cross-sections, the railway facilities, and the related design criteria for designing the cross-sections. In Fig. 12.27, the main procedures for applying the BIM are demonstrated. In fact, the key challenge of the modeling lies in the precise conversion from the 2D cross-sections to the 3D segments of RSS. Hence, the major steps of BIM should contain: 1) matching the components in the adjacent cross-sections, 2) matching the filler sections in the matched components, 3) matching the filler layers in the matched filler sections and performing the BIM based on the above hierarchical orders of matching.

The layout of facilities is utilized for determining the matching relations among the hierarchical elements of RSS. In this work, there are totally six facilities considered: track, platform, road, yard, wall and ditch. Based on these facilities, the matching relations of components and filler sections are determined. Next, the matching relations for filler layers should be confirmed.

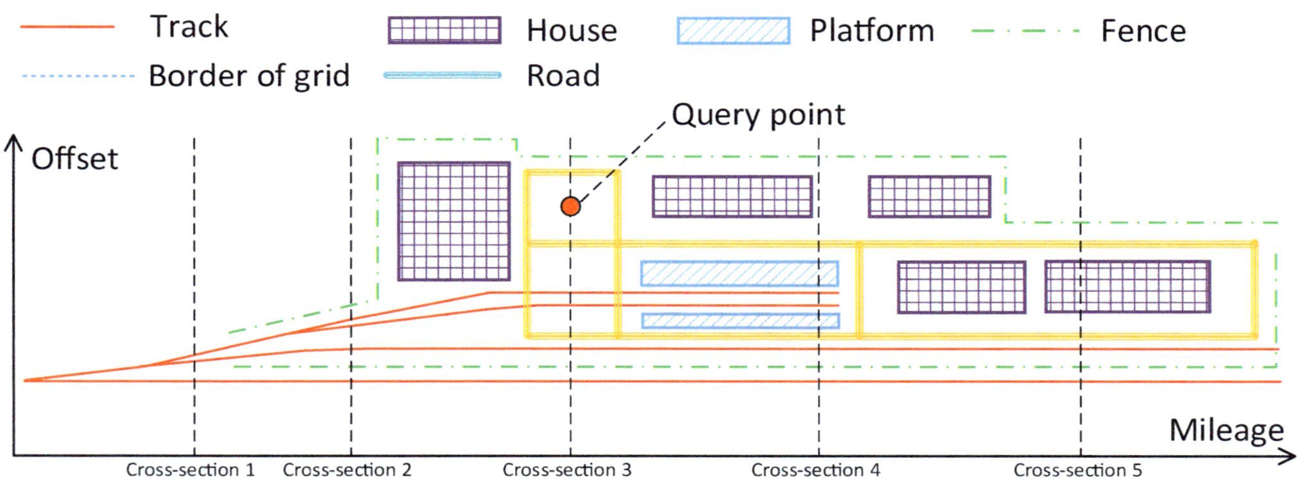

Figure 12.25 Demonstration of mileage-offset based query.

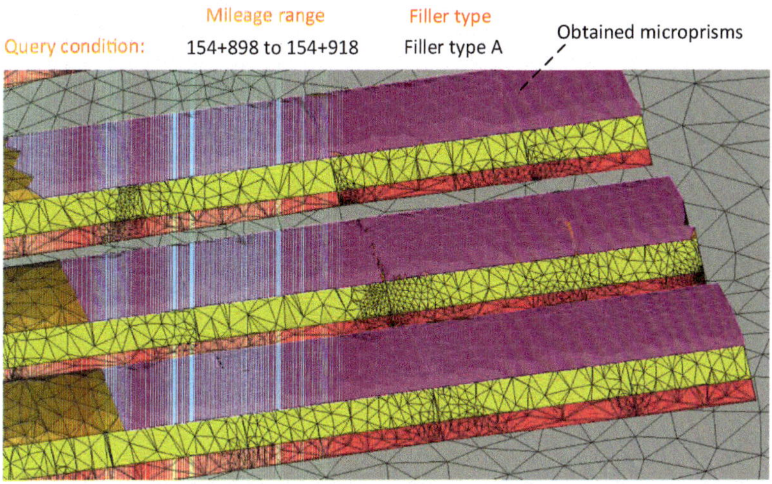

Figure 12.26 Result of retrieval based on a mileage range and a specific filler type, in which all obtained microprisms are highlighted.

At the filler section level, the numbers of layers may be different between two matched filler sections, due to which some layers should be vertically merged into one matched filler layer. For matching filler layers, two key attributes (level and filler type) of a filler layer are considered to determine the matching relations.

After the above procedures, all filler layers are processed and matched, by using which the 3D modeling process can be conducted.

Geometry Modeling Method. This step generates the 3D meshes based on the matching relations obtained from the previous step. The utilized 3D modeling method is generalized hexahedron, which can be used to build a closed 3D solid by using two closed 2D polylines. A 2D filler layer in this study consists of four polylines for expressing the upper, low, left and right contours. Hence, there can be eight polylines for a pair of matching filler layers, based on which the modeling method produces six sets of meshes to build a closed 3D solid.

12.2.3.2 Partially updating algorithm

Determination of Influenced Domain. In a railway station project, an RSS BIM is usually modified by a specific model change of a design criterion or facility, which may produce a requirement of partially updating the model. To perform a partial update, an influenced domain should be determined for securing all the influenced elements of the model.

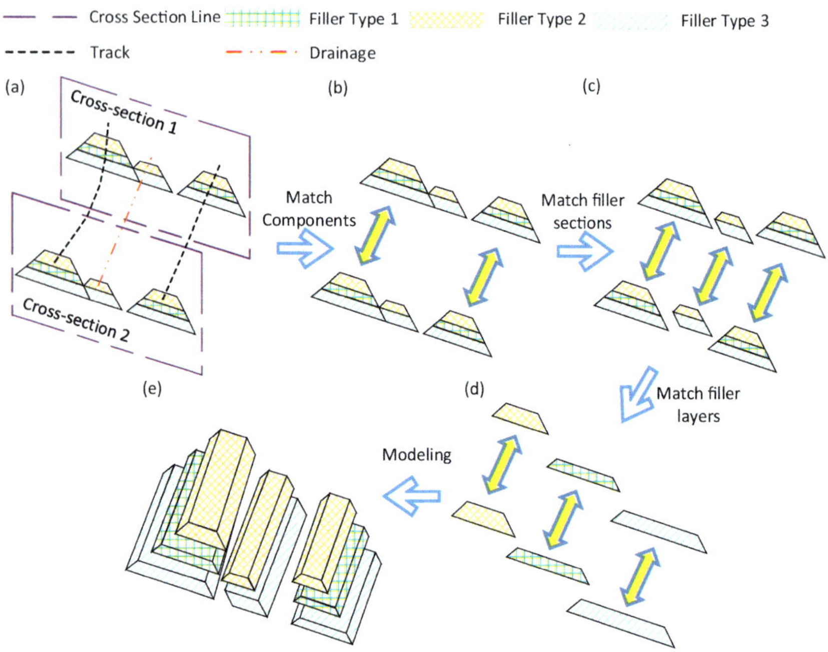

Figure 12.27 Procedures for modeling a segment of RSS: (a) the original two cross-sections; (b) matching the components in cross-sections; (c) matching the filler sections; (d) matching the filler layers; (e) performing the modeling.

Once a model change is made, a progressive search for locating the influenced 3D solids is performed, which is actually based on the acquisition of a cross-sectional influenced domain (2DIFD) and the corresponding three-dimensional influenced domain (3DIFD).

To confirm a 2DIFD, a triggering element (a facility or design criterion) should be first secured, which is the initial element influenced by a model change. Next, a set of cross-sections is found according to the triggering element. Then, in every cross-section, the influenced components, filler sections and layers are located in the hierarchical order. In the end, the 2DIFD is obtained.

After the acquisition of 2DIFD, the next step secures the 3DIFD by locating the related 3D solids from the original BIM model. Then, based on the type of model change, these 3D solids are directly updated or removed from the BIM for the next updating procedure.

Updating Model based on Influenced Domain. The procedure for updating the relevant 3D solids in the influenced domain is shown in Fig. 12.28.

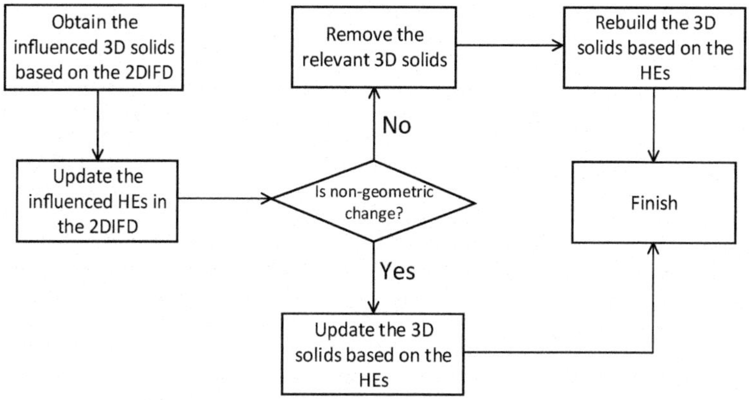

Figure 12.28 The steps for updating the RSS model according to the influenced domain.

The main procedures in Fig. 12.28 are as follows:

1. Obtain the influenced 3D solids based on 2DIFD.
2. Update the influenced elements in 2DIFD: after determining a 2DIFD, the next step is to modify the influenced elements in the domain, based on which the original BIM model can be finally updated.
3. Remove the relevant 3D solids in 3DIFD: the relevant 3D solids specify the exact parts to be modified of original RSS BIM, which are removed first for the next rebuilding procedure.
4. Rebuild the 3D solids in 3DIFD based on 2DIFD: The next step is to rebuild the 3D filler solids based on the updated elements in the 2DIFD. The modeling process includes two procedures.
   a) Mapping relations for elements: this step aims to match two fillers with layers in a pair of adjacent cross-sections, according to which a 3D solid is generated.
   b) Building 3D solids based on the matching relations: we employ a method called a generalized hexahedron to generate a 3D solid with two cross-sectional elements, by which the 2D elements of an RSS are converted to the 3D solids.

12.2.4 BIM for railway station facilities

Railway stations serve as crucial nodes in a railway system, handling important tasks such as passenger and cargo inflows and outflows, train yielding, overtaking, marshalling, disassembly, and vehicle maintenance [2]. To fulfill these complex operational functions, a railway station is commonly segmented into multiple yards according to their different functions. For instance, a district station contains a set of yards: a train departure and arrivals yard for passenger boarding and alighting, a marshalling yard for dispatching and coordinating, and freight yards for cargo handling, sorting, and storage. Each yard operates independently while also collaborating with others through tracks, turnouts, and signal facilities.

In a yard, numerous facilities are used collaboratively to accomplish various tasks [22–24]. For example, in the routine train arrival and departure process, a train is initially guided from the main track, then guided through multiple turnouts to reach an arrival-departure track. Upon arrival, the train halts on the arrival-departure track, whose effective length should exceed the length of the stopped train. The effective length is influenced by various facilities, including clearance posts and signal machines. Therefore, railway station facilities play a crucial role throughout the entire lifecycle of a railway station, making their BIM-based modeling highly valuable and meaningful.

12.2.4.1 Semantic analysis of railway station facilities

Railway station facilities can be categorized into three types based on their geometric features, namely point facilities, linear facilities and faceted facilities, as shown in Fig. 12.29. Point facilities such as turnouts, clearance posts, signal machines, and car bumpers, are typically described by specific three-dimensional (3D) coordinates and rotation angles to indicate their position and orientation. The geometric features of a linear facility include a designated baseline and a cross-section, based on which the 3D solid can be obtained by extruding the cross-section along the baseline, such as tracks and ditches. Faceted facilities, including platforms, level crossings, cargo and passenger yards, are usually designed based on a planar contour (a closed polyline).

12.2.4.2 Modeling with IFC

Based on the existing IFC, a corresponding BIM method for constructing railway station facilities with complex associative relations is proposed. The workflow of the proposed modeling method is shown in Fig. 12.30. First, the IFC file is generated according to the IFC. Next, the file is loaded for fetching the geometric and semantic information of the related railway station facilities. According to the geometric information, the geometric shapes of the facilities are constructed.

The geometric modeling method for three types of facilities (i.e., point, linear, and faceted facilities) is presented here. The construction of point

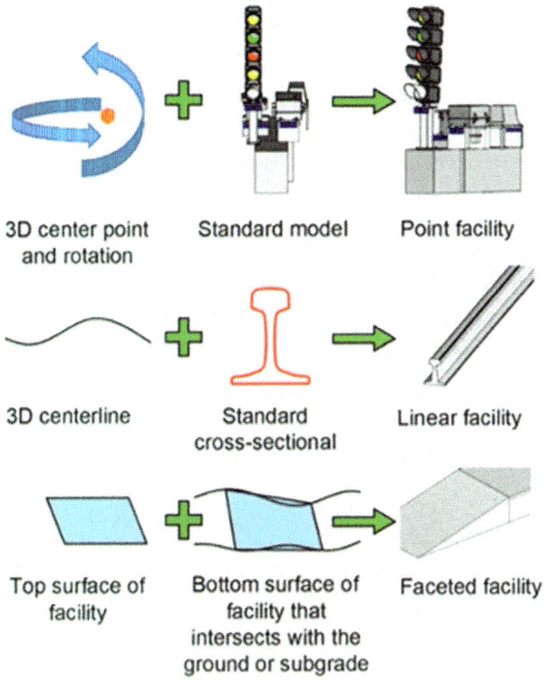

Figure 12.29 Geometric features of the point, linear and faceted facilities.

facilities primarily includes determining the geometric insertion center point and a 3D rotation matrix. This is achieved by selecting a standard model of point facilities that matches the requirements, inserting it at the center point, and performing a three-dimensional rotation. The establishment of linear facilities primarily focuses on establishing the centerline of the facility, achieved by placing and connecting standard cross-sections along this path. The faceted facilities are needed to determine the various face domains of the facility, or to determine the top face domain through corner points, and extrude the face domain vertically with reference to the height of the facility. Following this, the extruded body is intersected with the terrain surface to pinpoint the bottom face region. Finally, the formation of the body is undertaken, incorporating both the top and bottom face domains.

12.2.5 Combination BIM of railway station facilities and subgrades

The above sections have detailed the construction methods and technical specifics for station subgrade BIM models and railway station equipment

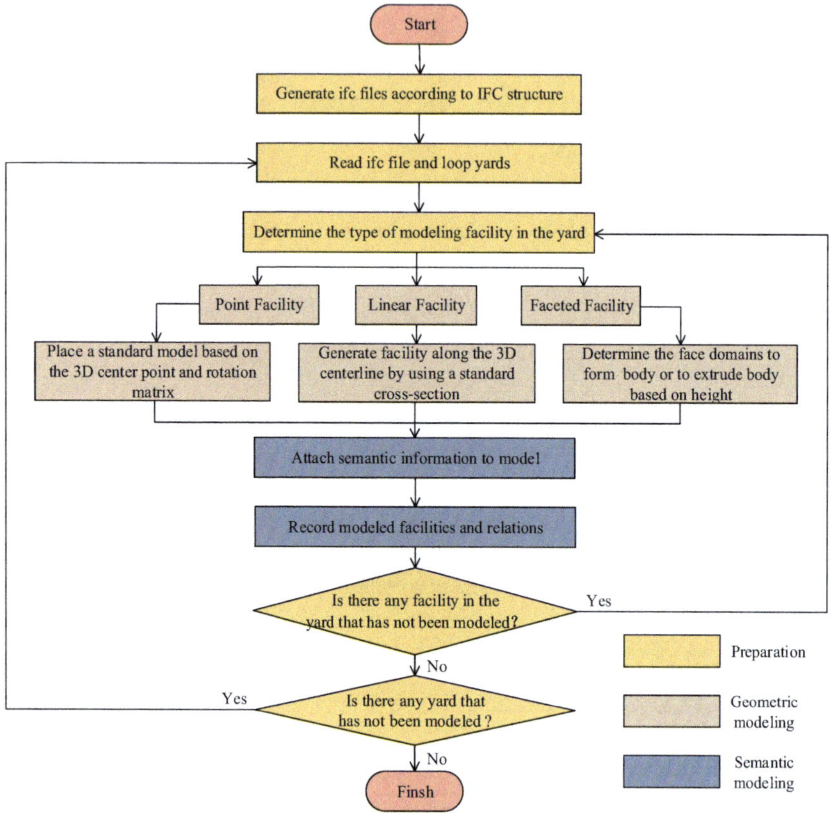

Figure 12.30 Workflow for constructing a BIM of railway station facilities.

BIM models. As independent digital entities, they already contain precise geometric information and rich attribute data. However, the core value of BIM lies not only in digitizing individual components, but in integrating and coordinating models across specialties and elements to form a comprehensive, data-driven digital twin of the station. This section focuses on combining the trackside equipment models with the subgrade models and demonstrates the resulting integrated BIM applications for railway stations.

12.2.5.1 Model integration

Model integration is not simply geometric merging, but a deep integration process based on a unified coordinate system and standard. Precise positioning and assembly are required. Using the same project reference point and coordinate system established during model creation, the completed station subgrade model serves as the spatial reference and foundational support. Through BIM collaboration platforms' "link" or "workset" functionality, station equipment models—including rails, turnouts, signals, and platforms—are accurately aligned and installed. Rails are precisely laid at the designed track-top elevation of the subgrade, turnouts seamlessly connect with rails, and signals or catenary supports are firmly embedded in the subgrade slopes or designated foundations, achieving full integration from underground foundations to above-ground facilities.

In addition to geometric integration, the combined model supports interconnected data interactions. For example, selecting a segment of rail allows viewing not only its type and material information, but also quickly locating the underlying subgrade layers, compaction, and drainage layout via established associations. Conversely, querying data for a subgrade settlement monitoring point can simultaneously display whether the overlying track and equipment are affected, providing a complete data chain for structural safety analysis.

12.2.5.2 Overall application effects after integration

The successful integration of trackside equipment models with subgrade models produces a digital twin of the station, as shown in Fig. 12.31. Its application benefits are particularly evident in the following aspects:

(a) Global visualization and walkthrough review. The integrated model provides an unprecedented overall perspective. Project stakeholders - including owners, designers, contractors, and operations teams, can perform virtual walkthroughs from first-person or free viewpoints to intuitively examine the station layout. Both macro-scale elements (e.g., track alignment and station layout) and micro-scale details (e.g., spatial relations between subgrade, equipment, and inter-equipment spacing) are clearly visible. This significantly enhances the presentation of design results and supports faster, more informed decision-making.

(b) Cross-disciplinary clash detection and conflict resolution. Hidden conflicts that were previously difficult to detect in three-dimensional space are now fully visible. The system can automatically perform collision detection, efficiently identifying issues such as: interference between catenary foundations and underground cable trenches, conflicts between drainage channels and signal cable pits, and insufficient clearance between platforms and track envelopes. Resolving these

Figure 12.31 BIM model for a railway station.

conflicts prior to construction prevents rework, saving both time and cost.

(c) Integrated quantity takeoff. With the complete integrated model, the BIM platform can generate precise and comprehensive quantity reports at the click of a button. The computation is no longer limited to individual specialties; instead, combined engineering quantities, such as subgrade earthwork, track materials, and concrete for equipment foundations, can be computed, providing accurate data for cost estimation and material procurement.

(d) Unified information carrier for construction and operations. The integrated model serves as a consistent data source throughout the project lifecycle. During construction, it supports simulation, technical briefings, and progress management. During operations, the model can interface with asset management systems, enabling full traceability of each piece of equipment, its foundation, and surrounding environment. This establishes a solid foundation for lean and intelligent facility management.

In summary, integrating station subgrade and facility BIM models represents a critical leap from fragmented design to fully integrated design for railway stations. It not only achieves visual integration but also fuses data and information, greatly enhancing design quality, construction efficiency, and paving the way for digital, intelligent station operations.

References

[1] C. Eastman, An Outline of the Building Description System, architectural drafting, 23, 1974. Research Report No. 50.

[2] Y. Xu, W. Li, J. Wang, et al., A 3D digital modelling method for railway station subgrade based on generalized hexahedron, Int. J. Rail Transp. 9 (2021) 79–99.
[3] N. Anwar, F. Najam, Response and design of column cross-sections, in: Structural Cross-Sections, Butterworth-Heinemann, 2017, pp. 329–390.
[4] B. Lin, Y. Chen, Transportation infrastructure and efficient energy services: A perspective of China's manufacturing industry, Energy Econ. 89 (2020).
[5] A. Costin, A. Adibfar, H. Hu, et al., Building information modeling (BIM) for transportation infrastructure - literature review, applications, challenges, and recommendations, Autom. ConStruct. 94 (2018) 257–281.
[6] Y. Tang, N. Xia, Y. Lu, et al., BIM-based safety design for emergency evacuation of metro stations, Autom. ConStruct. 123 (2021).
[7] G. Gao, Y. Liu, M. Wang, et al., A query expansion method for retrieving online BIM resources based on industry foundation classes, Autom. ConStruct. 56 (2015) 14–25.
[8] buildingSmart, Industry Foundation Classes (IFC) - An introduction, https://technical.buildingsmart.org/standards/ifc/.
[9] International Organization for Standardization, ISO 10303-21:2016, Industrial automation systems and integration - Product data representation and exchange - Part 21: Implementation methods: Clear text encoding of the exchange structure, https://www.iso.org/standard/63141.html.
[10] Data storage standard of railway engineering information model (Version 1.0), https://www.cnki.com.cn/Article/CJFDTotal-TJCX201601002.htm.
[11] Z. Ma, Z. Wei, W. Song, et al., Application and extension of the IFC standard in construction cost estimating for tendering in China, Autom. ConStruct. 20 (2011) 196–204.
[12] M. Huang, H. Zhu, J. Ninic, et al., Multi-LOD BIM for underground metro station: Interoperability and design-to-design enhancement, Tunn. Undergr. Space Technol. 119 (2022).
[13] M. Fang, T. Hu, J. Rose, Geometric composition, structural behavior and material design for asphalt trackbed: A review, Constr. Build. Mater. 262 (2020).
[14] A. Woszuk, L. Bandura, W. Franus, Fly ash as low cost and environmentally friendly filler and its effect on the properties of mix asphalt, J. Clean. Prod. 235 (2019) 493–502.
[15] M. Khan, I. Kim, J. Seo, A boundary and voxel-based 3D geological data management system leveraging BIM and GIS, Int. J. Appl. Earth Obs. Geoinf. 118 (2023).
[16] Y. Xu, L. Liu, C. Gotsman, et al., Capacity-constrained delaunay triangulation for point distributions, Comput. Graph. 35 (2011) 510–516.
[17] W. Lin, P. Lin, Intelligent generation of indoor topology (i-GIT) for human indoor pathfinding based on IFC models and 3D GIS technology, Autom. ConStruct. 94 (2018) 340–359.
[18] R. Graham, An efficient algorithm for determining the convex hull of a finite planar set, Inf. Process. Lett. (1972) 132–133.
[19] H. Pu, X. Fan, P. Schonfeld, et al., Extending IFC for multi-component subgrade modeling in a railway station, Autom. ConStruct. 141 (2022).
[20] F. Mortari, S. Zlatanova, L. Liu, et al., Improved geometric network model' (IGNM): A novel approach for deriving connectivity graphs for indoor navigation, ISPRS annals of the photogrammetry, in: Remote Sensing and Spatial Information Sciences, II-4, 2014, pp. 45–51.
[21] J. Shewchuk, Delaunay refinement algorithms for triangular mesh generation, CGTA or Comput. Geom. 22 (2002) 21–74.
[22] China Ministry of Railways, Code for Design of Railway Station and Terminal, China Planning Press, Beijing, China, 2017.

[23] J. Qi, L. Yang, Y. Gao, et al., Integrated multi-track station layout design and train scheduling models on railway corridors, Transp. Res. C Emerg. Technol. 69 (2016) 91–119.

[24] J. Lin, X. Hu, J. Dang, et al., Traffic model of machine-type communication for railway signal equipment based on MMPP, IET Microw. Antennas Propag. 13 (2019) 1072–1079.

CHAPTER THIRTEEN

System development and applications

Abbreviations and corresponding full texts

3D-DT Three-dimensional Distance Transform
DEM Digital Elevation Model
JD Intersection Point (of horizontal alignment)
PSO Particle Swarm Optimization
RAOS Railway Alignment Optimization System
TIN Triangulated Irregular Network

13.1 Introduction

Based on the theories, models, and methods presented in the preceding chapters, the *Railway Alignment Optimization System* (RAOS V1.0) was developed, as shown in Fig. 13.1. Its creation was supported by more than 10 national, provincial, and collaborative research projects—including the

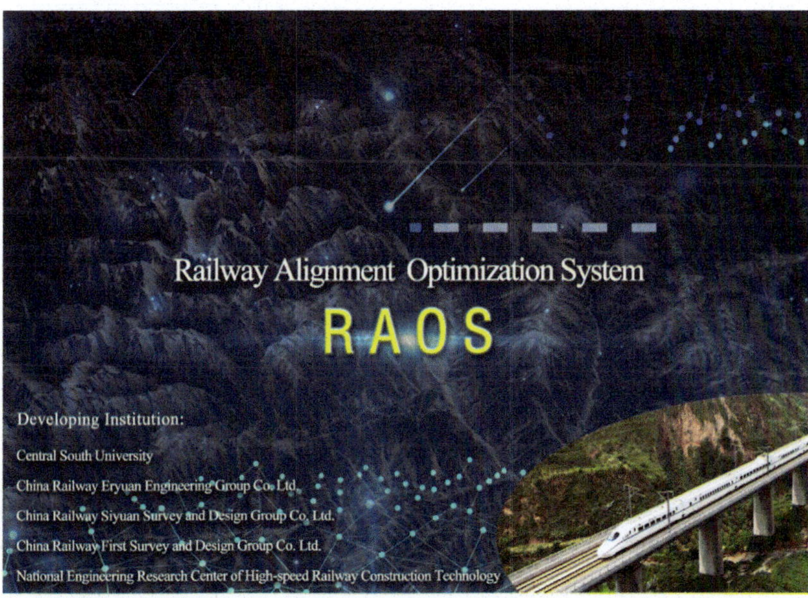

Figure 13.1 Railway alignment optimization system.

Intelligent Railway Alignment Design
ISBN: 978-0-443-45923-8
https://doi.org/10.1016/B978-0-443-45923-8.00006-6

National Key R&D Program, the National Natural Science Foundation of China, major projects of the China State Railway Group, and key research topics of the China Railway Engineering Corporation—together with nearly 30 years of railway alignment software development expertise. The functions of the software are shown in Fig. 13.2.

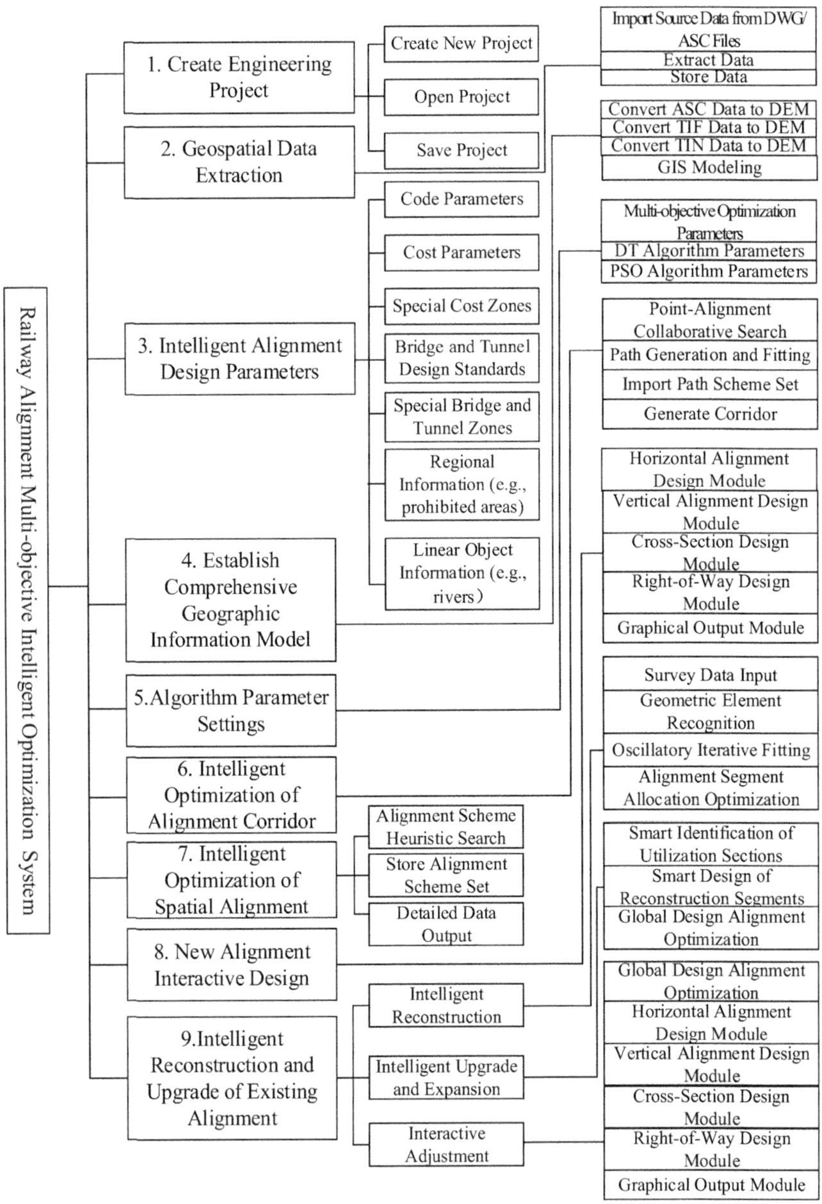

Figure 13.2 Framework diagram of the railway alignment optimization system.

13.2 Software functions

The software consists of nine modules, which are (1) project document management, (2) numerical model construction, (3) geographic information extraction, (4) alignment design information input, (5) alignment optimization parameters input, (6) alignment corridor optimization, (7) alignment spatial position optimization, (8) alignment interaction design, and (9) intelligent reconstruction of existing railway alignments.

13.2.1 Project document management

(1) New projects

Project folders are established with four subfolders: 3D, containing three-dimensional visualization data; DATA, containing design parameters and process data files; PMAP, containing basic geographic information data; and RESULT, containing design output files.

(2) Open a project

Select an existing project folder and open the process design file database in the DATA subfolder to continue the unfinished optimization design work.

(3) Save a project

Select the current planar, longitudinal or three-dimensional alignment design result and save it to the DATA subfolder.

13.2.2 Numerical model construction

(1) ASC data to DEM

Import grid elevation data in the standard ASC format and convert it to the grid DEM format required by the system.

(2) TIF data to DEM

Similarly, import elevation data in the standard TIF format and convert it to the grid DEM format required by the system.

(3) TIN data to DEM

Similarly, import elevation data in the standard TIN (Triangulation Network Numerical Model) format and convert it to the grid DEM format required by the system.

13.2.3 Geographic information extraction

(1) Read the source file data

Select the required source file (e.g., DWG, ASC) and load the basic geographic information data from it.

(2) Data extraction

According to the data type, geographic information such as land type, ground elevation, and land values is extracted by category (e.g., by DWG layers) (Fig. 13.3).

(3) Information input

Store the extracted geographic information data in the ACCESS database in the DATA subfolder.

13.2.4 Alignment design information input

Input all required design information for railway alignment design, including railway major technical standards (Fig. 13.4) and information on land prices, standards for bridges and tunnels, prohibited areas, disaster-prone regions, and other relevant datasets (Fig. 13.5). Based on the above information, a comprehensive geographic information model is created and stored in the PMAP subfolder.

13.2.5 Alignment optimization parameters input

Enter the multiobjective optimization parameters, 3D-DT algorithm search parameters, and PSO algorithm search parameters, respectively, as shown in Fig. 13.6.

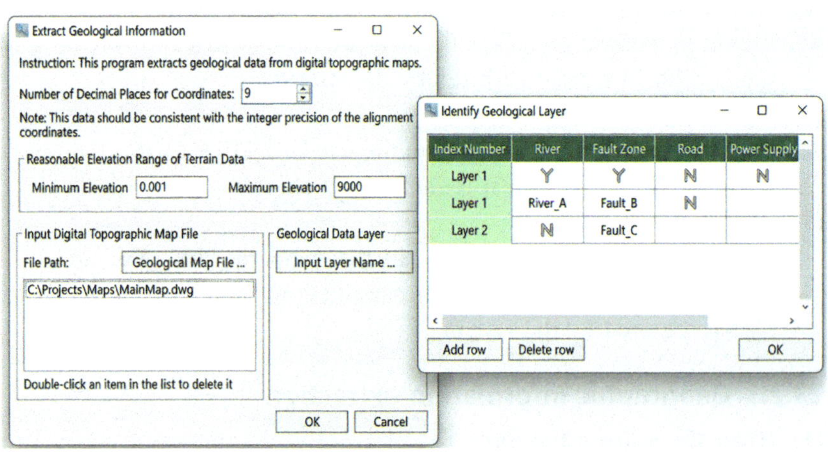

Figure 13.3 Geographic information data extraction interface.

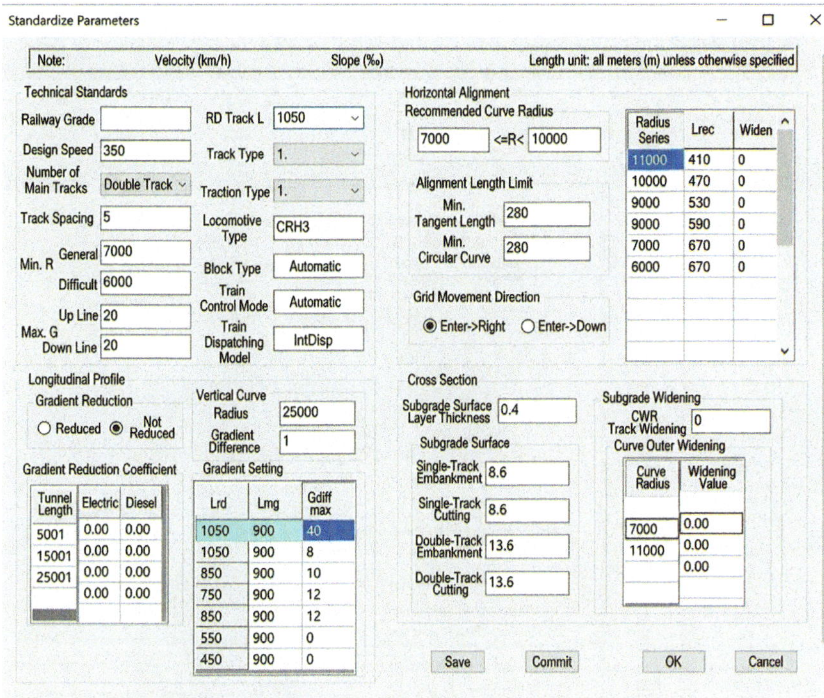

Figure 13.4 Interface for standard design parameter input.

13.2.6 Alignment corridor optimization

The 3D-DT algorithm is applied to search for the near-optimized railway path and associated structures. The resulting DT path is then fitted, and a near-optimized railway alignment corridor is constructed based on the fitted path, as shown in Fig. 13.7.

13.2.7 Alignment spatial position optimization

The PSO algorithm is applied to perform multiobjective optimization of the alignment spatial location, and the resulting optimized alignments are saved in the RESULT subfolder, as shown in Fig. 13.8.

13.2.8 Alignment interaction design

Using VizRail—a widely used modern railroad alignment design software—designers can fine-tune the alignment optimization results in the horizontal, longitudinal, and transverse dimensions, and produce the final design file.

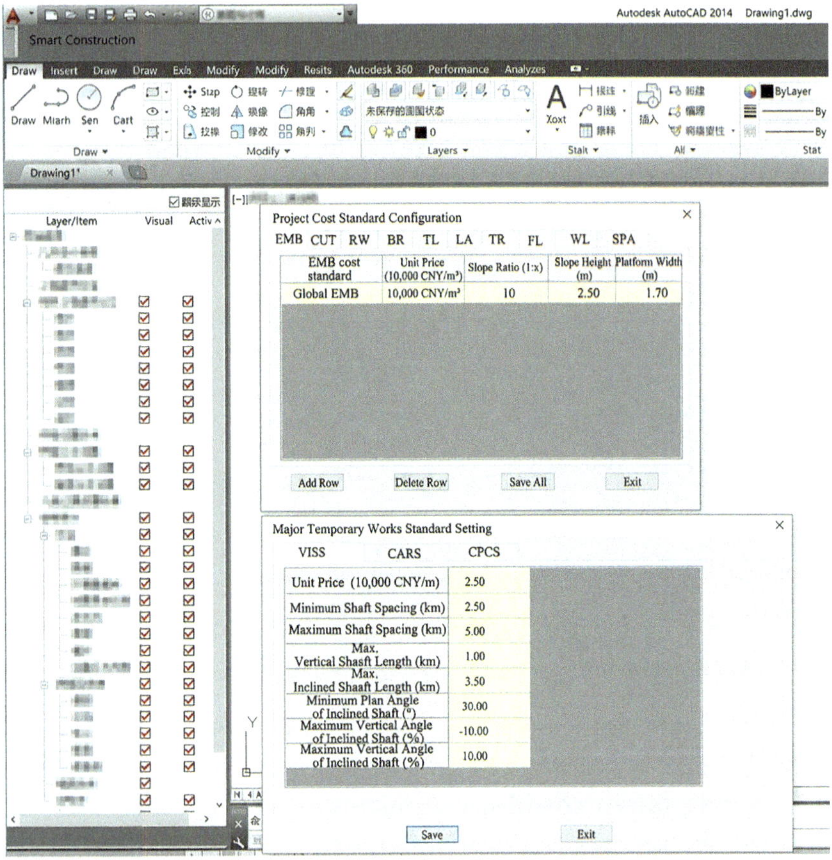

Figure 13.5 Comprehensive alignment parameter input interface.

13.2.9 Intelligent reconstruction of existing railway alignments

(1) Intelligent reconfiguration

The horizontal alignment is fitted using the full-line coordinate method. The process begins with importing the field-measured centerline coordinate data. The entire alignment is then automatically reconstructed through three steps: *automatic segmentation of tangents and curves* → *horizontal alignment fitting* → *saving of actual offset adjustments*. The specific operation dialog box is shown in Fig. 13.9.

(2) Automatic additions and alterations

For various scenarios, such as adding a second track or upgrading an existing double track, multiple horizontal alignment reconstruction

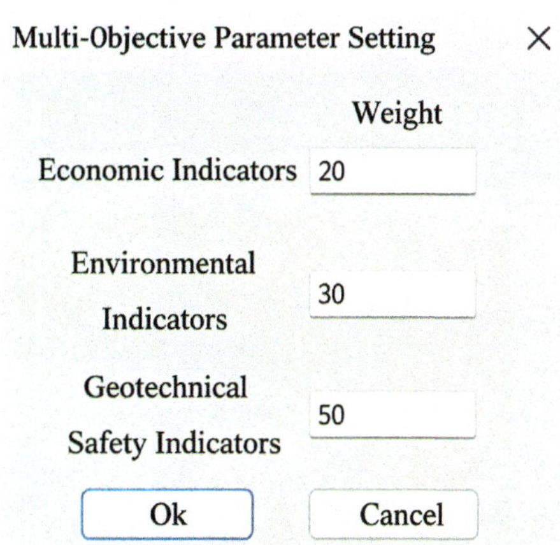

(a) Multi-object parameter setting

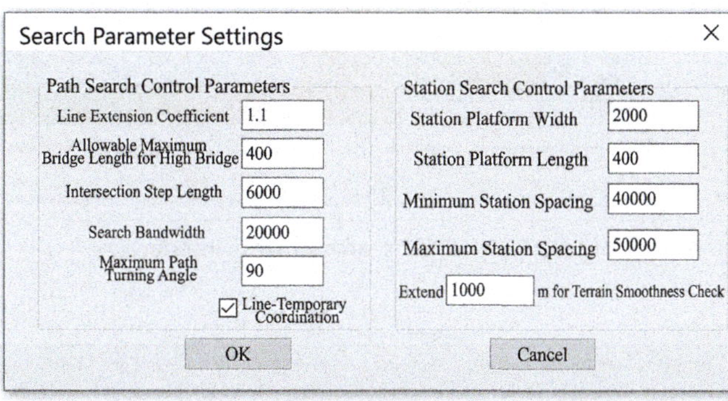

(b) 3D-DT algorithm

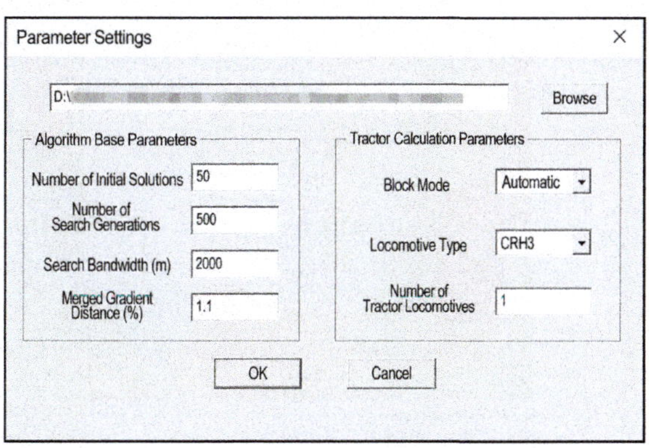

(c) PSO algorithm

Figure 13.6 The algorithm parameter input screen.

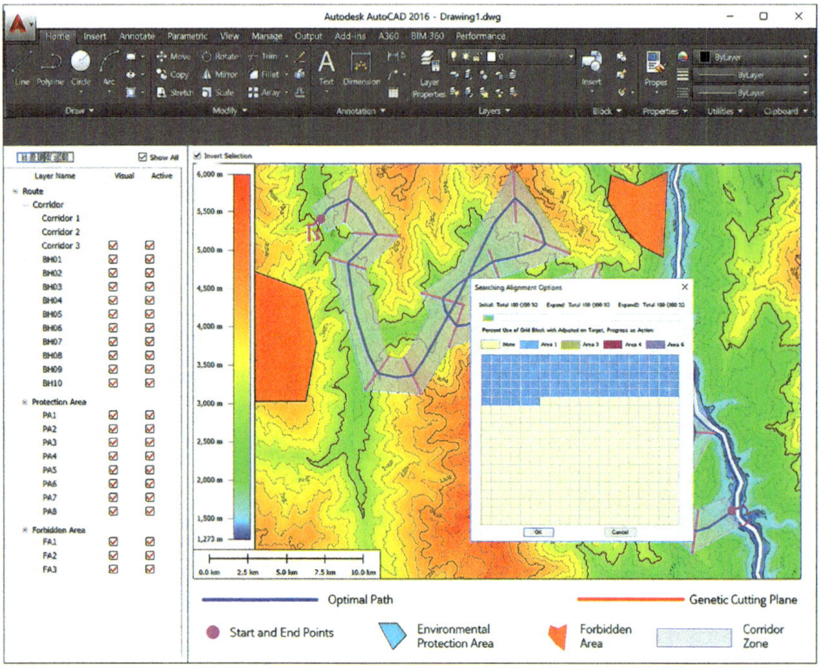

Figure 13.7 3D-DT search process interface.

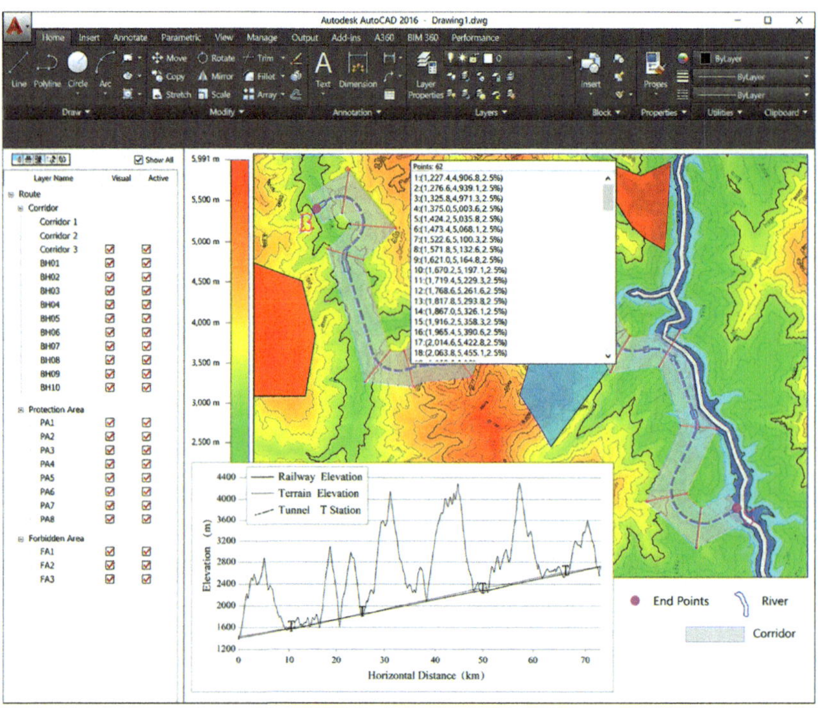

Figure 13.8 PSO optimization process display interface.

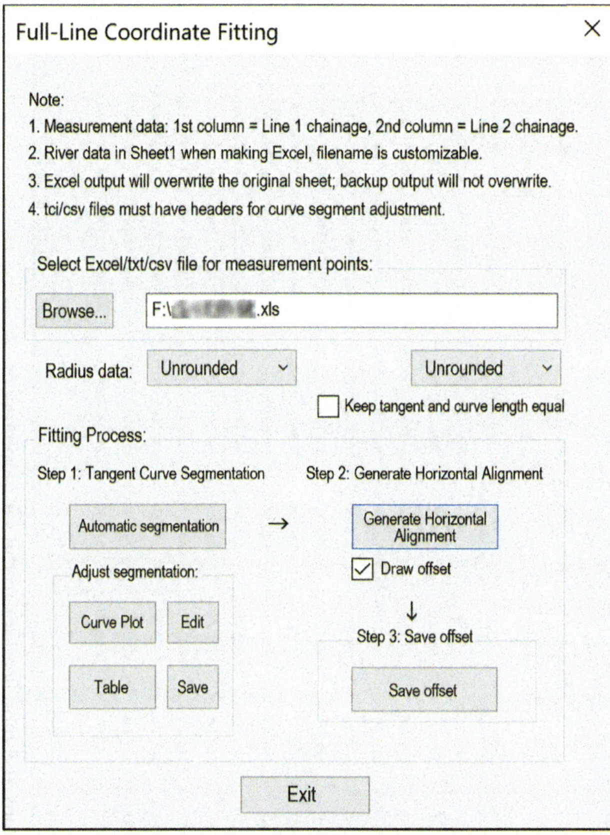

Figure 13.9 Full line coordinate fitting.

functions are provided. These include parallel reconstruction, bypass reconstruction, S-curve reconstruction, and straightening of curves. The system automatically marks the start and end points of the reconstructed sections, as well as chainage break points (Fig. 13.10).

Meanwhile, multiple methods for redesigning longitudinal gradients are available. Data from reusable sections of the existing longitudinal profile can be automatically extracted and applied to the reconstructed longitudinal profile.

(3) Interactive adjustment

As illustrated in Fig. 13.11, the existing baseline, the design baseline, and the second design line are treated as existing alignments with different attributes. These alignments are imported into the existing-line digital design software, which provides comprehensive tools for

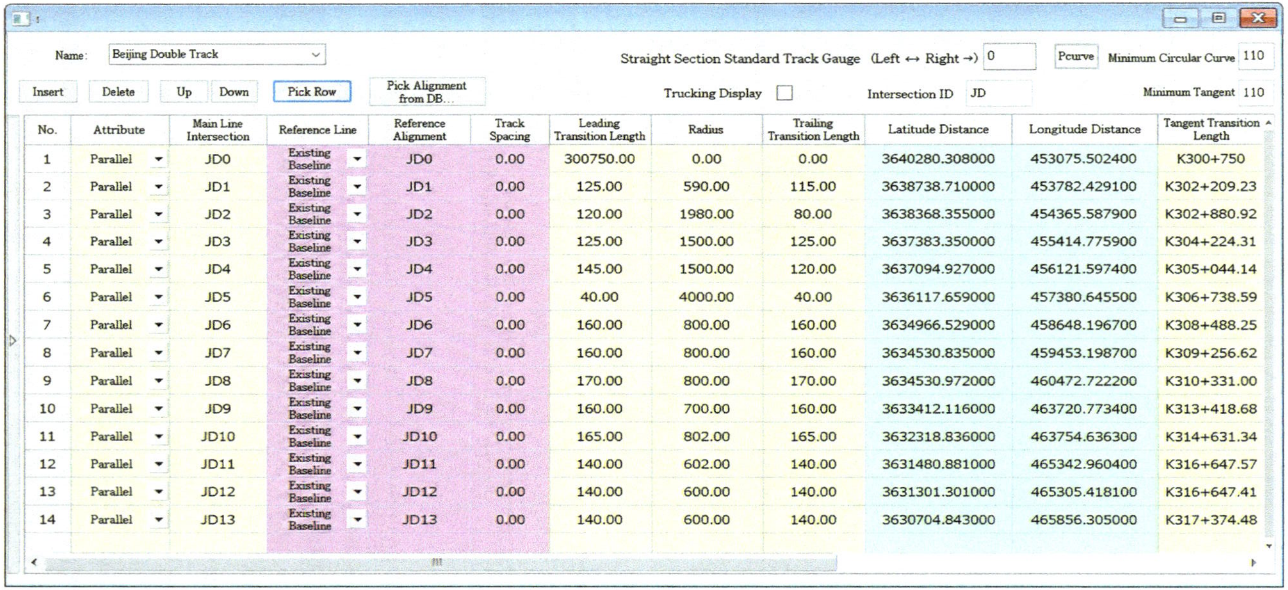

No.	Attribute	Main Line Intersection	Reference Line	Reference Alignment	Track Spacing	Leading Transition Length	Radius	Trailing Transition Length	Latitude Distance	Longitude Distance	Tangent Transition Length
1	Parallel	JD0	Existing Baseline	JD0	0.00	300750.00	0.00	0.00	3640280.308000	453075.502400	K300+750
2	Parallel	JD1	Existing Baseline	JD1	0.00	125.00	590.00	115.00	3638738.710000	453782.429100	K302+209.23
3	Parallel	JD2	Existing Baseline	JD2	0.00	120.00	1980.00	80.00	3638368.355000	454365.587900	K302+880.92
4	Parallel	JD3	Existing Baseline	JD3	0.00	125.00	1500.00	125.00	3637383.350000	455414.775900	K304+224.31
5	Parallel	JD4	Existing Baseline	JD4	0.00	145.00	1500.00	120.00	3637094.927000	456121.597400	K305+044.14
6	Parallel	JD5	Existing Baseline	JD5	0.00	40.00	4000.00	40.00	3636117.659000	457380.645500	K306+738.59
7	Parallel	JD6	Existing Baseline	JD6	0.00	160.00	800.00	160.00	3634966.529000	458648.196700	K308+488.25
8	Parallel	JD7	Existing Baseline	JD7	0.00	160.00	800.00	160.00	3634530.835000	459453.198700	K309+256.62
9	Parallel	JD8	Existing Baseline	JD8	0.00	170.00	800.00	170.00	3634530.972000	460472.722200	K310+331.00
10	Parallel	JD9	Existing Baseline	JD9	0.00	160.00	700.00	160.00	3633412.116000	463720.773400	K313+418.68
11	Parallel	JD10	Existing Baseline	JD10	0.00	165.00	802.00	165.00	3632318.836000	463754.636300	K314+631.34
12	Parallel	JD11	Existing Baseline	JD11	0.00	140.00	602.00	140.00	3631480.881000	465342.960400	K316+647.57
13	Parallel	JD12	Existing Baseline	JD12	0.00	140.00	600.00	140.00	3631301.301000	465305.418100	K316+647.41
14	Parallel	JD13	Existing Baseline	JD13	0.00	140.00	600.00	140.00	3630704.843000	465856.305000	K317+374.48

Figure 13.10 Additions and modifications to the set of references.

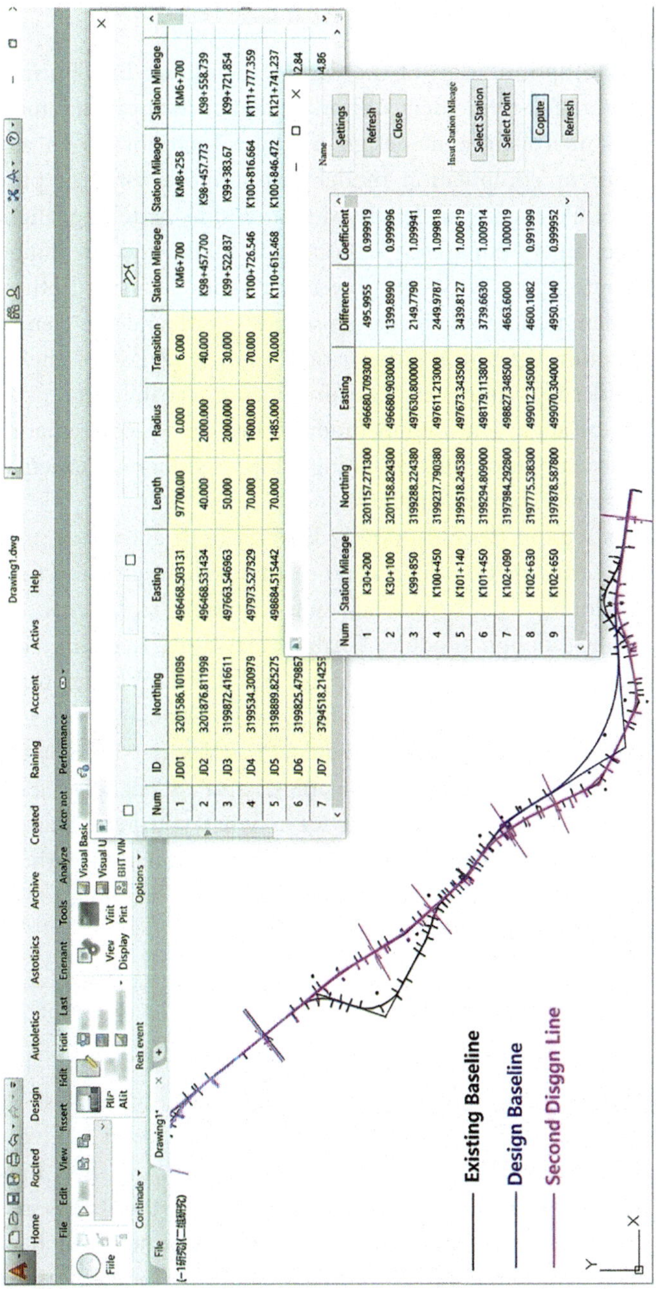

Figure 13.11 Detailed design output diagram.

detailed horizontal, vertical, and cross-sectional design of alignments derived from intelligent reconstruction or upgrading processes. The main functions include:

① Overall optimization of the design alignment: Enables integrated and iterative refinement of horizontal and vertical alignments.

② Horizontal alignment design module: Incorporates an advanced alignment computation model that supports efficient processing of double-track parallel, bypass, and side-switching alignments. Implements extensive, customized solid-entity techniques to generate dynamic design objects for Lines I and II (both existing and designed), gradients, cross-sections, and land-use elements. By integrating graphical information with data layers, the module provides real-time, drag-and-drop design capabilities.

③ Vertical alignment design module: Provides a convenient grade-adjustment tool incorporating slope-reduction principles and automated warning functions.

④ Cross-section design module: Enables both automatic and interactive cross-section design and earthwork computation for single-line, double-line, bypass, and side-switching alignments.

⑤ Land-use design module: Enables real-time, dynamic editing of land-use stakes (dragging, adding, and deleting) and automatically calculates land-use areas by land type and segment.

⑥ Drawing and chart output module: Provides user-friendly export of plan views, longitudinal profiles, and drainage maps, complemented by a "what-you-see-is-what-you-get" flag-insertion tool. Features include automatic plan segmentation with straightened display, and visualized elevation-break handling for longitudinal profiles.

13.3 Extension of applications

The above "Railway Alignment Optimization System" has been adopted by several major railway survey and design institutes, such as the China Railway First Survey and Design Institute Group Co., Ltd., the China Railway Eryuan Engineering Group Co., Ltd., and the China Railway Fourth Survey and Design Institute Group Co., Ltd. It has supported the design of thousands of kilometers of railway alignments, such as the X High-Speed Railway, the YR High-Speed Railway, the ST Railway, and the XT Railway.

According to application certificates from these institutes, the system can rapidly generate high-quality alignments tailored to design requirements, seamlessly integrate with detailed alignment design software, and enable refined design for any selected alignment. This establishes a complete workflow from concept to refinement, which significantly shortens the alignment design cycle, improves design efficiency by more than 8 times, and reduces project investment and labor costs by approximately 5%–10%. These outcomes yield significant social and economic benefits.

CHAPTER FOURTEEN

Research prospects

Research on railway alignment optimization has been conducted since the 1960s, evolving in recent years from theoretical studies to widespread practical application, markedly enhancing the quality and efficiency of alignment design. Driven by the national *Belt and Road Initiative*, *Transportation Power Strategy*, and *Western Development Strategy*, China's railway development focus has shifted westward and outward to countries along the Silk Road Economic Belt. The expansion encounters severe natural conditions, including complex terrain and geology, harsh climates, frequent natural hazards, and fragile ecosystems. Simultaneously, operational demands, such as high-speed passenger services, heavy-haul freight, and dense rail traffic, compound these challenges. Consequently, the three multiplicities of railway alignment design—multifactor environmental influences, multi-objective evaluation, and multiple coupled constraints—have become more prominent than ever.

The intelligent generation and optimization of railway alignments that ensure safety, sustainability, and cost-efficiency throughout the entire life cycle has become a critical bottleneck in the industry's progression toward green and intelligent transportation. At the same time, breakthroughs in artificial intelligence, including big data, foundation models, brain-inspired intelligence, and large-scale parallel computing, are providing new research paradigms posited to accelerate both theoretical and technological innovation in intelligent alignment optimization.

Future research should address increasingly complex construction and operational environments, stringent design standards, and stage-specific requirements across the entire alignment life cycle. Building upon existing optimization models, efforts should be extended to encompass multi-objective optimization for green and low-carbon design, disaster mitigation, and riding comfort. It will be essential to establish representation models and treatment methods for dense and widely distributed constraints, develop machine learning-based intelligent search methods, and design next-generation intelligent alignment systems capable of faster, more comprehensive, and more accurate computations. Ultimately, a new "model–method–system" framework for intelligent alignment optimization

Intelligent Railway Alignment Design
ISBN: 978-0-443-45923-8
https://doi.org/10.1016/B978-0-443-45923-8.00003-0

should be constructed, enabling optimization for highly complex combinations of objectives and constraints.

14.1 Intelligent alignment optimization with multi-level fidelity for stage-specific requirements

Railway alignment design follows a hierarchical process from concept to detail, including prefeasibility study, feasibility study, preliminary design, and construction design stages. Each stage presents distinct alignment problems: prefeasibility and feasibility studies emphasize macroscopic corridor selection and determination of major technical standards; preliminary design focuses on the alignment spatial position and coordinating major structures; construction design refines the alignment at local scales. Existing research largely targets mesoscopic spatial alignment search and develops stage-specific optimization models in isolation, paying insufficient attention to differentiated needs and cross-stage consistency. Consequently, a systematic, multilevel study of intelligent alignment—tailored to stage-specific objectives, constraints, and levels of detail—is essential.

14.1.1 Prefeasibility study: Integrated optimization of node–line–network at large spatiotemporal scales

Prefeasibility studies evaluate the strategic significance of proposed alignments within the broader rail network and the context of national transportation planning. Decisions on key nodes, junction points, and macroscopic routing have a decisive impact on the overall network configuration. By rationally integrating node selection, alignment routing, and network formation, alignments at this stage must remain consistent with national development strategies and regional economic objectives. Given the broad spatiotemporal scope of prefeasibility analysis, it is essential to examine the coupled interactions and co-evolution mechanisms among nodes, alignments, and networks, to formulate multi-objective models that balance economic efficiency, network performance, robustness, and coordination, and to develop intelligent search methods capable of solving such large-scale, integrated optimization problems.

14.1.2 Feasibility study: Data-driven optimization of key technical standards

The selection of major technical standards during the feasibility study stage is one of the most crucial early-stage decisions that fundamentally influences railway capacity, operational safety, transport efficiency, investment costs, and long-term benefits. Errors in these decisions can embed inherent design flaws, leading to significant construction and operational risks. The core of this optimization problem lies in modeling the dynamic coupling among multidimensional environmental factors and technical standards to establish accurate input–output relationships. Conventional methods based on predefined functional forms and statistical regression using limited samples are inadequate for capturing the full complex and nonlinear interactions inherent in these relationships.

Recent progress in big data analytics, machine learning, intelligent optimization, and parallel computing provides powerful new pathways for determining major technical standards. Big data technologies remove the need for predefined model assumptions by mining hidden patterns directly from massive datasets. Modern machine learning, especially deep learning, uses multilayer neural networks that emulate aspects of human cognition, enabling the extraction of complex and nonlinear relations from large-scale observational and simulated data. High-performance computing, through clusters architectures and multi-GPU parallelization, further accelerates processing, making it feasible to solve high-dimensional optimization problems that were previously computationally prohibitive. By establishing comprehensive, large-scale alignment datasets and applying deep learning to optimize major technical standards, the decision-making process can shift decisively from experience-driven qualitative decisions to data-driven quantitative optimization, enhancing accuracy, objectivity, and long-term design reliability.

14.1.3 Preliminary design: Integrated line–construction facility co-optimization

An optimal alignment must satisfy not only safety, cost, comfort, and environmental requirements but also ensure constructability and enable timely delivery. At the preliminary design stage, large temporary works, such as shafts, inclined tunnels, access roads, beam yards, and spoil disposal sites, are vital to construction feasibility and exhibit strong spatial coupling with the alignment and the distribution of bridges, tunnels, and stations. Effective planning requires that these temporary facilities be harmonized with terrain, geological conditions, and existing transport networks to meet schedule, quality, and investment requirements.

Traditionally, alignment and temporary works are planned sequentially: the main alignment is first fixed, and temporary works are subsequently arranged to support its construction. This "alignment-dominant" approach can be effective in relatively simple terrain. However, in complex environments characterized by high bridge–tunnel ratios, long-span structures, and deep tunnels, the number, scale, and difficulty of temporary works increase dramatically. Under such conditions, an otherwise optimal alignment from an operational or cost perspective may prove unbuildable due to the absence of feasible temporary works schemes. In these cases, the dependency reverses: the configuration and feasibility of temporary works directly constrain and even dictate alignment selection.

This necessitates a paradigm shift from sequential to integrated design. Temporary works must be embedded within the alignment process itself to enable the concurrent optimization of both the alignment and associated facilities. The alignment output should deliver not only an optimized spatial alignment but also the most compatible temporary works scheme, achieving full alignment–facility co-optimization.

14.1.4 Construction design: Intelligent fine-tuning under strong constraints

At the construction design stage, the structural design of bridges, tunnels, and stations is already basically determined, and coordination among different professional fields becomes critical. The alignment often requires localized fine-tuning in response to detailed geotechnical data, structure layouts, and construction requirements. These intersecting factors impose dense, tightly coupled constraints, creating a highly restrictive optimization problem. The central challenge is to implement minimal alignment adjustments that satisfy all technical and construction requirements while preserving overall design standards and performance objectives.

14.2 Intelligent alignment models for multi-objective and highly constrained optimization

Current alignment optimization models primarily focus on the minimization of construction and operating costs, with partial consideration of supplementary objectives such as geological hazard risk and environmental protection. However, as construction environments become increasingly complex and design standards more stringent, alignment design must expand its objective system to incorporate additional priorities, including green and low-carbon development, disaster risk reduction, and enhanced riding comfort, while addressing densely distributed constraints.

14.2.1 Disaster-mitigation-oriented intelligent alignment

As railway projects expand into rugged and geologically unstable regions, geological hazards pose escalating risk to both construction and operations over the entire life cycle. Alignment design fundamentally determines the macroscale exposure to such hazards. While substantial progress has been made in developing comprehensive evaluation methods using uncertainty analysis and accumulated extensive empirical knowledge, the direct integration of these principles into intelligent alignment optimization frameworks remains a significant challenge.

Future research should focus on formalizing and standardizing disaster-mitigation principles for alignment design, translating qualitative engineering expertise into quantitative evaluation models. Enhanced hazard risk assessment frameworks will enable intelligent alignment systems to make safer and better-informed decisions, mitigating vulnerability to geological hazards at the source.

14.2.2 Comfort-oriented intelligent alignment

In addition to external adaptation to terrain, geological conditions, and ecological environments, railway alignment must be internally harmonized with rolling-stock dynamic performance. Existing intelligent optimization methods largely focus on external environmental adaptation—minimizing construction cost, geological risk, and ecological impact—while giving limited attention to vehicle–track dynamic interactions. Incorporating vehicle–track dynamic performance evaluation into alignment optimization models would enable the integrated consideration of vehicle–track–environment coupling, thereby enhancing operational stability, passenger comfort while continuing to satisfy traditional objectives of economic efficiency and safety.

14.2.3 Suitability evaluation in densely constrained environments

Railway alignments increasingly face dense networks of above-ground and underground constraints in highly developed urban areas. Feasible alignment corridors are often fragmented and tightly restricted by existing transport networks, utilities, geohazards, and environmentally sensitive areas. The adoption of high design speeds and stricter geometric standards further reduces the alignment flexibility, while the growing prevalence of deep tunnels and long-span bridges dramatically expands both the vertical and horizontal search space.

Conventional "alignment-first–evaluate-later" optimization method, where candidate alignments are generated and subsequently iteratively adjusted to satisfy constraints, is inefficient in densely constrained environments. In such conditions, substantial computational resources are expended on unproductive exploration, and the optimization process frequently fails to converge to viable solutions within reasonable timeframes.

Inspired by manual design practice, where engineers preassess alignment suitability to focus on promising corridors, a "terrain-attracts-line" strategy should replace blind trial-and-error search. Preevaluating environmental suitability to identify high-quality corridors effectively prunes the search space, guiding alignment generation toward feasible regions and significantly improving computational efficiency and solution quality.

14.3 Other key research directions

Other possible research directions may include more detailed evaluation of right-of-way availability and costs based on detailed geographic information, explicit estimation of benefits and maximization of net benefits (benefits minus costs) rather than minimization of costs, more detailed estimation of potential demand and access costs to candidate station locations, more detailed probabilistic demand forecasts and analyses of the effects of demand levels on alignment optimization (e.g., how higher future demand may justify straighter and more expensive alignments), more analysis of how mixed train types and evolving train characteristics should influence alignment design, alignment optimization for rail transit lines, stations, and networks, and alignments for magnetic levitation trains.

Index

'*Note:* Page numbers followed by "f" indicate figures and "t" indicate tables.'

N

O

P

T

U

V

W

Printed and bound by CPI Group (UK) Ltd, Croydon, CR0 4YY
14/09/2026
02216409-0006